Damir Janigro
Editor

Mammalian Brain Development

Foreword by Denis Noble, Oxford University

Humana Press

Editor
Damir Janigro
Cleveland Clinic Lerner
College of Medicine
Cleveland, OH, USA
janigrd@ccf.org

ISBN 978-1-61779-648-7 e-ISBN 978-1-60761-287-2
DOI 10.1007/978-1-60761-287-2
Springer New York Dordrecht Heidelberg London

Softcover reprint of the hardcover 1st edition 2009

Cover illustration: Courtesy of Vince Fazio, Cleveland Clinic Lerner College of Medicine, NB-20 LRI, 9600 Euclid Avenue, Cleveland, OH 44195

Printed on acid-free paper

Springer is part of Springer Science+Business Media (www.springer.com)

To Kim, Mattia and Alice

Foreword

Denis Noble

Nearly a decade after completion of the first draft of the entire Human Genome sequence we are in a better position to assess the nature and the consequences of that heroic achievement, which can be seen as the culmination of the molecular biological revolution of the second half of the twentieth century. The achievement itself was celebrated at the highest levels (President and Prime Minister) on both sides of the Atlantic, and rightly so. DNA sequencing has become sufficiently common now, even to the extent of being used in law courts, that it is easy to forget how technically difficult it was and how cleverly the sequencing teams solved those problems in the exciting race to finish by the turn of the century [1, 2].

The fanfares were misplaced, however, in an important respect. The metaphors used to describe the project and its biological significance gave the impression to the public at large, and to many scientists themselves, that this sequence would reveal the secrets of life. DNA had already been likened to a computer program [3]. The "genetic program" for life was therefore to be found in those sequences: A kind of map that had simply to be unfolded during development. The even more colourful "book of life" metaphor gave the promise that reading that book would lead to a veritable outpouring of new cures for diseases, hundreds of new drug targets, and a brave new world of medicine.

Advances based on knowing the DNA sequences have certainly come, but they are fairly rare. We are far from using gene therapy generally in medical treatment. And, contrary to expectations, the pharmaceutical companies are producing fewer drugs at even greater expense than they did a decade ago. The most promising new drugs in my own field are treatments for angina, ivabradine (Servier) and ranolazine (CV therapeutics), that have relied on *pre-genomic* methods targeting receptors that were already known, while stem cells currently seem a more promising source of new treatments than gene therapy.

Why have the benefits been slow to materialise? The reason is simple. DNA is not the program of life [4, 5]. Nor is it the sole cause of phenotype characteristics [6, 7]. The relation between DNA and the phenotype is vastly more complicated

Denis Noble
Department of Physiology, Anatomy & Genetics, Parks Road, Oxford, OX1 3PT, UK

than many people thought. Moreover, the causal relationships go both ways. Hence my metaphor of the genome as a pipe organ (yes, there are pipe organs with as many pipes as genes in human DNA) played by the rest of the organism, and the environment. We are therefore moving away from the era of *The Selfish Gene* [8] and its simplistic "[genes] created us body and mind" to a more subtle twenty-first century biology that recognises that it is the system as a whole that is selected (selection depends on death, and it is only organisms that live or die) and which is the only means by which DNA can be read. The fertilised egg cell is also an "eternal replicator," without which DNA would lack meaning. This is the era of Systems Biology. And I believe that the nervous system will be its greatest challenge and probably its greatest triumph.

This volume on the development of the mammalian brain beautifully illustrates these principles. Damir Janigro and the contributors are to be congratulated on producing a book that will be of seminal importance.

The ability to dissect out the extent to which different sexual dimorphisms can be attributed to genetic and epigenetic factors is fundamental. Carrer and Cambiasso enlighten this particular form of "nature or nurture" question by showing that sex differences in spatial perception ability, in the density of TH immunoreactive neurons in the mesencephalon and of vasopressinergic fibers in the lateral septum are entirely controlled by the genome, whereas most other sexual dimorphisms in the brain of vertebrates are produced by the epigenetic action of gonadal hormones. There is considerable phenotype plasticity in the development of sexual dimorphisms.

Phenotype plasticity is even more evident in the work described by Weaver. He and his colleagues have revealed the remarkable degree to which maternal behaviour can mark the genome epigenetically in a way that then predisposes the young to display that behaviour when they in turn become adults. So, "the nature of the mother-offspring interaction influences the expression of genes that control the neural circuitry of behavioral responses in the offspring through life." This is an important development, but as his chapter points out, it is restricted to one gene and one region of the brain. That underlines the importance of further research in this area. The implications of this form of epigenetic influence are profound since it can transmit patterns of epigenetic marking through the generations even if all such marks are removed in the germline. In my view, it constitutes a form of inheritance of acquired characteristics.

If maternal behaviour can be so influential, what of other early life experiences, such as seizures in the infant? This is the question taken up by Marsh and Brooks-Kayal. As they emphasise, this question is difficult since it is hard to distinguish the effects of the seizures themselves from the effects of medication to treat them. Their work brings out the practical clinical importance of this problem. Other chapters on seizure show that susceptibility is the highest during development (Veliskova, Vezzani and Nehlig) and that both seizures and antiepileptic drugs may alter learning and behaviour (Galanopoulu, Velisek and Moshé). "The effects of seizures and antiepileptic drugs are further modified by genetic and epigenetic factors, biological and metabolic underlying conditions, or environmental influences."

Organisms, the human in particular, are clearly not Turing machines following a prescribed logic written down in the DNA, or even elsewhere in the organism. They are interaction machines dependent on the open system that we call the environment as well as on the "internal wiring" of the genes, proteins, cellular machinery and the nervous system. As Craig Venter has admitted in his autobiography, appropriately called *A life Decoded* (since it was his own DNA that was used): "One of the most profound discoveries I have made in all my research is that you cannot define a human life or any life based on DNA alone…." Why? Because "An organism's environment is ultimately as unique as its genetic code" [2]. John Sulston is also cautious: "The complexity of control, overlaid by the unique experience of each individual, means that we must continue to treat every human as unique and special, and not imagine that we can predict the course of a human life other than in broad terms" [1].

Disorders of development of the brain form the focus of the chapter by Batra, Carlton and Franco, a development that they describe as "nothing short of a miracle." In my little book on Systems Biology, *The Music of Life*, I replaced Richard Dawkins' "Selfish gene" metaphor (which lays emphasis on individual gene-selection in evolution) with "genes as prisoners" (which lays emphasis on the co-operativity of genes in generating functions for which they may be selected). But what do we know of this co-operativity? We have broken humpty-dumpty into his billions of pieces, his genes, proteins and what-have-you, but putting him back together again is going to prove difficult in an unimaginably greater way. Integration is much harder than reduction in biology. So much so that, in referring to development, I wrote "through which [the interactions], blindly, as if by magic, function emerges." I wondered long and hard about the phrase "as if by magic" since magic is usually seen as the opposite of science (and therefore impossible). But, as a description of our present state of ignorance, words like "miracle" and "magic" are natural resources for expressing a sense of awe and wonder at what happens during development. Sure, we know that this process has been refined and perfected by evolution over billions of years. Still, it seems a wonder that it happened at all. The human brain may well be the most complicated thing in the universe. Batra, Carlton and Franco take us through what is known of the errors of its development and how to tackle these problems in the future.

Cucullo reveals the processes involved in the development of the Blood-Brain Barrier, a structure without which the development of the mammalian brain would be impossible. This process is also multifactorial "regulated by a complex pattern of cell signaling, soluble factors, and local environment." This is a special feature, in the case of the brain, of a process that occurs in the development of all organs: Angiogenesis.

There is another barrier of vital importance of course in development: The placental barrier, which is the subject of the chapter by Ghosh and Marchi. As they show, the placental barrier performs tasks later taken over by the Blood-Brain barrier.

Gluncic writes about the use of imaging techniques: "Live imaging is essential to determine dynamic parameters or stability of a structure," and this is particularly true of the brain where we would like to have fine (ideally mm) 3D resolution capable also of temporal resolution at the level (1 ms) of neuronal activity. In practice of course we have to compromise and balance the resolutions in the different dimensions against each other.

Finally Bianchi and D'Adamo tackle the problem of chromosome abnormalities, long known to have great importance in the development of the nervous system, particularly in relation to conditions of mental retardation. As they point out, these abnormalities are not by any means limited to inherited genetic defects. It is important therefore to distinguish between congenital and hereditary. Congenital disease can occur without there being a genetic mutation.

I see this book, therefore, as charting the way forward in a form of biological science that will, I suspect, dominate the field in the twenty-first century. This is the form in which we recognise that the development of an individual is fundamentally an *interaction* between its genes and environment, and in which the integration of biological function can occur at all levels between the gene and the organism itself, with no level having a privileged form of causation [7, 9].

References

1. Sulston J, Ferry G (2002) The common thread. Bantam Press, London
2. Venter C (2007) A life decoded. Allen Lane, Penguin books, London
3. Monod J, Jacob F (1961) Teleonomic mechanisms in cellular metabolism, growth and differentiation. Cold Spring Harb Symp Quant Biol 26:389–401
4. Atlan H, Koppel M (1990) The cellular computer DNA: Program or data? Bull Math Biol 52:335–348
5. Noble D (2006) The music of life. OUP, Oxford
6. Keller EF (2000) The century of the gene. Harvard University Press, Cambridge, MA
7. Noble D (2008) Genes and causation. Philos Trans R Soc A 366:3001–3015
8. Dawkins R (1976) The selfish gene. OUP, Oxford
9. Noble D (2008) Mind over molecule: Activating biological demons. Ann NY Acad Sci 1123:xi–xix

Contents

Sexual Differentiation of the Brain: Genetic, Hormonal and Trophic Factors 1
Hugo F. Carrer and María J. Cambiasso

Life at the Interface Between a Dynamic Environment and a Fixed Genome: Epigenetic Programming of Stress Responses by Maternal Behavior 17
Ian C.G. Weaver

Effects of Early Life Seizures and Anti-epileptic Drug Treatment on Human Brain Development in Human Models 41
Eric D. Marsh and Amy R. Brooks-Kayal

Prenatal Development of the Human Blood-Brain Barrier 53
Luca Cucullo

Seizure Propensity and Brain Development: A Lesson from Animal Models 77
Jana Velíšková, Annamaria Vezzani, and Astrid Nehlig

Seizures and Antiepileptic Drugs: Does Exposure Alter Normal Brain Development in Animal Models? 105
Aristea S. Galanopoulou, Libor Velíšek, and Solomon L. Moshé

Overview of Neural Mechanisms in Developmental Disorders 133
Ayush Batra, Erin Carlton, and Kathleen Franco

Drug Permeation Across the Fetal Maternal Barrier 153
Chaitali Ghosh and Nicola Marchi

In Vivo Imaging of Brain Development: Technologies, Models, Applications, and Impact on Understanding the Etiology of Mental Retardation 171
Vicko Gluncic

Congenital, Non-inheritable Chromosomal Abnormalities Responsible for Neurological Disorders 193
Riccardo Bianchi and Patrizia D'Adamo

Index 219

Contributors

Ayush Batra Cleveland Clinic Lerner College of Medicine, 9500 Euclid Avenue, Cleveland, OH 44195, USA
Cerebrovascular Research, 9500 Euclid Avenue, Cleveland, OH 44195, USA

Riccardo Bianchi Robert F. Furchgott Center for Neural and Behavioral Science, and Department of Physiology and Pharmacology, State University of New York Downstate Medical Center, Box 29, 450 Clarkson Avenue, Brooklyn, NY, USA
rbianchi@downstate.edu

Amy R. Brooks-Kayal Division of Child Neurology, Children's Hospital of Philadelphia, Philadelphia, PA 19104, USA
brooks-kayal.amy@tchden.org
Departments of Neurology and Pediatrics, University of Pennsylvania, Philadelphia, PA 19104, USA

María J. Cambiasso Instituto de Investigación Médica Mercedes y Martin Ferreyra, Casilla de Correo 389, 5000 Córdoba, Argentina

Erin Carlton Departments of Psychiatry & Psychology, 9500 Euclid Avenue, Cleveland, OH 44195, USA
Cerebrovascular Research, 9500 Euclid Avenue, Cleveland, OH 44195, USA

Hugo F. Carrer Instituto de Investigación Médica Mercedes y Martin Ferreyra, Casilla de Correo 389, 5000 Córdoba, Argentina
hfcarrer@immf.uncor.edu

Luca Cucullo Cleveland Clinic, Cerebrovascular Research, 9500 Euclid Avenue, NB-20, Cleveland, OH 44195, USA
cuculll@ccf.org

Patrizia D'Adamo Dulbecco Telethon Institute at DIBIT-HSR, Molecular Genetic of Mental Retardation, via Olgettina 58, 20132 Milan, Italy

Kathleen Franco Departments of Psychiatry & Psychology, 9500 Euclid Avenue, Cleveland, OH 44195, USA
francok@ccf.org

Aristea Galanopoulou Saul R. Korey Department of Neurology, Dominick P. Purpura Department of Neuroscience, Einstein/Montefiore Comprehensive Epilepsy Management Center, Bronx, NY, USA

Chaitali Ghosh Cleveland Clinic, Cerebrovascular Research, Lerner Research Institute, 9500 Euclid Avenue, NB2-137, Cleveland, OH 44195, USA

Vicko Gluncic Section of Neurosurgery, The University of Chicago Medical Center, University of Chicago, 5841 South Maryland Avenue, Chicago, IL 60637, USA
vicko.gluncic@gmail.com

Nicola Marchi Cleveland Clinic, Cerebrovascular Research, Lerner Research Institute, 9500 Euclid Avenue, NB2-137, Cleveland, OH 44195, USA
marchin@ccf.org

Eric D. Marsh Division of Child Neurology, Children's Hospital of Philadelphia, Philadelphia, PA 19104, USA

Solomon L. Moshé Saul R. Korey Department of Neurology, Dominick P. Purpura Department of Neuroscience, Einstein/Montefiore Comprehensive Epilepsy Management Center, Bronx, NY, USA
Department of Pediatrics, Albert Einstein College of Medicine, Bronx, NY, USA

Astrid Nehlig INSERM U 666, Faculté de Médecine, 11 rue Humann, Strasbourg, France

Libor Velíšek Saul R. Korey Department of Neurology, Dominick P. Purpura Department of Neuroscience, Einstein/Montefiore Comprehensive Epilepsy Management Center, Bronx, NY, USA
velisek@aecom.yu.edu

Jana Velíšková The Saul R. Korey Department of Neurology and Dominick P. Purpura Department of Neuroscience, Albert Einstein College of Medicine, 1410 Pelham Parkway South, Bronx, NY, USA

Annamaria Vezzani Department of Neuroscience, Mario Negri Institute for Pharmacological Research, Milano, Italy
Vezzani@marionegri.it

Ian C.G. Weaver Developmental Biology Program, Hospital for Sick Children, Toronto Medical Discovery East Tower, Medical & Related Sciences (MaRS) Centre, 101 College Street, Toronto, ON, Canada M5G 1X8
ian.weaver@utoronto.ca

Sexual Differentiation of the Brain: Genetic, Hormonal and Trophic Factors

Hugo F. Carrer and María J. Cambiasso

Abstract The current hypothesis to explain the sexual dimorphism of structure and function in the brain of vertebrates maintains that these differences are produced by the interaction of genetic mechanisms and gonadal hormones. In this chapter we summarize the evidence from our laboratory, as well as other laboratories analyzing the mechanisms that control sexually differentiated growth of axons in hypothalamic neurons in vitro.

Keywords Ventromedial hypothalamus • Axon growth • Membrane receptor • Intracellular signaling • ERK

Verification of differences between males and females has progressed from the comparison of form (gross anatomy) to the application of refined analytical tools, identifying sex differences at genetic, systemic and behavioral levels. One of the main targets of this exploration has obviously been the brain, focusing on the substrate of behavioral, functional and most recently, neuropathological differences between males and females. It has become apparent that the abundant sexual dimorphisms described to-date represents only a fraction of those likely to exist in the brain. The number of neurons, ratio of gray to white matter, neurochemical composition, glial function, genetic, metabolic, and neurotransmitter systems are examples of a rapidly growing list of sexual differences being discovered thanks to the application of sophisticated techniques, such as genetic modification and imaging analysis. These differences exist not only in the hypothalamus, where the first brain differences between the sexes were identified, but also in structures such as the neocortex, hippocampus, amygdala and olfactory system, and a set of functions not obviously related to reproduction, such as verbal fluency [1], anxiety levels [2], and motor skills [3]. Moreover, it is important to keep in mind that the process of sexual

---

H.F. Carrer
Instituto de Investigación Médica Mercedes y Martin Ferreyra, Instituto Ferreyra,
Casilla de Correo 389, 5000, Córdoba, Argentina
e-mail: hfcarrer@immf.uncor.edu

D. Janigro (ed.), *Mammalian Brain Development*, Contemporary Neuroscience,
DOI 10.1007/978-1-60761-287-2_1,

differentiation is not "all or none." Individuals are a mosaic of male and female characteristics expressed in varying relative amounts, and for every trait analyzed there is a whole spectrum of values across individuals of a given species. For example, Kimchi et al. [120] have elegantly demonstrated the potential bisexuality of individual mice: following incapacitation of the accessory olfactory system, female mice show many components of male reproductive behavior. It is most interesting that several studies report sexual differences in neural activity despite no behavioral difference between the sexes.

A recent report by Grabowski et al. [5] examined the neural correlates of naming images. Men and women performed the task equally well, but the patterns of brain activity associated with their performance differed significantly. Findings such as these indicate that isomorphic performance of a certain function between the sexes does not necessitate isomorphic neural mechanisms.

The process of brain sexual differentiation results from the interaction of genetic determinants and gonadal hormones. The relative contribution of each of these influences has been summarized in several excellent reviews [6–14]. The thorough and updated review by Wilson and Davies [15] deserves particular mention, as well as Cahill's work [16], calling attention to the medical importance of brain sexual differences. Becker et al. [17] are to be commended on their work by charting the way to define, describe, and explain these differences.

1 Brain Sex Depends on Hormones and Genes

Although it is clear that most sexual dimorphisms in the brain of vertebrates (and differentiated functions thereof) are produced by the epigenetic action of gonadal hormones, genetic mechanisms preceding, or in parallel with, hormonal effects are also important [18, 19]. The initial evidence on this matter came from Pilgrim's laboratory [20, 21], showing that in mesencephalic neuron cultures collected on gestational day (GD) 15 more cells express dopamine in cultures from female rat embryos as compared to cultures from male rat embryos. This and other sexual differences [22] could not be attributed to differences in hormonal environment, because the donor embryos were obtained when steroid gonadal secretion is just beginning [23], that is to say, before the perinatal surge of testosterone that determines the development of a male brain, which begins on GD17 or GD18 [24].

Work from our laboratory [25] also revealed sexual differences in the growth and differentiation of neurons in vitro. In neuronal cultures taken from embryos on GD16, (i.e., before the surge of gonadal steroids), we found an unexpected sexual difference: axons from male neurons (but not from female neurons) were longer when cultures were treated with estradiol (E2) [26]. The establishment of connections between neurons within and between Central Nervous System (CNS) structures contributes to the conformation of neuronal networks that are function-specific and therefore sexually differentiated. These differences are due to the effect of estrogen/testosterone on programmed cell death, neurite growth, axon guidance, and synaptogenesis.

On this basis, the regulation and control of neurite growth is decisive. Similarly, in cultures of hypothalamic neurons taken from GD16 embryos, treatment of cultures with E2 increased the levels of tyrosine kinase type B (TrkB) and insulin-like growth factor I (IGF-I) receptors in male, but not in female, neurons [26]. Similar sexual differences in the response of neurons to E2 in vitro have been corroborated in other systems. For example, aromatase activity in hypothalamic neurons of embryonic mice, the level of tyrosine hydroxylase (TH) message per cell, or the number of TH positive cells, are cases where sexual differences have been found to precede steroid-dependent differentiation [27–30].

Apart from regulating neuritogenesis, E2 contributes to brain differentiation regulating neurogenesis, cell death, terminal differentiation, and migration [31, 32]. However, a genetically-dependent bias was recently demonstrated in the movement of a cohort of cells located in the preoptic area [4]. There is at least one neuritic parameter, namely, the number of primary neurites, which is not affected by the hormonal treatment in the hippocampus [33], the amygdala [34], or the hypothalamus [35]. Treatment in situ with E2, did not modify the number of primary dendrites or branch points of ventromedial hypothalamic neurons [36]. These findings are suggestive that genetic, rather than epigenetic, determinants may control neuron soma shape and number of primary dendrites.

The difference in the response to E2 between males and females suggests that a temporal pattern of sexual differentiation exists with respect to the effects of E2 on two related but different processes: cellular polarization and axon elongation. In the male brain, hypothalamic neurons may prematurely polarize and differentiate axons, which when exposed to E2 are capable of faster axon elongation. On the other hand, even on GD19, neurons from a female hypothalamus have not yet defined a cellular machinery for polarization; however, this process is accelerated by E2. This would explain why cultures of neurons taken only from males at GD16 show axon elongation whereas neurons from female GD19 embryos respond to E2 with axon polarization. If male neurons define their axons earlier than female neurons, this difference should be reflected in vivo as well. The existence of sexual differences in the amount of GAP-43, a protein associated with the establishment of polarity, has been demonstrated in several brain structures [37]. The level of GAP-43 hybridization signal was significantly higher in the male cortex, bed nucleus, and medial preoptic nucleus, but not at the ventromedial and arcuate nuclei, as compared to postnatal females [38]. These differences were suppressed by hormonal manipulations after birth, suggesting epigenetic, sex steroid-dependent differentiation [37]. However, these studies were performed on postnatal rats, so the doubt persists on whether genetically determined sexual differences may exist during embryonic development.

In our studies, we analyzed how treatment with E2 affects the frequency distribution of axon length. We invariably found that cultured cells respond as a single homogenous population, with most or all of the neurons reacting to the presence of E2 in the medium [33, 34, 39]. Given that the number of cells carrying estrogen receptors (ER) is scarce both in vitro [20] and in vivo [40–46], this was an unexpected result, as it implied that E2 can exert its effects on neurons devoid of the classical nuclear receptors. Therefore, the issue of

ER density in cultured neurons may need to be reconsidered. As more sensitive immunocytochemical methods are developed, more evidence indicates that almost all neurons in hippocampal cultures stain positive for ERα or ERβ [47, 48]. Alternatively, the explanation for this phenomenon could be related to the fact that E2 exerts its effects through a novel, and yet unidentified, receptor (see below).

In most mammals, the SRY gene found on the Y chromosome determines the development of the testis through the regulation of downstream genes, including those of Mullerian inhibiting hormone and aromatase. It is thought that the presence of the SRY protein is necessary and sufficient for the initiation of testis development. The finding of SRY transcripts in the brain of humans, rats, and a marsupial [49–53] suggests the possibility that a diffusible substance of local origin could establish a sexually dimorphic environment in the brain, preceding the stage of sexual differentiation induced by gonadal secretions. If the SRY gene is deleted from male mice (XY^-), the resulting phenotype is female but if a SRY transgene is inserted into an autosome in these mice (XY^- SRY), they develop testes and become fertile males. Mating them with normal females results in four different progenies: XY^- and XX^- mice with ovaries (females) and XX and XY mice with the SRY gene that have testis (males) [54]. Comparison of these four types of mice can separate chromosome effects from hormone effects on sexual differentiation of structure and function. Using this model, sex differences in spatial perception ability [55] and in the density of TH immunoreactive neurons in the mesencephalon [56] and vasopressinergic fibers in the lateral septum [57] have been shown to be entirely controlled by the genome. It is worth noting that the sexual difference in response to E2 disappears with development: in cultures from GD19 embryos both male and female neurons increase their axon length in response to E2 in the medium [35]. These results are reminiscent of observations made by Reisert et al. [21] on the neuritogenic effect of E2 on dopaminergic diencephalic cells. They also found a sexual difference that leveled off at later stages of development, although in their system the difference pointed to precocious maturation in females. It has been demonstrated that SRY is specifically expressed in the substantia nigra [58] of the adult male brain rodent in tyrosine-expressing neurons and that down regulation of SRY during development causes a decrease of tyrosine hydroxylase expression, which leads to motor deficits, independently of hormonal effects.

Several additional genes, such as SF-1, DAX-1, SOX-9, WT-1 and AMH, show a temporal and spatial pattern of expression that closely overlaps with SRY, and it is believed that these genes may be involved in the genetic cascade of sexual determination [59]. Pompolo and Harvey [60] have found no sexual difference in the expression of SOX-9 in the brain of rats and mice, but activation and presumptive participation in the generation of brain sexual differentiation for the other sex-associated genes remains to be studied. The orphan nuclear receptor steroidogenic factor 1 (SF-1, also called Ad4BP and NR5A1) plays an essential role in regulating not only endocrine differentiation and function [57] but also controls the genesis of sex specific nuclear groups in the ventromedial hypothalamus [61]. Using microarrays and RT-PCR, Dewing et al. [62] have identified 50 genes that exhibited male/female differences in expression, of which at least seven are expressed in the brain. It remains to be seen which, if any, of these genes participate in the control of brain differentiation.

Male-type brain circuitry results from exposure to androgens during a "critical period" of brain development, whereas the female brain develops in the absence of testicular secretions. The relative contribution of ovarian secretions to the full development of female brain organization and function is still a matter of debate [63]. Many of the masculinizing actions of androgen on the brain appear to be exclusively the result of aromatization of testosterone to estrogen [64–66]. More recently, data has been obtained indicating that during puberty there exists a second "critical period," coincident with an increase in circulating gonadal steroids. It has been suggested that structural changes in the brain (some of which are testosterone dependent) occurring at this time, are important for full brain differentiation leading to normal behavior and neuroendocrine physiology [67].

2 Neurotrophic Factors Are Important or Sexual Dimorphism

The mechanisms responsible for the differentiating effects of E2 remain largely undisclosed. Still, we have shown that on GD16 the axogenic effect of E2 depends on the interaction between neurons and glia from a target region, and that neurons from fetal male donors appear to mature earlier than neurons from females, a differentiated response that appears to take place prior to differential exposure to gonadal secretions. Numerous observations indicate that neurotrophic factors and E2 interact to bring about their effects on neuron survival, growth and differentiation [68–75]. E2 could increase the amount of trophic factors released to the medium by glial cells, the number and/or sensitivity of receptors or a combination of these mechanisms. The capacity of glial cells from a variety of brain regions to produce the appropriate growth factors has been documented [76].The fact that in our experiments both E2 and co-culture with target glia were necessary for the axogenic response to appear argues in favor of a triple interaction among E2, neurons equipped with the necessary receptors and the appropriate glia releasing to the medium some factor(s) which produce(s) the observed effect in all or most neurons present in the culture.

Findings from several laboratories indicate that structural and functional characteristics of astroglia are sexually dimorphic and can be altered by experimental modifications of gonadal steroid levels (for review see (77)). Our experiments using sexually segregated neuron-glia co-cultures posed an additional question, which is to what extent the effect of E2 was exerted on glial cells. Bearing this question in mind, we studied the effects of target and non-target glia-conditioned media (CM) on the E2-induced growth of neuronal processes of hypothalamic neurons obtained from sexually segregated fetal donors. Glia cells, mainly astrocytes, were cultured without addition of E2, to isolate the effects of the hormone on neurons themselves.

The growth and differentiation of hypothalamic neurons was differentially affected by E2 and the astroglia-CM, depending on the genetic sex of the neurons, and confirming the results observed in co-cultures with glia [39]. Hypothalamic neurons from males were particularly sensitive to the presence of astroglia-CM, since media conditioned by target (ventral mesencephalon) and non-target (striatum or cerebral cortex) glia modified the number of primary neurites and the growth of axons, whereas in female cells, only target-derived CM affected axonal growth. The lack of an E2 effect on female-derived neurons was not due to a generalized deficit in growth capacity in these cultures, as demonstrated by the fact that nerve growth factor and target derived-CM did induce additional axon growth.

The evidence obtained support the hypothesis that the developmental actions of estrogen on neurite growth and differentiation can result from modulatory interactions with endogenous growth factors, rather than from direct estrogenic action alone. In fact, the most obvious explanation of our results is that the axogenic effect of E2 depends on the participation of a soluble factor released by glia to the culture medium.

This interpretation agrees with evidence indicating that astrocytes or their precursors have a secretory capacity which can modulate the growth of neurons [78–80]; this modulatory effect is particularly evident for glia taken from target regions. There is growing consensus that the production of these factors by cells from target regions conforms to a gradient that serves as a guidance mechanism for axons to find their synaptic destination [81], apart from the fact that interactions between neurons, glia and their long processes orient extending growth cones [82]. The modulatory effect of glia cells on neuronal proliferation and differentiation is far from simple, and steroids influence this relationship [83]. Moreover, Sohrabji et al. [71] demonstrated that E2 could modulate the concentrations of $p75^{NGFR}$ (low-affinity nerve growth factor receptor) and TrkA (tyrosine kinase type A) receptors for nerve growth factor. Studies using gene localization and immunochemical staining have shown that mRNA for neurotrophins and neurotrophins themselves are localized in glial cells and that their amounts are modulated by E2 [84, 85]. Also, it has been shown that E2 can modulate both neurotrophins and their receptors in a region-specific manner [70, 86, 87] and this regulation can be reciprocal [73]. Indeed, Toran-Allerand et al. [68] have made a compelling case for the proposition that E2 may act synergistically with trophic factors to affect neuronal differentiation (including neurite growth), survival and functional control.

Apart from the sexual differences in response to E2 made evident in our sexually segregated cultures, it is important to bear in mind that gonadal steroid hormones have dynamic effects on neuritogenesis that are regionally specific. Whereas in hypothalamic and hippocampal neurons an axogenic effect has been observed [33, 35], in neurons from the amygdala the effect was exerted on dendritic arborization [34]. On the other hand, in the arcuate nucleus E2 causes a *decrease* of axosomatic synapses [88, 89] whereas cerebellar cells fail to respond to variations in gonadal hormone levels [90]. In males astroglia-CM from areas that are not targets for hypothalamic neurons (cortex or striatum), had an *inhibitory* effect on axon growth that was not demonstrated in female-derived neurons. Our results and those from other laboratories [91] underline the highly dynamic and mutually interactive nature of neuron-glia interactions. These interactions are age-dependent and region specific, meaning that generalizations about their influence and outcome should be made with caution, particularly if during experimental procedures separation of neurons and glia may have modified the growth of certain surviving cell lineages or cell subtypes.

Although the presence of ERα mRNA was demonstrated in the donor tissues as well as in the neurons cultured with or without E2, neither the Type III steroidal receptor blocker tamoxifen nor Type I antiestrogen ICI182780 – a nonsteroidal receptor antagonist – prevented the axogenic effects of the hormone [92]. The efficacy of these and other ER antagonists has been characterized on the basis of their ability to prevent the transcriptional activation of genes containing an estrogen response element (ERE). Therefore, it is possible that this "non genomic" effect of estradiol, escaped the blockade by acting upstream of any potential interaction with ERE [93] or cross-coupling with neuritotrophic signaling cascades such as mitogen-activated protein kinase (MAPK) or protein kinase A (PKA) [75, 94–96] in the cytoplasm.

The insertion of the receptor protein in the cell membrane could equally explain the inability of Tamoxifen and ICI182780 to block the effect of estradiol, a possibility we investigated using a membrane-impermeable E2-albumin construct (E2-BSA) [97]. Estradiol conjugated to a protein of high molecular weight preserved its axogenic capacity, suggesting the possibility of a membrane effect responsible for the action of E2. In this instance, the simultaneous activation of the estrogenic and trophic signaling pathways through membrane bound specific receptors parsimoniously explains the results. This interpretation coincides with the mechanism of steroid action characterized as non-genomic or membrane signaling [98], which requires the activation of a membrane-associated binding site, in the form of a different, albeit unknown, type of ER linked to an intracellular signaling pathway (mER).

Direct membrane effects of E2 have repeatedly been observed [99], but the identity of the receptor responsible for this effect remains elusive. Although the presence of ERα anchored to the membrane of cultured hippocampal neurons has been demonstrated [47], the existence of a hitherto unidentified isoform of ER has gained ground [100]. Singh et al. [75] postulate that a membrane bound ER may be part of a multimeric complex including also hsp90 and B-Raf. Evidence has been obtained that both cytosolic and membrane bound ERα as well as ERβ originate from a single transcript [101], suggesting that a post-translational modification of the amino acid sequence allows the insertion in the cell membrane [102]. The existence of a putative, novel, estradiol-sensitive receptor, designated ER-X placed within plasma membrane caveolae was proposed [103]. In addition, the E2 binding capacity of a seven-transmembrane receptor (7TMR) has been demonstrated in several cell lines [104, 105], including gonadotrophin releasing hormone neurons [106]. This receptor, also identified as GPR30 (one of the G protein coupled receptors), has been localized in the plasma membrane in vitro [107] as well as in vivo [108], and mediates estrogen-elicited rapid responses, including activation of a stimulatory G protein, activation of adenylyl cyclase and calcium mobilization. Although less compelling, evidence has been obtained for a host of alternative mERs (see (109, 110)).

The results described above suggested a dynamic interaction between some neurotrophic factor(s) present in the astroglia-CM and neurons cultured in the presence of E2. This interaction may be instrumented through the regulation of the respective receptors, thus modulating the effect of the specific ligands. In agreement with this possibility, we showed that in cultures of male derived neurons grown with E2 and CM from target glia, the amounts of TrkB had increased notably. Densitometric quantification showed that these cultures had more TrkB than cultures with CM alone or E2 alone.

On the contrary, in cultures of female derived neurons, the presence of CM alone induced maximum levels of TrkB, which were not further increased by E2.

Electrophoretic analyses of these extracts showed that they also contained a protein migrating in SDS-PAGE at 97 kD, characteristic for the insulin-like growth factor receptor (IGF-I R_{β}). This band reacted with the polyclonal antibody against the β subunit of this receptor. Western-blot analyses showed that these cells expressed the IGF-I_{β} subunit receptor only in cultures of male derived neurons grown with both CM and E2, whereas controls without E2 or without astroglia-CM or female-derived neurons in all conditions did not. The close correlation between an increase in receptors for the neurotrophic factor IGF-I and TrkB, but not TrkA and TrkC [26], supports the conclusion that E2 and the trophic factors released by astrocytes may be acting simultaneously and concurrently to induce axonal growth. The finding that in neurons from GD16 males neither astroglia-CM nor estradiol alone are able to promote the neuritogenic response supports this possibility.

To further investigate the interaction between E2 and the neurotrophin receptors, we used a specific antisense oligonucleotide (AS) to prevent the estradiol-induced increase of TrkB in cultures of hypothalamic neurons from males. The effect of estradiol was suppressed in cultures in which TrkB was down-regulated by the AS, showing decreased axonal elongation when compared with neurons treated with E2 without AS or with sense TrkB. It is important to note that in cells grown with sense TrkB, neurite length as well as number of minor processes was not different from cells grown with vehicle, and the axogenic effect of estradiol was preserved. The simplest interpretation of our results would be that the TrkB antisense prevented the axogenic response to estradiol, because increased TrkB levels normally observed in neurons treated with estradiol are necessary to accelerate axon growth. This explanation agrees with our previous results [26] indicating that estradiol and trophic factors released by astrocytes from a target region may act concurrently to induce axon elongation. In fact, it has been shown that estrogen and nerve growth factor may regulate receptor and/or ligand availability through reciprocal regulation at the level of gene transcription (for review see (68, 97, 111–113)). This convergence could explain why for the axogenic effect to occur, the concurrent action of both estrogen and soluble trophic factor(s) from target astroglia is necessary.

Even if E2 can activate a signaling cascade independent of any classic genomic regulation, changes in protein transcription at some level are indispensable to explain the hormonal effects. Among the signal cascades studied so far in neural cells, estrogen has been shown to stimulate the formation of cyclic adenosine monophosphate (cAMP) [114], the phosphorylation of the cAMP response element binding protein CREB [115], the formation of inositol triphosphate (IP3) [116], the transient increase of intracellular Ca^{2+} levels [117], and to activate the MAPK signaling pathway in a neuroblastoma cell line [94] and in cerebral cortical explants [75]. Also, in organotypic slice cultures of the developing mouse cerebral cortex, cells that respond to estradiol treatment by phosphorylation of extracellular regulated kinases ERK1 and ERK2 have been identified. The requirement for Hsp90 in estrogen-induced activation of ERK1 and ERK2 by MEK2 was demonstrated, providing an insight into the mechanisms by which estradiol may influence cytoplasmic and nuclear events in responsive neurons via the MAP kinase cascade [118].

The putative participation of the MAP kinase cascade was investigated in our model evaluating activation of ERK1/2. After 48 h in culture with E2 and astroglia-CM, a 3 h washout period was performed. The cultures were pulsed with estradiol 10 nM for 30 and 60 min or with BDNF 50 ng/ml for 60 min and harvested for Western blot analysis of ERK phosphorylation. An increase of phosphorylated ERK1/2 was observed in cultures treated with E2 for 1 h. As expected, BDNF also increased phosphorylated ERK1/2. Gorosito and Cambiasso [119] have shown that treatment with an intracellular Ca^{2+} chelator, a Ca^{2+}-dependent PKC inhibitor, or two specific inhibitors of ERK completely inhibited the E2-induced axogenesis. E2 and E2-BSA rapidly induced phosphorylation of ERK, which was blocked by the specific inhibitor of the ERK pathway UO126 but not by the ER antagonist ICI182,780. Decrease of intracellular free Ca^{2+} or disruption of PKC activation by Ro32-0432 attenuated ERK activation, indicating the confluence of signals in the MAPK pathway. Sub-cellular analysis of ERK demonstrated that the phospho-ERK signal is augmented in the nucleus after 15 min of E2 stimulation. It was also shown that the activation of the ERK pathway by E2 leads to an increase in the phosphorylation of CREB, suggesting the participation of this transcription factor in the neuritogenic effect of E2.

3 Conclusion

Differences between the sexes in brain structure and function are due to differences in chromosome complement of neurons, as well as differences in exposure to sexual steroids and trophic factors. Estrogen availability to the brain cells at particular times and places affects the growth, differentiation, and selective survival of neurons and glia ordaining the sex specific synaptic connections and their functional profile. This complex process requires the participation of specific receptors for estrogen as well as for neurotrophic factors, some of them from glial origin. Apart from the genomic effects resulting from the translocation of the liganded nuclear estrogen receptor, there are second messenger effects mediated through a membrane-associated ER. This pathway would follow the MAPK signaling cascade (ERK1/2) to activate genes controlling synthesis of neurotrophin receptors and other growth associated proteins. This signaling cascade may be one of the putative sites of cross talk between the estrogen and neurotrophic signals resulting in additive, potentiating or interdependently complementary effects on growth and differentiation. In view of the multiple and varied alternatives succinctly described above, it seems superfluous to state that the precise mechanisms involved from E2 activation of receptors to neurite growth are still not definitely established.

Acknowledgments Work from the authors' laboratory was supported by grants from Agencia Córdoba Ciencia, Agencia Nacional de Promoción Científica y Tecnológica, Consejo Nacional de Investigaciones Científicas y Técnicas (CONICET) of Argentina and the European Commission. H.F.C. and M.J.C are career members of CONICET.

References

1. Herlitz A, Nilsson LG, Backman L (1997) Gender differences in episodic memory. Mem Cognit 25:801–811
2. Zimmerberg B, Farley MJ (1993) Sex differences in anxiety behavior in rats: role of gonadal hormones. Physiol Behav 54:1119–1124
3. Kimura D (1996) Sex, sexual orientation and sex hormones influence human cognitive function. Curr Opin Neurobiol 6:259–263
4. Knoll JG, Wolfe CA, Tobet SA (2007) Estrogen modulates neuronal movements within the developing preotic area-anterior hypothalamus. Eur J Neurosci 26:1091–1099
5. Grabowski TJ, Damasio H, Eichhorn GR, Tranel D (2003) Effects of gender on blood flow correlates of naming concrete entities. Neuroimage 20:940–954
6. McCarthy MM (1994) Molecular aspects of sexual differentiation of the rodent brain. Psychoneuroendocrinology 19:415–427
7. Segovia S, Guillamon A, del Cerro MCR, Ortega E, Pérez-Laso C, Rodriguez-Zafra M et al (1999) The development of brain sex differences: a multisignaling process. Behav Brain Res 105:69–80
8. Arnold AP, Gorski RA (1984) Gonadal steroid induction of structural sex differences in the central nervous system. Annu Rev Neurosci 7:413–442
9. Breedlove SM, Cooke BM, Jordan CL (1999) The orthodox view of brain sexual differentiation. Brain Behav Evol 54:8–14
10. Becu Villalobos D, Gonzalez Iglesias A, Diaz Torga G, Hockl P, Libertun C (1997) Brain sexual differentiation and gonadotropins secretion in the rat. Cell Mol Neurobiol 17:699–715
11. Kow LM, Florea C, Schwanzel-Fukuda M, Devidze N, Kami KH, Lee A et al (2007) Development of a sexually differentiated behavior and its underlying CNS arousal functions. Curr Top Dev Biol 79:37–59
12. Davies W, Wilkinson LS (2006) It is not all hormones: alternative explanations for sexual differentiation of the brain. Brain Res 1126:36–45
13. Morris JA, Jordan CL, Breedlove SM (2004) Sexual differentiation of the vertebrate nervous system. Nat Neurosci 7:1034–1039
14. Arnold A, Burgoyne P (2004) Are XX and XY brain cells intrinsically different? Trends Endocrinol Metab 15:6–11
15. Wilson CA, Davies DC (2007) The control of sexual differentiation of the reproductive system and brain. Reproduction 133:331–359
16. Cahill L (2006) Why sex matters for neuroscience. Nat Rev Neurosci 7:477–484
17. Becker JB, Arnold AP, Berkley KJ, Blaustein JD, Eckel LA, Hampson E et al (2005) Strategies and methods for research on sex differences in brain and behavior. Endocrinology 146:1650–1673
18. Arnold AP (1996) Genetically triggered sexual differentiation of brain and behavior. Horm Behav 30:495–505
19. Xu J, Burgoyne PS, Arnold AP (2002) Sex differences in sex chromosome gene expression in mouse brain. Hum Mol Genet 11:1409–1419
20. Engele J, Pilgrim C, Reisert I (1989) Sexual differentiation of mesencephalic neurons in vitro – effects of sex and gonadal hormones. Int J Dev Neurosci 7:603–6203
21. Reisert I, Engele J, Pilgrim C (1989) Early sexual differentiation of diencephalic dopaminergic neurons of the rat in vitro. Cell Tissue Res 255:411–417
22. Beyer C, Pilgrim C, Reisert I (1991) Dopamine content and metabolism in mesencephalic and diencephalic cell cultures: sex differences and effects of sex steroids. J Neurosci 11:1325–1333
23. Picon R (1976) Testosterone secretion by foetal rat testes in vitro. J Endocrinol 71:231–238
24. Weisz J, Ward IL (1980) Plasma testosterone and progesterone titers of pregnant rats, their male and female fetuses, and neonatal offspring. Endocrinology 106:306–316

25. Carrer HF, Diaz H, Lorenzo A, Cáceres A (eds) (1990) Sexual differences and effects of estrogen on neuronal growth and differentiation. In: Garcia-Austt E, Macadar O, Trujillo-Cenóz O, Velluti R (eds) Fundamental Neurobiology. Universidad de la República, Montevideo, Uruguay, pp 27–38.
26. Cambiasso MJ, Colombo JA, Carrer HF (2000) Differential effect of oestradiol and astroglia-conditioned media on the growth of hypothalamic neurons from male and female rat brains. Eur J Neurosci 12:2291–2298
27. Beyer C, Wozniak A, Hutchison JB (1993) Sex-specific aromatization of testosterone in mouse hypothalamic neurons. Neuroendocrinology 58:673–681
28. Beyer C, Green SJ, Barker PJ, Huskisson NS, Hutchison JB (1994) Aromatase-immunoreactivity is localised specifically in neurones in the developing mouse hypothalamus and cortex. Brain Res 638:203–210
29. Hutchison JB, Beyer C (1994) Gender-specific brain formation of oestrogen in behavioural development. Psychoneuroendocrinology 19:529–541
30. Raab H, Pilgrim C, Reisert I (1995) Effects of sex and estrogen on tyrosine hydroxylase mRNA in cultured embryonic rat mesencephalon. Mol Brain Res 33:157–164
31. Tobet SA (2002) Genes controlling hypothalamic development and sexual differentiation. Eur J Neurosci 16:373–376
32. Simerly RB (2002) Wired for reproduction: organization and development of sexually dimorphic circuits in the mammalian forebrain. Annu Rev Neurosci 25:507–536
33. Blanco G, Diaz H, Carrer HF, Beauge L (1990) Differentiation of rat hippocampal neurons induced by estrogen in vitro – effects on neuritogenesis and Na, k-ATPase activity. J Neurosci Res 27:47–54
34. Lorenzo A, Diaz H, Carrer HF, Caceres A (1992) Amygdala neurons in vitro – neurite growth and effects of estradiol. J Neurosci Res 33:418–435
35. Diaz H, Lorenzo A, Carrer HF, Caceres A (1992) Time lapse study of neurite growth in hypothalamic dissociated neurons in culture – sex differences and estrogen effects. J Neurosci Res 33:266–281
36. Frankfurt M, Gould E, Woolley CS, McEwen BS (1990) Gonadal steroids modify dendritic spine density in ventromedial hypothalamic neurons – a golgi study in the adult rat. Neuroendocrinology 51:530–535
37. Lustig RH, Hua P, Wilson MC, Federoff HJ (1993) Ontogeny, sex dimorphism, and neonatal sex hormone determination of synapse-associated messenger RNAs in rat. Mol Brain Res 20:101–110
38. Shughrue PJ, Dorsa DM (1994) The ontogeny of GAP-43 (neuromodulin) mRNA in postnatal rat brain: evidence for a sex dimorphism. J Comp Neurol 340:174–184
39. Cambiasso MJ, Diaz H, Caceres A, Carrer HF (1995) Neuritogenic effect of estradiol on rat ventromedial hypothalamic neurons co-cultured with homotopic or heterotopic glia. J Neurosci Res 42:700–709
40. Gerlach JL, McEwen BS, Toran-Allerand CD, Friedman WJ (1983) Perinatal development of estrogen receptors in mouse brain assessed by radioautography, nuclear isolation and receptor assay. Brain Res 313:7–18
41. Brown TJ, MacLusky NJ, Toran-Allerand CD, Zielinski JE, Hochberg RB (1989) Characterization of 11B-methoxy 16a-iodoestradiol binding: neuronal localization of estrogen binding sites in the developing rat brain. Endocrinology 124:2074–2088
42. Funabashi T, Kleopoulos SP, Brooks PJ, Kimura F, Pfaff DW, Shinohara K et al (2000) Changes in estrogenic regulation of estrogen receptor alpha mRNA and progesterone receptor mRNA in the female rat hypothalamus during aging: an in situ hybridization study. Neurosci Res 38:85–92
43. Shughrue PJ, Merchenthaler I (2001) Distribution of estrogen receptor beta immunoreactivity in the rat central nervous system. J Comp Neurol 436:64–81
44. Scott CJ, Tilbrook AJ, Simmons DM, Rawson JA, Chu S, Fuller PJ et al (2000) The distribution of cells containing estrogen receptor-alpha (ER alpha) and ER beta messenger ribonucleic acid in the preoptic area and hypothalamus of the sheep: comparison of males and females. Endocrinology 141:2951–2962

45. Pasterkamp RJ, Yuri K, Visser DTM, Hayashi SJ, Kawata M (1996) The perinatal ontogeny of estrogen receptor- immunoreactivity in the developing male and female rat hypothalamus. Brain Res Dev Brain Res 91:300–303
46. Donahue JE, Stopa EG, Chorsky RL, King JC, Schipper HM, Tobet SA et al (2000) Cells containing immunoreactive estrogen receptor-alpha in the human basal forebrain. Brain Res 856:142–151
47. Clarke CH, Norfleet AM, Clarke MSF, Watson CS, Cunningham ML, Thomas ML (2000) Perimembrane localization of the estrogen receptor alpha protein in neuronal processes of cultured hippocampal neurons. Neuroendocrinology 71:34–42
48. Su JD, Qiu J, Zhong YP, Li XY, Wang JW, Chen YZ (2001) Expression of estrogen receptor (ER)-alpha and-beta immunoreactivity in hippocampal cell cultures with special attention to GABAergic neurons. J Neurosci Res 65:396–402
49. Harley VR, Goodfellow PN (1994) The biochemical role of SRY in sex determination. Mol Reprod Dev 39:184–193
50. Clepet C, Schafer AJ, Sinclair AH, Palmer MS, Lovell B, Goodfellow PN (1993) The human SRY transcript. Hum Mol Genet 2:2007–2012
51. Harry JL, Koopman P, Brennan FE, Graves JA, Renfree MB (1995) Widespread expression of the testis-determining gene SRY in a marsupial. Nat Genet 11:347–349
52. Lahr G, Maxson SC, Mayer A, Just W, Pilgrim C, Reisert I (1995) Transcription of the Y chromosomal gene, Sry, in adult mouse brain. Brain Res Mol Brain Res 33:179–182
53. Mayer A, Lahr G, Swaab DF, Pilgrim C, Reisert I (1998) The Y-chromosomal genes SRY and ZFY are transcribed in adult human brain. Neurogenetics 1:281–288
54. Arnold AP (2004) Sex chromosomes and brain gender. Nat Rev Neurosci 5:701–708
55. Stavnezer AJ, McDowell CS, Hyde LA, Bimonte HA, Balogh SA, Hoplight BJ et al (2000) Spatial ability of XY sex-reversed female mice. Behav Brain Res 112:135–143
56. Carruth LL, Reisert I, Arnold AP (2002) Sex chromosome genes directly affect brain sexual differentiation. Nat Neurosci 5:933–934
57. Parker KL, Rice DA, Lala DS, Ikeda Y, Luo X, Wong M et al (2002) Steroidogenic factor 1: an essential mediator of endocrine development. Recent Prog Horm Res 57:19–36
58. Dewing P, Chiang CW, Sinchak K, Sim H, Fernagut PO, Kelly S et al (2006) Direct regulation of adult brain function by the male-specific factor SRY. Curr Biol 16:415–420
59. O'Neill MJ, O'Neill RJ (1999) Whatever happened to SRY? Cell Mol Life Sci 56(11–12):883–893
60. Pompolo S, Harley VR (2001) Localisation of the SRY-related HMG box protein, SOX9, in rodent brain. Brain Res 906:143–148
61. Ikeda Y, Luo X, Abbud R, Nilson JH, Parker KL (1995) The nuclear receptor steroidogenic factor 1 is essential for the formation of the ventromedial hypothalamic nucleus. Mol Endocrinol 9:478–486
62. Dewing P, Shi T, Horvath S, Vilain E (2003) Sexually dimorphic gene expression in mouse brain precedes gonadal differentiation. Brain Res Mol Brain Res 118:82–90
63. Fitch RH, Denenberg VH (1998) A role for ovarian hormones in sexual differentiation of the brain. Behav Brain Sci 21(3):311
64. Lephart ED, Lund TD, Horvath TL (2001) Brain androgen and progesterone metabolizing enzymes: biosynthesis, distribution and function. Brain Res Rev 37:25–37
65. Hutchison JB, Wozniak A, Beyer C, Karolczak M, Hutchison RE (1999) Steroid metabolising enzymes in the determination of brain gender. J Steroid Biochem Mol Biol 69:85–96
66. MacLusky NJ, Walters MJ, Clark AS, Toran-Allerand CD (1994) Aromatase in the cerebral cortex, hippocampus, and mid-brain: ontogeny and developmental implications. Mol Cell Neurosci 5:691–698
67. Baron-Cohen S, Knickmeyer RC, Belmonte MK (2005) Sex differences in the brain: implications for explaining autism. Science 310:819–823
68. Toran-Allerand CD, Singh M, Setalo G (1999) Novel mechanisms of estrogen action in the brain: new players in an old story. Front Neuroendocrinol 20:97–121

69. Garcia Segura LM, Chowen JA, Naftolin F, TorresAleman I (1999) Steroid effects on brain plasticity – role of glial cells and trophic factors. In: Baulieu EE, Robel P, Schumacher M (eds) Neurosteroids. Humana, Totowa, NJ, pp 255–268
70. Miranda RC, Sohrabji F, Toran-Allerand CD (1993) Presumptive estrogen target neurons express messenger RNAs for both the neurotrophins and neurotrophin receptors – A. Mol Cell Neurosci 4:510–525
71. Sohrabji F, Miranda RC, Toran-Allerand CD (1994) Estrogen differentially regulates estrogen and nerve growth factor receptor mRNAs in adult sensory neurons. J Neurosci 14:459–471
72. Sohrabji F, Greene LA, Miranda RC, Toran-Allerand CD (1994) Reciprocal regulation of estrogen and NGF receptors by their ligands in PC12 cells. J Neurobiol 25:974–988
73. Miranda RC, Sohrabji F, Singh M, Toran-Allerand D (1996) Nerve growth factor (NGF) regulation of estrogen receptors in explant cultures of the developing forebrain. J Neurobiol 31:77–87
74. Sohrabji F, Miranda RCG, Toran-Allerand CD (1995) Identification of a putative estrogen response element in the gene encoding brain-derived neurotrophic factor. Proc Natl Acad Sci USA 92:11110–11114
75. Singh M, Setalo G, Guan XP, Warren M, Toran-Allerand CD (1999) Estrogen-induced activation of mitogen-activated protein kinase in cerebral cortical explants: convergence of estrogen and neurotrophin signaling pathways. J Neurosci 19:1179–1188
76. Labourdette G, Sensenbrenner M (1995) Growth factors and their receptors in the central nervous system. In: Kettenmann H, Ransom BR (eds) Neuroglia. Oxford University Press, New York, NY, pp 441–459
77. Garcia Segura LM, Chowen JA, Dueñas M, Parducz A, Naftolin F (1996) Gonadal steroids and astroglial plasticity. Cell Mol Neurobiol 16:225–237
78. Joosten EAJ, Gribnau AAM (1989) Astrocites and guidance of outgrowing corticospinal tracts in the rat: an immunocytochemical study using anti-vimentin and anti-glial fibrillary acidic protein. Neuroscience 31:439–452
79. Smith GM, Rutishauser U, Silver J, Miller RH (1990) Changing role of forebrain astrocytes during development, regenerative failure, and induced regeneration upon transplantation. J Comp Neurol 251:23–43
80. Engele J, Shubert D, Bohn MC (1991) Conditioned media derived from glial cell lines promote survival and differentiation of dopaminergic neurons in vitro: role of mesencephalic glia. J Neurosci Res 30:359–371
81. Jessell TM (1991) Cell migration and axon guidance. In: Kandel ER, Schwartz JH, Jessell TL (eds) Principles of neural science. Prentice-Hall, London, pp 908–928
82. Hidalgo A, Booth GE (2000) Glia dictate pioneer axon trajectories in the Drosophila embryonic CNS. Development 127:393–402
83. Chowen JA, Azcoitia I, Cardona Gomez GP, Garcia Segura LM (2000) Sex steroids and the brain: lessons from animal studies. J Pediatr Endocrinol Metab 13(8):1045–1066
84. Ma YJ, Junier MP, Costa ME, Ojeda SR (1992) Transforming growth factor-alpha gene expression in the hypothalamus is developmentally regulated and linked to sexual maturation. Neuron 9:657–670
85. Dueñas M, Luquin S, Chowen JA, Torres Aleman I, Naftolin F, Garcia Segura LM (1994) Gonadal hormone regulation of insulin-like growth factor-I- like immunoreactivity in hypothalamic astroglia of developing and adult rats. Neuroendocrinology 59:528–538
86. Jezierski MK, Sohrabji F (2000) Region- and peptide-specific regulation of the neurotrophins by estrogen. Mol Brain Res 85:77–84
87. Miranda RC, Sohrabji F, Toran-Allerand D (1994) Interactions of estrogen with the neurotrophins and their receptors during neural development. Horm Behav 28:367–375
88. Párducz A, Perez J, Garcia Segura LM (1993) Estradiol induces plasticity of gabaergic synapses in the hypothalamus. Neuroscience 53:395–401
89. Naftolin F, Garcia Segura LM, Keefe D, Leranth C, MacLusky NJ, Brawer JR (1990) Estrogen effects on the synaptology and neural membranes of the rat hypothalamic arcuate nucleus. Biol Reprod 42:21–28

90. Luquin S, Naftolin F, Garcia Segura LM (1993) Natural fluctuation and gonadal hormone regulation of astrocyte immunoreactivity in dentate gyrus. J Neurobiol 24:913–924
91. Garcia Segura LM, Dueñas M, Busiguina S, Naftolin F, Chowen JA (1995) Gonadal hormone regulation of neuronal-glial interactions in the developing neuroendocrine hypothalamus. J Steroid Biochem Mol Biol 53:293–298
92. Cambiasso MJ, Carrer HF (2001) Nongenomic mechanism mediates estradiol stimulation of axon growth in male rat hypothalamic neurons in vitro. J Neurosci Res 66:475–481
93. Sukovich DA, Mukherjee R, Benfield PA (1994) A novel, cell-type-specific mechanism for estrogen receptor-mediated gene activation in the absence of an estrogen-responsive element. Mol Cell Biol 14:7134–7143
94. Watters JJ, Campbell JS, Cunningham MJ, Krebs EG, Dorsa DM (1997) Rapid membrane effects of steroids in neuroblastoma cells: effects of estrogen on mitogen activated protein kinase signalling cascade and c-fos immediate early gene transcription. Endocrinology 138:4030–4033
95. Beyer C, Karolczak M (2000) Estrogenic stimulation of neurite growth in midbrain dopaminergic neurons depends on cAMP/protein kinase A signalling. J Neurosci Res 59:107–116
96. Nethrapalli IS, Singh M, Guan XP, Guo QF, Lubahn DB, Korach KS et al (2001) Estradiol (E2) elicits Src phosphorylation in the mouse neocortex: the initial event in E2 activation of the MAPK cascade? Endocrinology 142:5145–5148
97. Zakon HH (1998) The effects of steroid hormones on electrical activity of excitable cells. Trends Neurosci 21:202–207
98. Hall JM, Couse JF, Korach KS (2001) The multifaceted mechanisms of estradiol and estrogen receptor signaling. J Biol Chem 276:36869–36872
99. Belcher SM, Zsarnovszky A (2001) Estrogenic actions in the brain: estrogen, phytoestrogens, and rapid intracellular signaling mechanisms. J Pharmacol Exp Ther 299:408–414
100. Shughrue PJ, Askew GR, Dellovade TL, Merchenthaler I (2002) Estrogen-binding sites and their functional capacity in estrogen receptor double knockout mouse brain. Endocrinology 143:1643–1650
101. Razandi M, Pedram A, Greene GL, Levin ER (1999) Cell membrane and nuclear estrogen receptors (ERs) originate from a single transcript: studies of ER alpha and ER beta expressed in Chinese hamster ovary cells. Mol Endocrinol 13:307–319
102. Levin ER (1999) Cellular functions of the plasma membrane estrogen receptor. Trends Endocrinol Metab 10:374–377
103. Toran-Allerand CD (2000) Novel sites and mechanisms of oestrogen action in the brain. Novartis Found Symp 230:56–69
104. Thomas P, Pang Y, Filardo EJ, Dong J (2005) Identity of an estrogen membrane receptor coupled to a G protein in human breast cancer cells. Endocrinology 146:624–632
105. Revankar CM, Cimino DF, Sklar LA, Arterburn JB, Prossnitz ER (2005) A transmembrane intracellular estrogen receptor mediates rapid cell signaling. Science 307:1625–1630
106. Jacobi JS, Martin C, Nava G, Jeziorski MC, Clapp C, Martinez de la Escalera G (2007) 17-beta-estradiol directly regulates the expression of adrenergic receptors and kisspeptin/GPR54 system in GT1-7 GnRH neurons. Neuroendocrinology 86(4):260–269
107. Filardo E, Quinn J, Pang Y, Graeber C, Shaw S, Dong J et al (2007) Activation of the novel estrogen receptor G protein-coupled receptor 30 (GPR30) at the plasma membrane. Endocrinology 148:3236–3245
108. Funakoshi T, Yanai A, Shinoda K, Kawano MM, Mizukami Y (2006) G protein-coupled receptor 30 is an estrogen receptor in the plasma membrane. Biochem Biophys Res Commun 346:904–910
109. Toran-Allerand CD (2004) Minireview: a plethora of estrogen receptors in the brain: where will it end? Endocrinology 145:1069–1074
110. Hasbi A, O'Dowd BF, George SR (2005) A G protein-coupled receptor for estrogen: the end of the search? Mol Interv 5:158–161
111. Kato S, Masuhiro Y, Watanabe M, Kobayashi Y, Takeyama KI, Endoh H et al (2000) Molecular mechanism of a cross-talk between oestrogen and growth factor signalling pathways. Genes Cells 5:593–601

112. Garcia Segura LM, Naftolin F, Hutchison JB, Azcoitia I, Chowen JA (1999) Role of astroglia in estrogen regulation of synaptic plasticity and brain repair. J Neurobiol 40:574–584
113. Jordan CL (1999) Glia as mediators of steroid hormone action on the nervous system: an overview. J Neurobiol 40:434–445
114. Gu Q, Moss RL (1996) 17 beta-estradiol potentiates kainate-induced currents via activation of the cAMP cascade. J Neurosci 16:3620–3629
115. Zhou Y, Watters JJ, Dorsa DM (1996) Estrogen rapidly induces the phosphorylation of the cAMP response element binding protein in rat brain. Endocrinology 137:2163–2166
116. Favit A, Fiore L, Nicoletti F, Canonico PL (1991) Estrogen modulates stimulation of inositol phospholipid hydrolysis by norepinephrine in rat brain slices. Brain Res 555:65–69
117. Beyer C, Raab H (1998) Nongenomic effects of oestrogen: embryonic mouse midbrain neurones respond with a rapid release of calcium from intracellular stores. Eur J Neurosci 10:255–262
118. Setalo G, Singh M, Guan XP, Toran-Allerand CD (2002) Estradiol-induced phosphorylation of ERK1/2 in explants of the mouse cerebral cortex: the roles of heat shock protein 90 (Hsp90) and MEK2. J Neurobiol 50:1–12
119. Gorosito SV, Cambiasso MJ (2008) Axogenic effect of estrogen in male rat hypothalamic neurons involves Ca(2+), protein kinase C, and extracellular signal-regulated kinase signaling. J Neurosci Res 86(1):145–157
120. Kimchi T, Xu J, Dulac C (2007) A functional circuit underlying male sexual behaviour in the female mouse brain. Nature 448:1009–1014

Life at the Interface Between a Dynamic Environment and a Fixed Genome: Epigenetic Programming of Stress Responses by Maternal Behavior

Ian C.G. Weaver

Abstract In evolutionary developmental biology, a common theme across all life forms is the ability to vary phenotype in response to environmental conditions, a process termed phenotypic plasticity. In humans and non-human primates, the nature of mother-infant interactions in early postnatal life has a profound role in mediating variation in offspring phenotype, including emotional and cognitive development, which is endured through life. Similarly, maternal behavior in rodents is associated with long-term programming of individual differences in behavioral and hypothalamic-pituitary-adrenal (HPA) responses to stress in the offspring. One critical question is how is such "environmental programming" established and sustained in the offspring?

In this chapter, we discuss a novel mechanism to explain how maternal licking/grooming behavior in the rat can alter the hippocampal glucocorticoid receptor (GR) expression in the offspring, which concomitantly alters the HPA axis and the stress responsiveness of these animals. Both in vivo and in vitro studies show that maternal behavior increases GR expression in the offspring via increased hippocampal serotonergic tone accompanied by increased histone acetylase transferase activity, histone acetylation and DNA demethylation mediated by the transcription factor NGFI-A.

In summary, this research demonstrates that an epigenetic state of a gene can be established through early in life experience, and is potentially reversible in adulthood. We suggest that epigenetic modifications of specific genomic regions in response to variations in environmental conditions might serve as a major source of variation in biological and behavioral phenotypes. Accordingly, we consider the relevance of our findings in relation to other known epigenetic mechanisms in neuroscience and discuss the clinical implications of these observations for future research.

Keywords Maternal care • Hippocampus • Glucocorticoid receptor • Chromatin • DNA methylation

I.C.G. Weaver
Developmental and Stem Cell Biology Program, Hospital for Sick Children, Toronto Medical Discovery East Tower, Medical & Related Sciences (MaRS) Centre, 101 College Street, Toronto, Ontario, Canada, M5G 1X8
e-mail: ian.weaver@utoronto.ca

D. Janigro (ed.), *Mammalian Brain Development*, Contemporary Neuroscience,
DOI 10.1007/978-1-60761-287-2_2,

1 Early in Life Environmental Influences on Brain Development and Behavioral Responses

In mammals, early postnatal life represents a period when external stimuli can influence the development of individual differences in cognitive and emotional development in the offspring [1–3]. The relation between early environment quality and such long-term neurodevelopmental programming appears, in part, to be mediated by the closeness or degree of positive attachment in parent-infant bonding and parental investment during early life [4]. Indeed, parental effects are apparent across all forms of life, ranging from mammals to reptiles to insects to birds and even plant species [5–8]. Variations in parental investment in mammals, such as nutrient supply provided by the parent and behavioral interactions, affect the development of (1) defensive responses and (2) reproductive strategies in the progeny [9–11].

There is substantial evidence for the importance of maternal influences on subsequent performance both in terms of neurophysiological functions and behavioral responses of the offspring [9–11]. Importantly, the nature of the mother-offspring interaction influences the expression of genes that control the neural circuitry of behavioral responses in the offspring through life. Exposure of the mother to environmental adversity alters the nature of mother-offspring interaction, which, in turn, influences the development of defensive responses to threat as well as reproductive strategies in the offspring [7, 12]. Such developmental phenotypic plasticity is thought to have evolved to harmonize the development of specific biological systems to enhance the match between phenotype and environmental demand (stressors) [12]. Therefore, the variations in mother-offspring interactions are, in effect, the forecast of environmental quality (e.g., nutrient availability, predation, infection, population density) for the offspring.

Although enhancing the offspring's capacity to respond according to environmental cues early in life can have immediate adaptive value, such as improved resilience to stress, and favored by natural selection; the cost is that these adaptations serve as predictors of ill health in later life [13]. Epidemiological studies [14–20] and primate models [21] suggest that the developmental onset of chronic illness (including obesity, metabolic disorders and heart disease as well as affective disorders and drug abuse) in adolescence and adult life emerge as a function of: (1) altered responses to environmental stressors in conjunction with an increased level of prevailing adversity (i.e., allostatic overload) [22]; and, (2) when there is a mismatch between the environmental demand perceived by the mother and that which is encountered by the offspring later in life [14, 15, 20]. Although this may appear paradoxical that phenotypic plasticity increases a risk to health, it is important to consider that in the context of natural selection health is relevant only to the extent that it influences reproductive success, not longevity.

Thus, stress resilience (i.e., the ability to cope) plays an integral role in mediating the effects of changing environmental demand on health outcome. Consistent with this idea, stress-diathesis models suggest that maternal influences on the development of neuroendocrine systems that underlie hypothalamic-pituitary-adrenal (HPA) and behavioral responses to stress mediate the relation between early in life environment and health in the adult offspring [13, 23–27]. The enduring effects of maternal care on stress reactivity raises questions concerning the identity of the relevant genomic targets, the nature of the gene-environment interactions and their relation to phenotype and, perhaps most fundamental, the mechanism that fixes the transient exposure to environmental adversity in early life to a stable change in gene expression programming long after the exposure has gone.

The aim of this chapter is to present current progress in our molecular understanding of how early environment influences brain development in a manner that persists through life. The conserved influence of parental care on developmental programming between humans and other altricial species suggests that animal models could be used to further understand the molecular mechanisms involved. We present recent evidence mostly obtained in rodents that suggests that maternal care in the first week of postnatal life establishes diverse and stable phenotypes in the offspring through the epigenetic modification of genes expressed in the brain that regulate neuroendocrine and behavioral responses throughout life.

In particular, we will focus on maternal programming of the rat hippocampal glucocorticoid receptor (GR) and HPA stress responses in the offspring. We will then discuss the increasing evidence from animal studies that supports the mechanism for prenatal and early postnatal inheritance through environmentally induced epigenetic changes. We will conclude the chapter by exploring how these studies in animal models could be extended to determine the importance of environmental epigenetics in human disease susceptibility and discuss the clinical research implications of such studies that might ultimately lead to new diagnostic, prognostic and therapeutic strategies.

2 Maternal Care in the Rat and HPA and Behavioral Responses to Stress in Adulthood

Though the importance of maternal care in promoting the health, survival and cognitive development of offspring has been demonstrated across the mammalian kingdom [28–30], the most extensively researched animal model of mother-infant interactions is the rat. Maternal behavior in the rat during the first weeks of life provides the nurturing environment that is crucial for survival of the young and allows the dam to meet the physiological demands of prolonged care of the offspring. Mother-pup contact primarily occurs within the context of a nest-bout, which begins when the mother approaches the litter, licks and grooms her pups, and nurses while occasionally licking and grooming the pups [31]. Observational studies [for review see reference [32]] provide evidence for stable individual differences in two forms of maternal behavior, licking/grooming (LG) and arched-back nursing (ABN) posture, over the first week of lactation [29, 31, 33–35]. Exposure to different levels of maternal LG-ABN during the postnatal period is associated with HPA blunting and changes in forebrain GR expression levels that persist into adulthood [29, 34].

The magnitude of the HPA response to stress is a function of the neural stimulation of hypothalamic corticotropin-releasing factor (CRF) release, which activates the

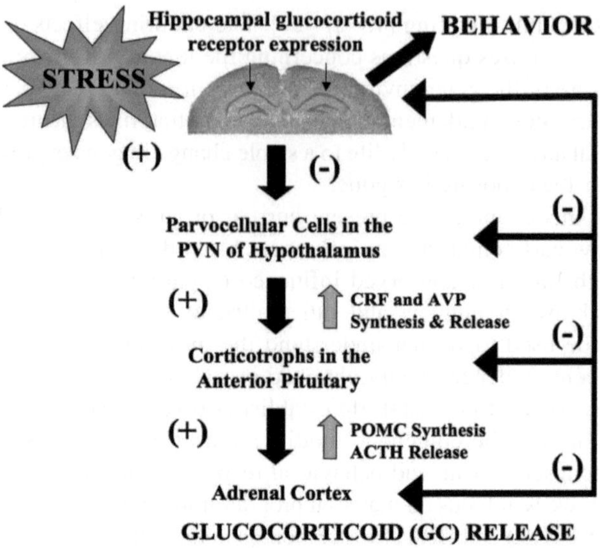

Fig. 1 Hypothalamic-pituitary-adrenal (HPA) axis. In response to a physical or psychological stressor, parvocellular cells in the paraventricular nucleus (PVN) of the hypothalamus produce hormone corticotrophin-releasing factor (CRF) and arginine-vasopressin (AVP). The hypothalamus co-secretes CRF/AVP and CRF targets specific receptors on corticotrophs in the anterior pituitary, resulting in the stimulation of the synthesis of the adrenocorticotroph (ACTH) precursor peptide proopiomelanocortin (POMC) and the secretion of ACTH. ACTH potently induces the adrenal cortex to secrete glucocorticoids (cortisol in humans, corticosterone in rodents), which regulate many metabolic activities, in addition to stress resistance. In a classical endocrine feedback manner, these steroids inhibit the synthesis and secretion of CRF and AVP within the hypothalamus and POMC-derived peptides in the pituitary, to prevent steroid-induced cellular damage. Increased stress levels stimulate GR activation in the hippocampus, which mediates the negative feedback system and behavioral adaptation

pituitary-adrenal system, as well as modulatory influences, such as glucocorticoid (GC) negative feedback in the hippocampus that inhibits CRF synthesis and release, and thus dampens HPA responses to stress [36] (Fig. 1). GCs bind to GR to mediate the peripheral stress response and also feed back to the central nervous system (CNS) to modulate further activation of the HPA response [37].

In adulthood, the offspring of mothers that exhibit increased levels of pup licking/grooming and arched-back nursing (i.e., High LG-ABN mothers) over the first week of life show increased hippocampal GR expression and enhanced GC feedback sensitivity by comparison to adult animals reared by Low LG-ABN mothers [29, 34]. Adult offspring of High LG-ABN mothers show decreased hypothalamic CRF expression and more modest HPA responses to stress [29]. Eliminating the difference in hippocampal GR levels abolishes the effects of early experience on HPA responses to stress in adulthood [38], suggesting that the difference in hippocampal GR expression serves as a mechanism for the effect of early experience on the development of individual differences in HPA responses to stress [39].

In support of this, mice with forebrain-specific disruption of the GR gene and loss of hippocampal GR expression show impairments to HPA axis regulation and an increase of anxiety-related behavior [40]. Accordingly, the adult offspring of High LG-ABN mothers are behaviorally less fearful under conditions of stress than are animals reared by Low LG-ABN dams [33]. Using a cross-fostering design we [34, 41] showed that pups born to Low LG-ABN mothers but reared by High LG-ABN dams were comparable in terms of hippocampal GR expression and HPA responses to the biological offspring of High LG-ABN mothers (and vice versa).

Together, these findings suggest that the developing rodent forebrain is exquisitely sensitive to tactile stimulation provided by the mother during the first week of life and that different frequencies of LG-ABN provided during this period program neurodevelopment with long-lasting consequences on hippocampal GR function and HPA responses to stress. However, since the rearing mother and not the biological mother defines the behavioral and neuroendocrine responses to stress, these studies support a non-genomic mechanism rather than a genetic (germ-line inheritance) mechanism for the long-term programming [29, 34, 39, 42].

3 Molecular Mechanisms for Maternal Effects on HPA Responses to Stress

To-date, most work has focused on the molecular mechanisms by which manipulations (e.g., handling) that alter the level of maternal LG-ABN induce and increase hippocampal GR expression, with more recent studies assessing whether similar changes are also observed in response to natural variations in maternal LG-ABN [for more detail see [43]]. Results from in vitro and in vivo studies [44, 45] suggest that maternal effects on hippocampal GR expression are mediated by increases in serotonin (5-HT) turnover and in hippocampal expression of the transcription factor nerve-growth-factor-inducible-protein-A [NGFI-A, also termed AT225, d2, EGR-1, ETR-103, G0S30, KROX-24, TIS8, ZENK, ZFP-6, ZIF-268 and ZNF225 [46–55]] (Fig. 2). In cultured rat hippocampal neurons, 5-HT increases expression of both NGFI-A and GR and the effect of 5-HT is blocked by concurrent treatment with an antisense oligonucleotide directed to NGFI-A mRNA [56].

The 5′ non-coding variable exon 1 region of the rat hippocampal GR gene contains multiple alternate sequences, including the exon 1_7 sequence, which appears to be a brain-specific promoter [57] (Fig. 3). In adult rats, hippocampal expression of GR mRNA splice variants containing exon 1_7 is increased by maternal LG-ABN behavior. The GR exon 1_7 promoter contains an NGFI-A-binding consensus sequence (Fig. 3). Maternal LG-ABN increases hippocampal NGFI-A expression and chromatin immunoprecipitation (ChIP) assays with hippocampal samples from post-natal-day (PND) 6 pups reveal that NGFI-A binding to the GR exon 1_7 promoter is dramatically increased in the offspring of High LG-ABN mothers compared with the offspring of Low LG-ABN mothers [56]. Co-transfection of an NGFI-A expression vector and a GR exon 1_7 promoter-luciferase construct into human embryonic

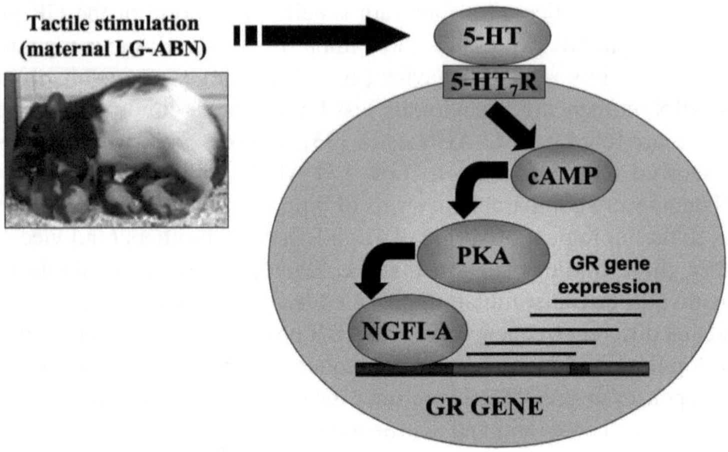

Fig. 2 Maternal programming of hippocampal glucocorticoid receptor (GR) gene expression. Maternal licking/grooming and arched-back nursing (LG-ABN) of the offspring increases hippocampal serotonin (5-HT) turnover and activation of a 5-HT_7 receptor, which is positively coupled to cyclic adenosine 3′,5′ monophosphate (cAMP). Increased cAMP activity results in activation of protein kinase-A (PKA) and cAMP response element-binding protein (CREB). Subsequent phosphorylated-CREB (pCREB) activity drives expression of the transcription factor NGFI-A, which, when bound to the GR promoter (shaded *dark grey*) may drive the observed increase in hippocampal GR expression in the offspring

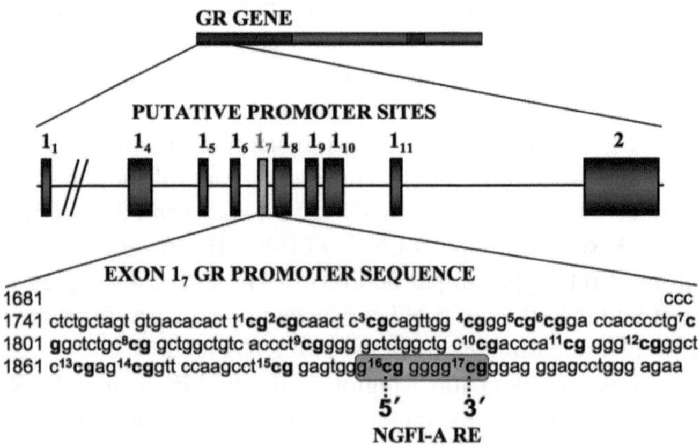

Fig. 3 Sequence map of the glucocorticoid receptor (GR) exon 1_7 promoter including the 17 CpG dinucleotides (bold) and the NGFI-A response element (RE, encircled) [adapted from [57]]

kidney (HEK) 293 cells increases luciferase activity, reflecting NGFI-A-induced activation of transcription through the GR exon 1_7 promoter [56].

These findings suggest that NGFI-A might increase GR expression in hippocampal neurons and provide a mechanism for the effect of maternal care over the first week of life. However, although there are striking differences in NGFI-A expression in the neonatal offspring of High LG-ABN and Low LG-ABN mothers, there is no effect of maternal care on hippocampal NGFI-A expression in the adult animals [44, 56]. This raises the question of whether the increased NGFI-A protein and exon 1_7 promoter sequence interaction in the PND 6 pups of High LG-ABN mothers result in a modification of the GR promoter that alters NGFI-A binding and maintains the maternal effect into adulthood.

4 Maternal Care Epigenetically Programs HPA Stress Responses in the Offspring

In mammals, gene expression profiles are controlled by the epigenome, which is comprised of DNA methylation [58] and chromatin structure, including histones and their modifications [59]. Chromatin structure and DNA methylation patterns are unique to each type of cell. The specific combination of epigenetic modifications determines the conformation of the chromatin fiber into which the DNA and histones are packaged, and can thereby regulate the transcriptional potential of the underlying genes. Active chromatin is generally associated with hypomethylated DNA, whereas inactive chromatin is associated with hypermethylated DNA [60]. Importantly, epigenetic changes are inherited mitotically in somatic cells, providing a potential mechanism by which environmental effects on the epigenome can have stable long-term effects on gene expression in the developing brain.

In mammalian cells, DNA methylation occurs at the carbon-5 position of cytosine residues of 5′-cytosine-phosphodiester-guanine (CpG)-3′ dinucleotide sequences [61]. We initially found evidence for greater methylation across the GR exon 1_7 promoter sequence in the hippocampus in adult offspring of Low LG-ABN mothers than in those of High LG-ABN mothers [41]. These findings suggested maternal effects on DNA methylation patterns of the offspring. High and Low LG-ABN mothers differ in the frequency of pup LG-ABN only during the first postnatal week. We examined the methylation status of individual CpG dinucleotides in the GR exon 1_7 promoter sequence during this period using sodium bisulfite mapping, focusing on the NGFI-A response element [41]. On embryonic day (ED) 20 (i.e., 24–48 h before birth), the entire exon 1_7 region of the GR promoter is completely unmethylated. On PND 1 both the 5′ and 3′ CpG dinucleotides within the NGFI-A response element are heavily methylated in both the High and Low LG-ABN offspring, suggesting a comparable postnatal wave of de novo methylation. The differences in methylation emerge between PND 1 and 6, which is precisely when differences in the maternal behavior of High and Low LG-ABN dams are apparent. By PND 6, the 5′ CpG dinucleotide of the NGFI-A response element is demethylated in the High LG-ABN but not in the Low LG-ABN group, and the maternal effect persists through to adulthood. These findings suggest that the group difference in DNA methylation involves a process of demethylation. The 5′ CpG dinucleotide appears to be always methylated in the offspring Low

LG-ABN mothers, and rarely methylated in those of High LG-ABN dams. Cross-fostering reverses the differences in the methylation of the 5′ CpG dinucleotide and suggests a direct relationship between maternal behavior and changes in DNA methylation of the GR exon 1_7 promoter [41].

The difference in methylation within the 5′ CpG dinucleotide of the NGFI-A response element suggests an alteration of NGFI-A binding. In vitro binding of increasing concentrations of purified recombinant NGFI-A protein to the NGFI-A response element was examined using an electrophoretic mobility shift assay (EMSA) with ^{32}P-labeled synthetic oligonucleotide sequences bearing the NGFI-A-binding site differentially methylated at the 5′ or 3′ CpG dinucleotides [56]. The results indicate that methylation of the cytosine within the 5′ CpG dinucleotide in the NGFI-A response element on the GR exon 1_7 promoter inhibits NGFI-A binding, a finding consistent with the role for cytosine methylation [41]. Differences in NGFI-A expression between the offspring of High and Low LG-ABN mothers in early postnatal life are no longer apparent in adulthood [56]. Instead, it appeared that the methylation of the NGFI-A response element interfered with NGFI-A binding to the GR exon 1_7 promoter in the offspring of Low LG-ABN mothers; ChIP assays indicated that binding of NGFI-A protein to the hippocampal GR exon 1_7 promoter in adult pups is threefold lower in offspring of Low LG-ABN mothers than in offspring of High LG-ABN dams.

NGFI-A activates the genes by recruiting cyclic adenosine 3,5 monophosphate (cAMP) response element binding protein-binding protein (CBP), a histone acetyl transferase (HAT), to their promoter regions. HATs catalyze the acetylation of selected positively charged amino acids (e.g., lysine and arginine) on the protruding histone tails, most commonly histone (H–) 3 or H4. In particular, histone acetylation of lysine (K–) 9 residues on the ε-NH_2 tail of H3 neutralizes the positive charge of the histone tail and decreases its affinity to negatively charged DNA and generates a more open DNA conformation resulting in breathing of the 146 bp DNA wrapped 1.65 turns around the octamer of histone residues [62, 63]. Transcription factors and the transcription apparatus then have access to the DNA, and expression of the corresponding genes is facilitated. Thus, H3K9 acetylation is a marker of active gene transcription.

ChIP assays using the same tissue samples used in the NGFI-A studies and an antibody against the acetylated form of H3 show increased association of acetylated H3K9 with the GR exon 1_7 promoter in offspring of the High LG-ABN mothers. Because histone acetylation is associated with active states of gene expression, these findings are consistent with the idea of increased NGFI-A binding to the GR exon 1_7 promoter, recruiting HATS resulting in increased transcriptional activation. Moreover, transient transfection studies show that DNA methylation reduces the ability of NGFI-A to activate the GR exon 1_7 promoter in HEK 293 cells [56].

Taken together these findings suggest that an epigenetic positional modification at a single cytosine residue within the NGFI-A response element alters NGFI-A binding and might explain the sustained effect of maternal care on hippocampal GR expression and HPA responses to stress.

5 Epigenetic Programming by Maternal Care is Reversible in the Adult Animal

Unlike the genetic code which is static, epigenetic programs are dynamic and potentially reversible even later in life and therefore amenable to therapeutic intervention [64]. Since neurons do not divide, epigenetic modifications in the brain could be reversible only if the DNA demethylation and methylation machinery remains present in the cell. Szyf and colleagues have previously proposed that the DNA methylation pattern is maintained through life by a dynamic equilibrium of methylation and demethylation [64]. We therefore tested whether the methylation state of the GR exon 1_7 promoter could be modified by pharmacological manipulations that disrupt the equilibrium that exists between the epigenetic machinery.

Histone deacetylase (HDAC) inhibitors can trigger active, replication-independent DNA demethylation [65]. Although the mechanism of this activation is unclear it has been proposed that increased acetylation results in increased accessibility of a gene to the demethylation machinery [65] and that transcription is required for active DNA demethylation, in methylation silenced genes [66]. We therefore tested whether this approach could reverse the epigenetic state of the GR exon 1_7 promoter in the adult offspring of Low LG-ABN mothers. Central infusion of adult rats with the HDAC inhibitor trichostatin A (TSA) significantly increased H3K9 acetylation, cytosine demethylation and NGFI-A binding to the GR exon 1_7 promoter in the offspring of Low LG-ABN mothers to levels comparable with those observed in the offspring of High LG-ABN dams [41].

Subsequent expression profiling of TSA-treated rats reveal specific effects of TSA on the hippocampal transcriptome [67]. The enhanced NGFI-A binding to the exon 1_7 promoter is associated with increased hippocampal GR expression in the offspring of Low LG-ABN mothers, to levels that are indistinguishable from those of the High LG-ABN offspring [41]. More importantly, TSA infusion also eliminates the effect of maternal care on HPA responses to acute stress [41]. Because there is considerable H3K9 acetylation and NGFI-A binding to the exon 1_7 promoter in the vehicle-treated offspring of High LG-ABN mothers, TSA is presumably less likely to effect NGFI-A binding and GR expression in these animals [41].

These results suggest causal relationships between maternal care, histone acetylation, DNA methylation of the exon 1_7 promoter, GR expression and HPA responses to stress. This was the first illustration of reversal of early in life behavioral programming by pharmacological modulation of the epigenome during adulthood and suggests that the enzyme(s) required for DNA demethylation are involved.

We then argued that if DNA methylating and demethylating enzymes dynamically maintain the DNA methylation pattern in adult neurons, then it should also be possible to reverse the demethylated state of the GR exon 1_7 promoter. The methyl donor *S*-adenosyl methionine (SAM) inhibits the demethylation reaction either by stimulating DNA methylation enzymes [68] or by inhibiting demethylases [64]. Systemic injection of the methyl-donor L-methionine was previously shown to increase the levels of SAM and DNA methylation [69]. Chronic central infusion of adult offspring of High or Low LG-ABN mothers with methionine increased DNA methylation within the NGFI-A response element and reduced NGFI-A binding to the exon 1_7 promoter selectively in the offspring of High LG-ABN dams, eliminating group differences in both hippocampal GR expression and HPA responses to stress [70].

These studies illustrate that maternal epigenetic programming early in life can be reversed later in life, suggesting that DNA methylation patterns are dynamic and potentially reversible even in adult neurons, which presumably contain the machinery required for de novo DNA demethylation or methylation and mediate CpG methylation rheostasis in the mature mammalian brain.

6 Mechanisms Leading from Maternal Care to Chromatin Plasticity

Our working hypothesis is that maternal care stimulates a signaling pathway, which activates certain transcription factors directing the epigenetic machinery (chromatin and DNA modifying enzymes) to specific targets within the genome. Maternal LG in early life elicits a thyroid hormone-dependent increase in 5-HT activity at 5-HT_7 receptors, and the subsequent activation of cAMP and cAMP-dependent protein kinase A (PKA) [44, 45]. This is accompanied by increased hippocampal expression of the transcription factor NGFI-A. However, evidence for a causal relation between NGFI-A binding and epigenetic reprogramming of the GR exon 1_7 promoter expression comes from using two tissue culture systems.

In HEK 293 cells transiently transfected with a methylated GR exon 1_7 promoter-luciferase vector, NGFI-A overexpression ultimately leads to demethylation of the 5′ CpG dinucleotide within the NGFI-A response element on the exon 1_7 promoter construct [56]. Because the non-integrated plasmid does not bear an origin of replication and does not replicate in HEK 293 cells, the assay measures the effects of NGFI-A on active, replication independent demethylation [65]. To demonstrate that DNA demethylation requires direct contact between NGFI-A and its cognate response element on the GR exon 1_7 promoter, we performed site directed mutagenesis of the 5′ and 3′ CpG dinucleotides of the NGFI-A response element. Mutation of the 3′ CpG dinucleotide blocks the ability of NGFI-A to bind the exon 1_7 promoter, and abolishes the ability of NGFI-A to demethylate and activate the GR promoter [56]. These data demonstrate that physical interaction between NGFI-A and the methylated GR exon 1_7 promoter is required for NGFI-A to target demethylation. Sodium bisulfite mapping confirmed that GR exon 1_7 promoter sequences that interact with NGFI-A are demethylated.

Next, using hippocampal cell cultures, we showed that treatment with either 5-HT or 8-bromo-cAMP results in increased GR expression and hypomethylation of the 5′ CpG dinucleotide within the NGFI-A response element on the GR exon 1_7 promoter, with no effect at the 3′ CpG dinucleotide, again in the absence of cell replication [56]. Strikingly, 5-HT treatment also increased expression of CBP and CBP binding to the exon 1_7 promoter, which is increased in the neonatal offspring of High LG-ABN

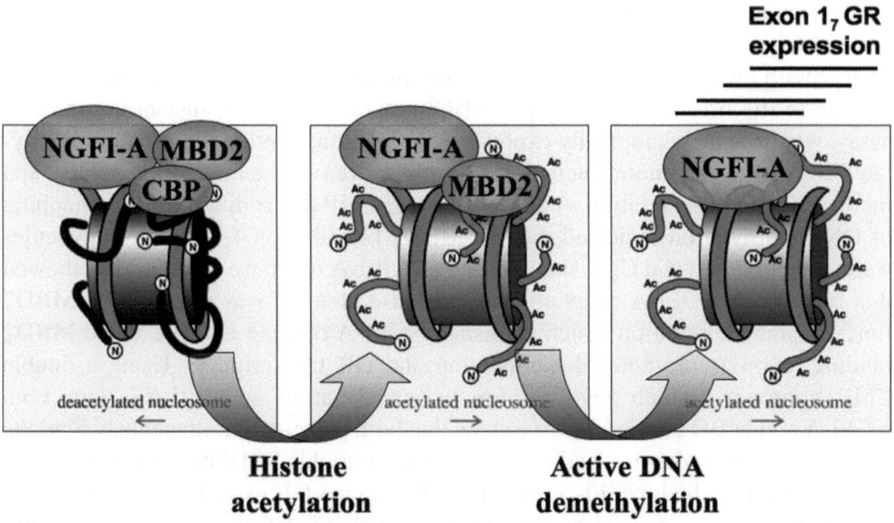

Fig. 4 Regulation of glucocorticoid receptor (GR) gene expression. NGFI-A recruits a CREB binding protein (CBP) that increases acetylation and accessibility to the DNA demethylase MBD2 and stable GR promoter activation (see text for details)

mothers [56]. We hypothesize that increased NGFI-A levels enhance the probability of low affinity NGFI-A binding to the methylated exon 1_7 promoter, recruitment of HATs such as CBP, histone acetylation and accessibility of demethylating enzyme(s) to the GR promoter, resulting in DNA demethylation (Fig. 4).

NGFI-A can actively target methylated-DNA-binding proteins to genomic targets [71] and we previously showed that a pharmacological increase in acetylation in vivo using TSA resulted in demethylation of the GR exon 1_7 promoter in the hippocampus [41]. Thus, we suggest that the role NGFI-A plays in regulation of GR expression is bimodal. During early development, high physiological levels of NGFI-A induced by maternal care interact with the methylated GR exon 1_7 promoter and trigger demethylation of the sequence, whereas later in life physiological levels of NGFI-A discriminate between the methylated and unmethylated exon 1_7 promoters and selectively activate the unmethylated sequences. Therefore, the different methylation states of the exon 1_7 promoters from the offspring of High and Low LG-ABN result in different levels of hippocampal GR expression. This suggests that the neonatal brain of altricial species such as the rat is not an immature version of the adult brain but is uniquely designed to optimize epigenetic programming by the mother.

The remaining question concerns the identity of the DNA demethylating enzyme(s). Though methyl-CpG binding domain protein (MBD)-2 silences methylated genes [72], the protein has also been reported to trigger active DNA demethylation and induce gene expression in mammalian cells [73–76]. Although this assignment of demethylase activity to MBD2 was contested [72], depletion of MBD2 results in hypermethylation of unmethylated genes in metastatic cancer [77, 78]. Consequently, as an enzyme, MBD2 might carry the potential for bidirectional activity and is highly expressed in the hippocampus of the adult rat brain [79, 80].

We examined whether MBD2 mediated active demethylation of the GR exon 1_7 promoter in the PND 6 offspring. Our results indicate increased association of MBD2 with the GR exon 1_7 promoter in offspring of High LG-ABN dams in comparison to the offspring of Low LG-ABN mothers. Using a transient transfection assay, we show that ectopically expressed MBD2 transcriptionally activates methylated GR exon 1_7 promoter-luciferase plasmid, increases the interaction of CBP and increases histone acetylation. A combination of ChIP and sodium bisulfite mapping of DNA methylation indicated that the MBD2 bound exon 1_7 promoter molecules were demethylated at a CpG site within the NGFI-A response element. We showed that binding of NGFI-A to its cognate response element was required for MBD2 function since a mutation which abolished NGFI-A binding also prevented MBD2 binding, exon 1_7 promoter demethylation and GR transcription. Using a double ChIP approach, which involves immunoprecipitation sequentially with both NGFI-A and MBD2 antibodies, we show that both proteins simultaneously bind the same exon 1_7 promoter molecule (Weaver et al., unpublished data). While the exact mechanisms by which NGFI-A recruits MBD2 to the GR exon 1_7 promoter remain unknown, our data is consistent with the idea that NGFI-A facilitates the accessibility of the sequence to MBD2 leading to target-specific demethylation (Fig. 4). Indeed, this method of recruitment of transcription factors to a gene might be a general mechanism for facilitating site-specific DNA demethylation.

Interestingly, although MBD2 deficient mice are viable, the postpartum MBD2 null mothers are significantly slower at retrieving pups to their nests in comparison to the wild type dams [81]. This suggests that MBD2 might also have an important role in the behavioral transmission of epigenetic modifications across generations by the mother, which we have previously shown is associated with cytosine methylation of the estrogen receptor (ER)-α 1b promoter and ER-α expression in the medial preoptic (MPOA) area of female offspring [82].

7 Persistent Alterations in DNA Methylation and Phenotypic Plasticity

A defining question concerns the mechanism by which environmental effects, including social interactions such as parenting, in early life are biologically embedded and sustained into adulthood. We propose that epigenetic changes could be an intermediate process that imprints dynamic environmental experiences on the fixed genome, resulting in stable alterations in phenotype; i.e., by a process of experience-dependent chromatin plasticity. In support of this theory, increasing evidence from animal studies indicates that prenatal and early postnatal environmental factors, such as xenobiotic chemicals [83–85], heavy metal toxins [86], reproductive factors [87, 88], low-dose radiation [89] and most recently, chemical byproducts from the manufacture of polycarbonate plastics used in baby bottles and dental composites [90] can result in altered epigenetic programming through DNA methylation and subsequent changes in the risk of developing disease. Epigenetic reprogramming in adulthood can thus occur following exposure to environmental contaminants [91]. Together with our evidence for a behavioral mode of programming, these studies suggest that not only the magnitude but also the timing of exposure to environmental factors plays an important role in mediating expression of phenotype in mammals.

We examined the effects of TSA and methionine pharmacological manipulation on brain gene expression and physiological and behavioral responses to stress within the context of tactile stimulation early in life. However, maternal nurturing also influences growth and development of the offspring to life-long health through providing nutrition. It has long been proposed [92] that poor fetal and infant growth and the subsequent development of disease in later life emerge from nutritional programming early in life [93]. Because diet-derived methyl donors and cofactors are necessary for the synthesis of SAM, which serves as the donor of methyl groups for DNA methylation, environmental factors that alter early nutrition and/or SAM synthesis can potentially influence adult metabolism via persistent alterations in DNA methylation [94–98]. For example, fetal deficiency in the essential amino acid methionine used in our studies and dietary folate (found in fresh fruits and vegetables), as well as genetic variants in methylenetetrahydrofolate reductase (MTHFR, a regulatory enzyme in folate metabolism), have been shown to alter intracellular SAM levels [99, 100] and linked to the risk of many serious health conditions [101].

Importantly, availability of dietary methionine and folate alter the parent-of-origin effects on the methylation status of imprinted genes [94, 102] which mediate many of the actions of growth hormone on somatic growth and tissue maintenance [103]. Our microarray analysis [67] revealed that TSA and methionine treatment altered the expression of several growth-regulatory genes including insulin-like growth factor (IGF)-2, which is an imprinted gene and thus perhaps subject to multiple phases of epigenetic programming (reprogramming), suggesting that the effects of maternal care on fetal and postnatal growth and development can be manipulated by diet and nutrition.

The isoflavinoid genistein (a phytoestrogen found in soy) is present at high levels in infant formula and has previously been shown to reverse stable epigenetic programming in adult mice [104]. Using the viable yellow agouti (A^{vy}) mouse model, recent studies have demonstrated that maternal supplementation with methyl donors [97] or genistein [97] during gestation can produce offspring with increased methylation of transposable elements in or upstream of the agouti gene, respectively, decreased agouti gene expression, brown fur and protection from obesity and cancer in adulthood. Together with our findings, these studies raise the possibility that early-life nutrition has the potential to influence epigenetic programming in the brain, not only during early development but also in adult life, and thereby modulate health throughout the life course.

This may have important therapeutic implications, since a common characteristic of ageing is a time-dependent decline in responsiveness or adaptation to the environment, a form of loss of phenotypic plasticity [105, 106]. Indeed, a decline in cognitive plasticity (learning and memory) is commonly observed in aged rodents [107] and in most human neurodegenerative diseases [108] and is largely attributed to neuron loss within the hippocampus. Basal HPA activity increases with age, resulting in elevated adrenocorticotroph (ACTH) and corticosterone plasma levels. Chronic treatment (3 months) of young adult rats with high levels of GCs or prolonged exposure to stress produces a similar hippocampal neuropathology to that observed in the aged animals [109, 110]. In the classic model of hypoxic ischemia, where all blood flow to the brain is transiently blocked, the presence of high GC concentrations increase the amount of hippocampal damage or accelerate the onset of damage [110–112]. These findings, taken together with the maternal effects on hippocampal GR sensitivity, suggests a relation between maternal care, GC sensitivity and neuron survival in the developing rat hippocampus.

The pro-apoptotic B-cell lymphoma (BCL)-2 associated protein-X (BAX) is an etiologic factor in the pruning of hippocampal neurons [113] and the development of neural sex differences [114], providing a molecular bases of hormone-regulated cell death in the forebrain. We reported that, relative to offspring of High LG-ABN mothers, the offspring of Low LG-ABN mothers have decreased hippocampal expression of GR and neurotrophic factors [115] and increased expression of BAX and a greater number of terminal deoxynucleotidyl transferase-mediated biotin-dUTP nick end-labeling (TUNEL)-positive neurons [116]. These observations lead us to believe that the differences in BAX expression may be involved in differential aging of the hippocampus. Chronic psychological stress is associated with and accelerates cellular aging [117], possibly through epigenetic regulation of telomere length [118]. We examined the effect of maternal behavior on DNA methylation patterns across the entire hippocampal genome of the adult offspring [119]. The levels of cytosine methylation varied between the different hippocampal regions and mRNA levels of DNA methyltransferases exhibited similar regional specificity and were correlated with global DNA methylation levels, suggesting a causal relation among methyltransferase expression, epigenomic state and neuron survival in the adult offspring.

Moreover, a shift in the balance between DNA methylating and demethylating machinery might account for inter-individual differences in cellular and behavioral plasticity through life. This idea is supported by studies in humans showing greater variance of total DNA methylation and H3K9 acetylation in older monozygotic twins than in younger twins [120] and might explain the frequent discordance of diseases such as bipolar disorder between monozygotic twins [121]. Our microarray analysis using rat hippocampal tissue revealed that TSA and methionine treatment altered the expression of several genes, similar to those observed in human neuropsychiatric disorders including age-related dementia [67]. For example, we found naturally occurring difference in the γ-amino butyric acid (GABA) synthesis enzyme L-glutamate-1-carboxylase (glutamate decarboxylase, GAD) isoform GAD67 and reelin expression within the adult offspring as a function of early-life experience, which are vulnerability factors in humans for the etiology and pathophysiology of schizophrenia [122]. Reelin activates activity-regulated cytoskeletal protein (Arc) protein synthesis [123] and therefore might be involved in regulation of synaptic plasticity through life.

TSA treatment enhanced reelin expression within the adult offspring of low LG-ABN mothers to levels comparable with the adult offspring of High LG-ABN mothers [67]. TSA can enhance induction of long-term potentiation (LTP) through increased histone acetylation [124]. Furthermore, in transgenic mice that express CBP in which the HAT activity is eliminated, the impaired stabilization of short-term memory into long-term memory is rescued by administration of TSA in adult animals [125], suggesting that the epigenome is a critical component of memory consolidation.

Interestingly, we have previously shown that the adult offspring of High LG-ABN mothers show increased hippocampal CBP expression [41] and synaptic density, enhanced LTP and perform better in paradigms that test hippocampal-dependent memory and learning [115, 126, 127], compared with the adult offspring of Low LG-ABN mothers. However, we need to examine the stability of these findings. For example, do the levels of BAX remain the same in aged animals and whether TSA or methionine treatment reverses the maternal effects on neuron survival, synaptic plasticity and cognitive behavior in the High and Low LG-ABN offspring. Nevertheless, epigenetic changes in hippocampal neurons would provide an efficient mechanism of neuroplasticity without requiring a dramatic change in hippocampal morphology.

Genes involved in the epigenetic regulation of basic developmental processes are highly conserved among vertebrates. A recent remarkable study using methyl CpG binding protein 2 (MECP2) null mice, showed that some of the delayed onset of behavioral and LTP phenotypes associated with the childhood neurodevelopmental disorder Rett syndrome (a severe autism spectrum disorder) could be rectified by restoration of the MECP2 gene, which presumably resumes its canonical role in epigenetic regulation of brain function [128]. Therefore, the rat model of natural variations in maternal care could potentially be useful in the study of gene-environment-therapeutic interactions of genomic regions that are know to be involved in disease syndromes.

8 Modeling Variations in Maternal Care-Mediated Epigenetic Programming in Humans

A number of findings have implicated forebrain GR in the regulation of the HPA axis and the development of affective disorders and other sequelae [129–131]. Thus, a fundamental question concerns whether a similar mechanism as described here operates in generating inter-individual differences in human behavior and health outcome.

Interestingly, the long term effects of tactile stimulation on individual differences in cognitive development and stress responses in rodents is consistent with work in humans [132] and non-human primates [133] examining the effects of neonatal touch in mother-infant interactions. Classic work in macaque monkeys showed that tactile contact with an inanimate surrogate mother influences emotional development and fearful responses in the offspring [133]. In humans, premature babies provided with skin-to-skin (kangaroo) care (i.e., wrapping the baby on the mother's chest for 1 h everyday) during the first 2 weeks of life performed better on cognitive and motor tasks at 6 months follow-up in comparison to the control group [132]. Likewise, results from recent epidemiological studies have shown that randomized control interventions aimed at improving parental care demonstrate improved behavioral outcomes and cognitive performance in the offspring that persist for years [134–136]. This begs the question of whether there are comparable epigenetic labile regions to the GR exon 1_7 promoter in the human genome. Alignment of splice sites reveals that the distally located GR exon 1F promoter in man shows high homology to the GR exon 1_7 promoter in the rat, and contains an NGFI-A-binding consensus sequence [137].

However, a critical question concerns the method of how to determine whether modulation of the epigenome during early human brain development also plays a causal role in the formation of human behavior. Because of the important ethical, social, and legal implications arising from any attempt to harvest living cells from the CNS, an alternative approach would be to examine the effects of maternal intervention programs on epigenetic alterations in the periphery tissues. Possible candidates are the immunological cytokines secreted by leukocytes in the blood circulation. The cognate receptors of these hormone-like signaling molecules are expressed throughout the CNS, including the hippocampus [138, 139]. Cross-talk between the immune system and the nervous system in response to adversity is thought to influence individual differences in HPA and behavioral responses to stress [138, 139]. Cytokines have previously been shown to be regulated by DNA methylation [140–143]. Notably, expression of the potent pro-inflammatory cytokine interleukin (IL)-1β is induced by the DNA methylation inhibitor 5-azacytidine [144], suggesting that common disease phenotypes may emerge in part through environmentally-triggered epigenetic changes in gene expression. ChIP, ChIP-on-chip (ChIP-chip) and ChIP-based sequencing (ChIP-Seq) could be used to explore gene-specific patterns of histone modifications on a genomic scale, which could be further tested by incorporating an assessment of the epigenome into population epidemiological studies [145]. These studies may open up the possibility that methylation patterns of certain genomic regions in white blood cells may be a useful marker for monitoring health in prospective longitudinal studies, such as a barometer for examining the effects of intervention programs aimed at improving child health.

9 Concluding Remarks

Sensory input during early development plays an important role in brain development with long-term consequences on brain functioning in adulthood. The studies presented in this chapter provide support for the effect of maternal behavior on hippocampal development and HPA responses to stress in the offspring, and that these effects are biologically embedded throughout life by epigenomic programming, yet reprogramming can take place at several points throughout the life-span in response to changes in environmental conditions. However, these findings are restricted to the study of a single promoter in only one gene in one brain region; at this time, these results might be best thought of as a proof of principle. The degree to which this mechanism generalizes to environmental programming in other systems remains to be determined, and may well reveal an alternative mechanism for programming of the epigenome.

Nevertheless, the epigenome appears to be an important target of environmental modification and central to phenotypic plasticity and mammalian brain development. The challenge will be to find comparable epigenetic labile regions. In our studies, the neonatal programming is mediated, in part, by the transcription factor NGFI-A. The function of such immediate early genes provides a mechanism by which the environment may interact with the genome to shape neuronal function. Examination of these genes from a developmental perspective, especially of genes whose products are involved in active DNA demethylation, will provide information on how the environment shapes development and influences phenotypic variation.

Together, this work adds to the knowledge of how complex behavior interacts with the epigenome and, in particular, illustrates the dynamic nature of gene-environment interactions throughout life. Accordingly, we are only beginning to understand the mechanisms whereby early-life experience suppresses or enhances expression of biological defense systems that respond to environmental adversity.

Acknowledgments These studies were supported by a grant from the Canadian Institutes for Health Research (CIHR) to Michael J. Meaney and Moshe Szyf and from the National Cancer Institute of Canada (NCIC) to Moshe Szyf who supervised I.C.G.W's doctoral studies. I.C.G.W. is currently supported by a Postdoctoral Fellowship Award from the CIHR.

References

1. Ainsworth AJ, Brown JE (1991) Detection of in ovo derived idiotypic antibodies. I. A model for maternal-neonatal idiotype network studies. J Immunol 147:910–914
2. Suomi SJ (1997) Early determinants of behaviour: evidence from primate studies. Br Med Bull 53:170–184
3. Ammerman RT (1991) The role of the child in physical abuse: a reappraisal. Violence Vict 6:87–101
4. Canetti L, Bachar E, Galili-Weisstub E, De-Nour AK, Shalev AY (1997) Parental bonding and mental health in adolescence. Adolescence 32:381–394
5. Stearns SC (1991) The evolution of life histories. Oxford University Press, New York, NY
6. Qvarnstrom A, Price TD (2001) Maternal effects, paternal effects and sexual selection. Trends Ecol Evol 16:95–100
7. Rossiter MC (1999) Maternal effects as adaptations. Oxford University Press, London
8. Mousseau TA, Fox CW (1998) The adaptive significance of maternal effects. Trends Ecol Evol 13:403–407
9. Ressler RH, Anderson LT (1973) Avoidance conditioning in mice as a function of their mothers' exposure to shock. Dev Psychobiol 6:105–111
10. Denenberg VH (1999) Commentary: is maternal stimulation the mediator of the handling effect in infancy? Dev Psychobiol 34:1–3
11. Ottinger DR, Tanabe G (1969) Maternal food restriction: effects on offspring behavior and development. Dev Psychobiol 2:7–9
12. Agrawal AA (2001) Phenotypic plasticity in the interactions and evolution of species. Science 294:321–326
13. Repetti RL, Taylor SE, Seeman TE (2002) Risky families: family social environments and the mental and physical health of offspring. Psychol Bull 128:330–366
14. Lissau I, Sorensen TI (1994) Parental neglect during childhood and increased risk of obesity in young adulthood. Lancet 343:324–327

15. Felitti VJ, Anda RF, Nordenberg D et al (1998) Relationship of childhood abuse and household dysfunction to many of the leading causes of death in adults. The Adverse Childhood Experiences (ACE) Study. Am J Prev Med 14:245–258
16. Manuck SB, Bleil ME, Petersen KL et al (2005) The socio-economic status of communities predicts variation in brain serotonergic responsivity. Psychol Med 35:519–528
17. Thapar A, Langley K, Fowler T et al (2005) Catechol O-methyltransferase gene variant and birth weight predict early-onset antisocial behavior in children with attention-deficit/hyperactivity disorder. Arch Gen Psychiatry 62:1275–1278
18. Caspi A, Moffitt TE, Cannon M et al (2005) Moderation of the effect of adolescent-onset cannabis use on adult psychosis by a functional polymorphism in the catechol-O-methyltransferase gene: longitudinal evidence of a gene X environment interaction. Biol Psychiatry 57:1117–1127
19. Caspi A, Sugden K, Moffitt TE et al (2003) Influence of life stress on depression: moderation by a polymorphism in the 5-HTT gene. Science 301:386–389
20. McCauley J, Kern DE, Kolodner K et al (1997) Clinical characteristics of women with a history of childhood abuse: unhealed wounds. JAMA 277:1362–1368
21. Bennett AJ, Lesch KP, Heils A et al (2002) Early experience and serotonin transporter gene variation interact to influence primate CNS function. Mol Psychiatry 7:118–122
22. McEwen BS (2004) Protection and damage from acute and chronic stress: allostasis and allostatic overload and relevance to the pathophysiology of psychiatric disorders. Ann N Y Acad Sci 1032:1–7
23. Heim C, Newport DJ, Heit S et al (2000) Pituitary-adrenal and autonomic responses to stress in women after sexual and physical abuse in childhood. JAMA 284:592–597
24. Francis DD, Meaney MJ (1999) Maternal care and the development of stress responses. Curr Opin Neurobiol 9:128–134
25. Sroufe LA (1997) Psychopathology as an outcome of development. Dev Psychopathol 9:251–268
26. Seckl JR, Meaney MJ (1993) Early life events and later development of ischaemic heart disease. Lancet 342:1236
27. Nemeroff CB (1996) The corticotropin-releasing factor (CRF) hypothesis of depression: new findings and new directions. Mol Psychiatry 1:336–342
28. Gale CR, Martyn CN (1996) Breastfeeding, dummy use, and adult intelligence. Lancet 347: 1072–1075
29. Liu D, Diorio J, Tannenbaum B et al (1997) Maternal care, hippocampal glucocorticoid receptors, and hypothalamic- pituitary-adrenal responses to stress [see comments]. Science 277:1659–1662
30. Kraemer GW (1997) Psychobiology of early social attachment in rhesus monkeys. Clinical implications. Ann N Y Acad Sci 807:401–418
31. Stern JM (1997) Offspring-induced nurturance: animal-human parallels. Dev Psychobiol 31:19–37
32. Champagne FA, Francis DD, Mar A, Meaney MJ (2003) Variations in maternal care in the rat as a mediating influence for the effects of environment on development. Physiol Behav 79:359–371
33. Caldji C, Tannenbaum B, Sharma S, Francis D, Plotsky PM, Meaney MJ (1998) Maternal care during infancy regulates the development of neural systems mediating the expression of fearfulness in the rat. Proc Natl Acad Sci USA 95:5335–5340
34. Francis D, Diorio J, Liu D, Meaney MJ (1999) Nongenomic transmission across generations of maternal behavior and stress responses in the rat. Science 286:1155–1158
35. Myers MM, Brunelli SA, Shair HN, Squire JM, Hofer MA (1989) Relationships between maternal behavior of SHR and WKY dams and adult blood pressures of cross-fostered F1 pups. Dev Psychobiol 22:55–67
36. De Kloet ER, Vreugdenhil E, Oitzl MS, Joels M (1998) Brain corticosteroid receptor balance in health and disease. Endocr Rev 19:269–301
37. Herman JP, Figueiredo H, Mueller NK et al (2003) Central mechanisms of stress integration: hierarchical circuitry controlling hypothalamo-pituitary-adrenocortical responsiveness. Front Neuroendocrinol 24:151–180

38. Meaney MJ, Aitken DH, Viau V, Sharma S, Sarrieau A (1989) Neonatal handling alters adrenocortical negative feedback sensitivity and hippocampal type II glucocorticoid receptor binding in the rat. Neuroendocrinology 50:597–604
39. Meaney MJ (2001) Maternal care, gene expression, and the transmission of individual differences in stress reactivity across generations. Annu Rev Neurosci 24:1161–1192
40. Boyle MP, Kolber BJ, Vogt SK, Wozniak DF, Muglia LJ (2006) Forebrain glucocorticoid receptors modulate anxiety-associated locomotor activation and adrenal responsiveness. J Neurosci 26:1971–1978
41. Weaver IC, Cervoni N, Champagne FA et al (2004) Epigenetic programming by maternal behavior. Nat Neurosci 7:847–854
42. Fleming AS, O'Day DH, Kraemer GW (1999) Neurobiology of mother-infant interactions: experience and central nervous system plasticity across development and generations. Neurosci Biobehav Rev 23:673–685
43. Weaver IC, Diorio J, Seckl JR, Szyf M, Meaney MJ (2004) Early environmental regulation of hippocampal glucocorticoid receptor gene expression: characterization of intracellular mediators and potential genomic target sites. Ann N Y Acad Sci 1024:182–212
44. Meaney MJ, Diorio J, Francis D et al (2000) Postnatal handling increases the expression of cAMP-inducible transcription factors in the rat hippocampus: the effects of thyroid hormones and serotonin. J Neurosci 20:3926–3935
45. Meaney MJ, Aitken DH, Sapolsky RM (1987) Thyroid hormones influence the development of hippocampal glucocorticoid receptors in the rat: a mechanism for the effects of postnatal handling on the development of the adrenocortical stress response. Neuroendocrinology 45:278–283
46. Milbrandt J (1987) A nerve growth factor-induced gene encodes a possible transcriptional regulatory factor. Science 238:797–799
47. Milbrandt J (1988) Nerve growth factor induces a gene homologous to the glucocorticoid receptor gene. Neuron 1:183–188
48. Montminy MR, Sevarino KA, Wagner JA, Mandel G, Goodman RH (1986) Identification of a cyclic-AMP-responsive element within the rat somatostatin gene. Proc Natl Acad Sci USA 83:6682–6686
49. Tsukada T, Fink JS, Mandel G, Goodman RH (1987) Identification of a region in the human vasoactive intestinal polypeptide gene responsible for regulation by cyclic AMP. J Biol Chem 262:8743–8747
50. Hyman SE, Comb M, Lin YS, Pearlberg J, Green MR, Goodman HM (1988) A common trans-acting factor is involved in transcriptional regulation of neurotransmitter genes by cyclic AMP. Mol Cell Biol 8:4225–4233
51. Sheng M, Greenberg ME (1990) The regulation and function of c-fos and other immediate early genes in the nervous system. Neuron 4:477–485
52. Self DW, Nestler EJ (1995) Molecular mechanisms of drug reinforcement and addiction. Annu Rev Neurosci 18:463–495
53. Habener JF, Meyer TE, Yun Y, Waeber G, Hoeffler JP (1990) Characterization of a cAMP-regulated enhancer-binding protein. Metabolism 39:13–16
54. Mitchell JB, Rowe W, Boksa P, Meaney MJ (1990) Serotonin regulates type II corticosteroid receptor binding in hippocampal cell cultures. J Neurosci 10:1745–1752
55. Smythe JW, Rowe WB, Meaney MJ (1994) Neonatal handling alters serotonin (5-HT) turnover and 5-HT2 receptor binding in selected brain regions: relationship to the handling effect on glucocorticoid receptor expression. Brain Res Dev Brain Res 80:183–189
56. Weaver IC, D'Alessio AC, Brown SE et al (2007) The transcription factor nerve growth factor-inducible protein a mediates epigenetic programming: altering epigenetic marks by immediate-early genes. J Neurosci 27:1756–1768
57. McCormick JA, Lyons V, Jacobson MD et al (2000) 5′-heterogeneity of glucocorticoid receptor messenger RNA is tissue specific: differential regulation of variant transcripts by early-life events. Mol Endocrinol 14:506–517
58. Razin A (1998) CpG methylation, chromatin structure and gene silencing-a three-way connection. EMBO J 17:4905–4908

59. Kadonaga JT (1998) Eukaryotic transcription: an interlaced network of transcription factors and chromatin-modifying machines. Cell 92:307–313
60. Holliday R, Pugh JE (1975) DNA modification mechanisms and gene activity during development. Science 187:226–232
61. Razin A, Szyf M (1984) DNA methylation patterns. Formation and function. Biochim Biophys Acta 782:331–342
62. Hong L, Schroth GP, Matthews HR, Yau P, Bradbury EM (1993) Studies of the DNA binding properties of histone H4 amino terminus. Thermal denaturation studies reveal that acetylation markedly reduces the binding constant of the H4 "tail" to DNA. J Biol Chem 268:305–314
63. Sealy L, Chalkley R (1978) DNA associated with hyperacetylated histone is preferentially digested by DNase I. Nucleic Acids Res 5:1863–1876
64. Szyf M (2001) Towards a pharmacology of DNA methylation. Trends Pharmacol Sci 22:350–354
65. Cervoni N, Szyf M (2001) Demethylase activity is directed by histone acetylation. J Biol Chem 276:40778–40787
66. D'Alessio AC, Weaver IC, Szyf M (2007) Acetylation induced transcription is required for active DNA demethylation, in methylation silenced genes. Mol Cell Biol 27(21):7462–7474
67. Weaver IC, Meaney MJ, Szyf M (2006) Maternal care effects on the hippocampal transcriptome and anxiety-mediated behaviors in the offspring that are reversible in adulthood. Proc Natl Acad Sci USA 103:3480–3485
68. Pascale RM, Simile MM, Satta G et al (1991) Comparative effects of L-methionine, S-adenosyl-L-methionine and 5′-methylthioadenosine on the growth of preneoplastic lesions and DNA methylation in rat liver during the early stages of hepatocarcinogenesis. Anticancer Res 11:1617–1624
69. Tremolizzo L, Carboni G, Ruzicka WB et al (2002) An epigenetic mouse model for molecular and behavioral neuropathologies related to schizophrenia vulnerability. Proc Natl Acad Sci USA 99:17095–17100
70. Weaver IC, Champagne FA, Brown SE et al (2005) Reversal of maternal programming of stress responses in adult offspring through methyl supplementation: altering epigenetic marking later in life. J Neurosci 25:11045–11054
71. Carvin CD, Parr RD, Kladde MP (2003) Site-selective in vivo targeting of cytosine-5 DNA methylation by zinc-finger proteins. Nucleic Acids Res 31:6493–6501
72. Ng HH, Zhang Y, Hendrich B et al (1999) MBD2 is a transcriptional repressor belonging to the MeCP1 histone deacetylase complex. Nat Genet 23:58–61
73. Bhattacharya SK, Ramchandani S, Cervoni N, Szyf M (1999) A mammalian protein with specific demethylase activity for mCpG DNA. Nature 397:579–583
74. Detich N, Bovenzi V, Szyf M (2003) Valproate induces replication-independent active DNA demethylation. J Biol Chem 278:27586–27592
75. Detich N, Hamm S, Just G, Knox JD, Szyf M (2003) The methyl donor S-adenosylmethionine inhibits active demethylation of DNA: a candidate novel mechanism for the pharmacological effects of S-adenosylmethionine. J Biol Chem 278:20812–20820
76. Detich N, Theberge J, Szyf M (2002) Promoter-specific activation and demethylation by MBD2/demethylase. J Biol Chem 277:35791–35794
77. Pakneshan P, Tetu B, Rabbani SA (2004) Demethylation of urokinase promoter as a prognostic marker in patients with breast carcinoma. Clin Cancer Res 10:3035–3041
78. Shukeir N, Pakneshan P, Chen G, Szyf M, Rabbani SA (2006) Alteration of the methylation status of tumor-promoting genes decreases prostate cancer cell invasiveness and tumorigenesis in vitro and in vivo. Cancer Res 66:9202–9210
79. Francis J, Jung B, Zhang G et al (2002) Kindling induces the mRNA expression of methyl DNA-binding factors in the adult rat hippocampus. Neuroscience 113:79–87
80. Jung BP, Zhang G, Ho W, Francis J, Eubanks JH (2002) Transient forebrain ischemia alters the mRNA expression of methyl DNA-binding factors in the adult rat hippocampus. Neuroscience 115:515–524
81. Hendrich B, Guy J, Ramsahoye B, Wilson VA, Bird A (2001) Closely related proteins MBD2 and MBD3 play distinctive but interacting roles in mouse development. Genes Dev 15:710–723

82. Champagne FA, Weaver IC, Diorio J, Dymov S, Szyf M, Meaney MJ (2006) Maternal care associated with methylation of the estrogen receptor alpha 1b promoter and estrogen receptor alpha expression in the medial preoptic area of female offspring. Endocrinology 147(6): 2909–2915
83. Anway MD, Skinner MK (2006) Epigenetic transgenerational actions of endocrine disruptors. Endocrinology 147:S43–S49
84. Ho SM, Tang WY, Belmonte de Frausto J, Prins GS (2006) Developmental exposure to estradiol and bisphenol A increases susceptibility to prostate carcinogenesis and epigenetically regulates phosphodiesterase type 4 variant 4. Cancer Res 66:5624–5632
85. Li S, Hansman R, Newbold R, Davis B, McLachlan JA, Barrett JC (2003) Neonatal diethylstilbestrol exposure induces persistent elevation of c-fos expression and hypomethylation in its exon-4 in mouse uterus. Mol Carcinog 38:78–84
86. Sutherland JE, Costa M (2003) Epigenetics and the environment. Ann N Y Acad Sci 983:151–160
87. Rossignol S, Steunou V, Chalas C et al (2006) The epigenetic imprinting defect of patients with Beckwith-Wiedemann syndrome born after assisted reproductive technology is not restricted to the 11p15 region. J Med Genet 43:902–907
88. Niemitz EL, Feinberg AP (2004) Epigenetics and assisted reproductive technology: a call for investigation. Am J Hum Genet 74:599–609
89. Koturbash I, Baker M, Loree J et al (2006) Epigenetic dysregulation underlies radiation-induced transgenerational genome instability in vivo. Int J Radiat Oncol Biol Phys 66:327–330
90. Dolinoy DC, Huang D, Jirtle RL (2007) Maternal nutrient supplementation counteracts bisphenol A-induced DNA hypomethylation in early development. Proc Natl Acad Sci USA 104:13056–13061
91. Okoji RS, Yu RC, Maronpot RR, Froines JR (2002) Sodium arsenite administration via drinking water increases genome-wide and Ha-ras DNA hypomethylation in methyl-deficient C57BL/6J mice. Carcinogenesis 23:777–785
92. Hales CN, Barker DJ (1992) Type 2 (non-insulin-dependent) diabetes mellitus: the thrifty phenotype hypothesis. Diabetologia 35:595–601
93. Wells JC (2003) The thrifty phenotype hypothesis: thrifty offspring or thrifty mother? J Theor Biol 221:143–161
94. Waterland RA, Lin JR, Smith CA, Jirtle RL (2006) Post-weaning diet affects genomic imprinting at the insulin-like growth factor 2 (Igf2) locus. Hum Mol Genet 15:705–716
95. Waterland RA, Dolinoy DC, Lin JR, Smith CA, Shi X, Tahiliani KG (2006) Maternal methyl supplements increase offspring DNA methylation at Axin Fused. Genesis 44: 401–406
96. Dolinoy DC, Weidman JR, Waterland RA, Jirtle RL (2006) Maternal genistein alters coat color and protects Avy mouse offspring from obesity by modifying the fetal epigenome. Environ Health Perspect 114:567–572
97. Waterland RA, Jirtle RL (2003) Transposable elements: targets for early nutritional effects on epigenetic gene regulation. Mol Cell Biol 23:5293–5300
98. Wolff GL, Kodell RL, Moore SR, Cooney CA (1998) Maternal epigenetics and methyl supplements affect agouti gene expression in Avy/a mice. FASEB J 12:949–957
99. Miller JW, Nadeau MR, Smith J, Smith D, Selhub J (1994) Folate-deficiency-induced homocysteinaemia in rats: disruption of S-adenosylmethionine's co-ordinate regulation of homocysteine metabolism. Biochem J 298(Pt 2):415–419
100. Poirier LA, Wise CK, Delongchamp RR, Sinha R (2001) Blood determinations of S-adenosylmethionine, S-adenosylhomocysteine, and homocysteine: correlations with diet. Cancer Epidemiol Biomarkers Prev 10:649–655
101. Giovannucci E (2004) Alcohol, one-carbon metabolism, and colorectal cancer: recent insights from molecular studies. J Nutr 134:2475S–2481S
102. Luedi PP, Hartemink AJ, Jirtle RL (2005) Genome-wide prediction of imprinted murine genes. Genome Res 15:875–884
103. Feinberg AP (2007) An epigenetic approach to cancer etiology. Cancer J 13:70–74

104. Day JK, Bauer AM, DesBordes C et al (2002) Genistein alters methylation patterns in mice. J Nutr 132:2419S–2423S
105. Swynghedauw B (2006) Phenotypic plasticity of adult myocardium: molecular mechanisms. J Exp Biol 209:2320–2327
106. Stout RD, Suttles J (2005) Immunosenescence and macrophage functional plasticity: dysregulation of macrophage function by age-associated microenvironmental changes. Immunol Rev 205:60–71
107. Kadar T, Silbermann M, Brandeis R, Levy A (1990) Age-related structural changes in the rat hippocampus: correlation with working memory deficiency. Brain Res 512:113–120
108. Anderton BH (1997) Changes in the ageing brain in health and disease. Philos Trans R Soc Lond B Biol Sci 352:1781–1792
109. Kerr DS, Campbell LW, Applegate MD, Brodish A, Landfield PW (1991) Chronic stress-induced acceleration of electrophysiologic and morphometric biomarkers of hippocampal aging. J Neurosci 11:1316–1324
110. Sapolsky RM (1985) Glucocorticoid toxicity in the hippocampus: temporal aspects of neuronal vulnerability. Brain Res 359:300–305
111. Koide Y, Kimura S, Inoue S et al (1986) Responsiveness of hypophyseal-adrenocortical axis to repetitive administration of synthetic ovine corticotropin-releasing hormone in patients with isolated adrenocorticotropin deficiency. J Clin Endocrinol Metab 63:329–335
112. Morse JK, Davis JN (1990) Regulation of ischemic hippocampal damage in the gerbil: adrenalectomy alters the rate of CA1 cell disappearance. Exp Neurol 110:86–92
113. White FA, Keller-Peck CR, Knudson CM, Korsmeyer SJ, Snider WD (1998) Widespread elimination of naturally occurring neuronal death in Bax-deficient mice. J Neurosci 18:1428–1439
114. Forger NG, Rosen GJ, Waters EM, Jacob D, Simerly RB, de Vries GJ (2004) Deletion of Bax eliminates sex differences in the mouse forebrain. Proc Natl Acad Sci USA 101: 13666–13671
115. Liu D, Diorio J, Day JC, Francis DD, Meaney MJ (2000) Maternal care, hippocampal synaptogenesis and cognitive development in rats. Nat Neurosci 3:799–806
116. Weaver IC, Grant RJ, Meaney MJ (2002) Maternal behavior regulates long-term hippocampal expression of BAX and apoptosis in the offspring. J Neurochem 82:998–1002
117. Epel ES, Blackburn EH, Lin J et al (2004) Accelerated telomere shortening in response to life stress. Proc Natl Acad Sci USA 101:17312–17315
118. Blasco MA (2007) The epigenetic regulation of mammalian telomeres. Nat Rev Genet 8:299–309
119. Brown SE, Weaver IC, Meaney MJ, Szyf M (2008) Regional-specific global cytosine methylation and DNA methyltransferase expression in the adult rat hippocampus. Neurosci Lett 440:49–53
120. Francis D, Diorio J, LaPlante P, Weaver S, Seckl JR, Meaney MJ (1996) The role of early environmental events in regulating neuroendocrine development. Moms, pups, stress, and glucocorticoid receptors. Ann N Y Acad Sci 794:136–152
121. Kato T, Iwamoto K, Kakiuchi C, Kuratomi G, Okazaki Y (2005) Genetic or epigenetic difference causing discordance between monozygotic twins as a clue to molecular basis of mental disorders. Mol Psychiatry 10:622–630
122. Caruncho HJ, Dopeso-Reyes IG, Loza MI, Rodriguez MA (2004) A GABA, reelin, and the neurodevelopmental hypothesis of schizophrenia. Crit Rev Neurobiol 16:25–32
123. Dong E, Caruncho H, Liu WS et al (2003) A reelin-integrin receptor interaction regulates Arc mRNA translation in synaptoneurosomes. Proc Natl Acad Sci USA 100:5479–5484
124. Levenson JM, O'Riordan KJ, Brown KD, Trinh MA, Molfese DL, Sweatt JD (2004) Regulation of histone acetylation during memory formation in the hippocampus. J Biol Chem 279:40545–40559
125. Korzus E, Rosenfeld MG, Mayford M (2004) CBP histone acetyltransferase activity is a critical component of memory consolidation. Neuron 42:961–972
126. Bredy TW, Grant RJ, Champagne DL, Meaney MJ (2003) Maternal care influences neuronal survival in the hippocampus of the rat. Eur J Neurosci 18:2903–2909

127. Bredy TW, Humpartzoomian RA, Cain DP, Meaney MJ (2003) Partial reversal of the effect of maternal care on cognitive function through environmental enrichment. Neuroscience 118:571–576
128. Guy J, Gan J, Selfridge J, Cobb S, Bird A (2007) Reversal of neurological defects in a mouse model of Rett syndrome. Science 315:1143–1147
129. Holsboer F (2000) The corticosteroid receptor hypothesis of depression. Neuropsychopharmacology 23:477–501
130. Invitti C, Redaelli G, Baldi G, Cavagnini F (1999) Glucocorticoid receptors in anorexia nervosa and Cushing's disease. Biol Psychiatry 45:1467–1471
131. DeRijk R, Sternberg EM (1997) Corticosteroid resistance and disease. Ann Med 29:79–82
132. Feldman R, Eidelman AI, Sirota L, Weller A (2002) Comparison of skin-to-skin (kangaroo) and traditional care: parenting outcomes and preterm infant development. Pediatrics 110:16–26
133. Harlow HF, Zimmermann RR (1959) Affectional responses in the infant monkey; orphaned baby monkeys develop a strong and persistent attachment to inanimate surrogate mothers. Science 130:421–432
134. Olds DL, Robinson J, Pettitt L et al (2004) Effects of home visits by paraprofessionals and by nurses: age 4 follow-up results of a randomized trial. Pediatrics 114:1560–1568
135. Olds DL, Kitzman H, Cole R et al (2004) Effects of nurse home-visiting on maternal life course and child development: age 6 follow-up results of a randomized trial. Pediatrics 114:1550–1559
136. Olds D, Henderson CR Jr, Cole R et al (1998) Long-term effects of nurse home visitation on children's criminal and antisocial behavior: 15-year follow-up of a randomized controlled trial. JAMA 280:1238–1244
137. Turner JD, Muller CP (2005) Structure of the glucocorticoid receptor (NR3C1) gene 5′ untranslated region: identification, and tissue distribution of multiple new human exon 1. J Mol Endocrinol 35:283–292
138. Takao T, Culp SG, Newton RC, De Souza EB (1992) Type I interleukin-1 receptors in the mouse brain-endocrine-immune axis labelled with [125I]recombinant human interleukin-1 receptor antagonist. J Neuroimmunol 41:51–60
139. Takao T, Tracey DE, Mitchell WM, De Souza EB (1990) Interleukin-1 receptors in mouse brain: characterization and neuronal localization. Endocrinology 127:3070–3078
140. Falek PR, Ben-Sasson SZ, Ariel M (2000) Correlation between DNA methylation and murine IFN-gamma and IL-4 expression. Cytokine 12:198–206
141. Northrop JK, Thomas RM, Wells AD, Shen H (2006) Epigenetic remodeling of the IL-2 and IFN-gamma loci in memory CD8 T cells is influenced by CD4 T cells. J Immunol 177:1062–1069
142. Fitzpatrick DR, Shirley KM, Kelso A (1999) Cutting edge: stable epigenetic inheritance of regional IFN-gamma promoter demethylation in CD44highCD8+ T lymphocytes. J Immunol 162:5053–5057
143. Reid TR, Merigan TC, Basham TY (1992) Resistance to interferon-alpha in a mouse B-cell lymphoma involves DNA methylation. J Interferon Res 12:131–137
144. Kovacs EJ, Oppenheim JJ, Carter DB, Young HA (1987) Enhanced interleukin-1 production by human monocyte cell lines following treatment with 5-azacytidine. J Leukoc Biol 41:40–46
145. Callinan PA, Feinberg AP (2006) The emerging science of epigenomics. Hum Mol Genet 15(Spec No 1):R95–R101

Effects of Early Life Seizures and Anti-epileptic Drug Treatment on Human Brain Development in Human Models

Eric D. Marsh and Amy R. Brooks-Kayal

Abstract When seizures, abnormal electrical discharges in the brain, occur during infancy or early childhood, there is a belief amongst child neurologists and pediatricians that the seizures are harmful to a child's brain development. Anti-epileptic drugs are used to treat these seizures, yet these drugs may have an impact on brain development as well. Animal research has begun to address the questions "do seizures and/or anti-epileptic drugs affect the developing brain," but more work needs to be completed to guide the clinicians view on the subject. Therefore, clinicians have attempted to answer these questions with a variety of different clinical studies. This chapter will review the clinical literature on the effects of seizures and anti-epileptic drugs on brain development and cognition in infants and young children. The current literature suggests that seizures, at least frequent ones, may have an impact on brain development, but that multiple confounding factors makes this clinically very difficult to study. Anti-epileptic drugs, however, have been shown to affect brain development during fetal exposure, but the impact on postnatal exposure is much less clear. Many more clinical studies will be required to answer these questions in a way that would change clinical practice. Clearly, animal studies will be very important as too many variables make clinical research always difficult to interpret.

Keywords Anti-epileptic drugs • Seizures • Development • Cognitive studies

1 Introduction

The primary goal of child neurologists and pediatricians is to maximize the developmental potential of their patients. Many physicians believe that seizures have a negative impact on development (although this is somewhat controversial) and that treatment of seizures occurring during the neonatal through early childhood periods

A.R. Books-Kayal
Division of Child Neurology, Children's Hospital of Philadelphia, Philadelphia, PA, 19104, USA
e-mail: brooks-kayal.amy@tchden.org

D. Janigro (ed.), *Mammalian Brain Development*, Contemporary Neuroscience,
DOI 10.1007/978-1-60761-287-2_3,

improves later outcome [1]. However, physicians are also concerned that anti-epileptic medications (AED) may impact brain development during this critical period. Thus, there is still much discussion amongst child neurologists regarding which neonatal or early childhood seizures should be treated and with what medications. As the current standard of care is to treat children who have recurrent seizures, it is important to garner reliable clinical information about the effects of seizures on brain development. Any studies of the effects of seizures on brain development must look at the seizures in isolation from the effects of treatment or the underlying etiology. Studies that can successfully separate these overlapping effects have been difficult to obtain. In addition to the clinical seizure studies being difficult to interpret, there is a paucity of drug studies where the primary end point investigates the effects of anti-epileptic drugs on the developmental outcome in children. Therefore, the data available regarding both the long and short term effects of drug treatment on children's cognitive development need to be evaluated with caution. Fortunately, neuroscientists and pharmacologists have been attempting to answer many of these questions in animal models where more controlled studies can be performed to separate the effects of seizures and AED treatments (this is discussed in another chapter in this volume). Until the animal research influences clinical care, child neurologists and pediatricians must use the available clinical literature, discussed below, to guide the current clinical practice.

Animal research about cognitive effects of anti-epileptic drugs and seizures on the developing brain have been extensively reported and reviewed [2, 3] (including the chapter in this volume by Moshe et al.) Briefly, some seizures appear to have a permanent deleterious effect during development, but the changes are different qualitatively and quantitatively than those observed when adult animals experience similar seizures [4–7]. Over the last decade, increasing numbers of studies have shown that AEDs can also have considerable negative effects on brain development [8, 9]. Basic science literature is very informative and makes scientists and physicians begin to question if the effects reported in animals are also observed in humans. What then does the available literature state about the impact of seizures and AEDs on the developing human brain? In this chapter, we review the literature with the hope of generating an improved understanding of the effects of seizures and AEDs during human brain development.

2 Caveats to Clinical Research on Effects of Seizures and AEDs During Development

Researching the effect of any environmental agent on a child's cognitive development is a difficult task as multiple factors may have an impact on development. For example, many studies have shown that low socioeconomic status contributes to lower intelligence quotient (IQ) testing in children who have biological risk factors [10, 11]. How the low socioeconomic status impacts the development of children with epilepsy is not known and should be the focus of further study. Issues such as access to appropriate health care, exposure to a healthy diet and other related factors likely all play roles. Aside from low socioeconomic status, parental (particularly maternal) level of education and IQ also impact on the final cognitive outcome in children [10, 11]. As these systemic factors can impact children's development controlling for these factors is necessary, but difficult, in any study that is attempting to elucidate an effect of drugs or an illness.

In addition to the environmental systemic factors that can confound physicians understanding of the effects of seizures and AEDs on development, there is a major intrinsic confounder to this study. Children with seizures have various underlying etiologies, such as brain malformations, channelopathies, gene mutations, and remote insults (e.g., in utero strokes, congenital infections, meningitis, and toxin exposure). No two etiologies may give the same developmental outcome. Within any given cause for the seizures, the severity of the "insult" cannot be easily determined further complicating the picture. Indeed, there are studies that compare patients IQ with and without seizures for a given etiology. For patient's with Tuberous Sclerosis higher seizure burden and TSC2 mutations lead to lower cognitive outcome [12]. However, the study does not clarify if the TSC2 mutations or the seizures are the major factor. Most likely, the developmental abnormality created by the TSC2 mutation is more severe and causes lower IQ and intractable seizures. But as will be discussed in detail in this chapter, separating these factors is difficult.

Besides these issues, the child's genetic make-up, and any pre-existent learning and attentional problems add to potential contributors to cognitive dysfunction in children with epilepsy. Finally, determining the individual effects of seizures or AEDs on a child's development is complicated by the fact that children with seizures are typically treated with AEDs resulting in an interaction that makes it very difficult to separate the independent effects of either variable. With these caveats, the studies attempting to show the effects of seizures and AEDs on early development are presented.

3 Developmental Impact of Early Life Seizures

To assess the effects of seizures on cognitive development an ideal study would include a cohort of patients with a single etiology who undergo neuropsychological testing prior to medication treatment, then yearly as their epilepsy persists. There are several practical difficulties with this approach. First, obtaining enough patients with a single etiology can be difficult. Recently this approach has been undertaken with childhood absence epilepsy (CAE) in an ongoing National Institutes of Health (NIH) multi-site sponsored study but no published information regarding intellectual outcome has yet been released. Another major problem, which was discussed above and is an issue with the CAE study, is that the patients will be treated with AEDs. If there is a negative effect on cognitive performance, it will be difficult to discriminate the effects of the seizures themselves or the medications used to treat the seizures. Of course a placebo controlled study with only half of the patients treated could answer this question, but with the danger of not treating recurrent seizures, this study design would not be ethically acceptable for most seizure types. A possible solution for this problem is being attempted in the NIH CAE study. These researchers are collecting neuropsychological data on the children treated with three different AEDs. This may address the issue by showing an effect of one medication and not another, arguing for an AED effect, or by reporting an issue with all medications, arguing for a seizure effect. Unfortunately, any shared effect between the medications will still make distinguishing seizure from AED effects difficult. With these caveats in mind, the prospective cohort study is the most compelling way to approach this difficult area of research.

4 Clinical Studies on Epilepsy and Cognition

A number of prospective cohort studies have been performed. A series of early studies reviewed by Dodrill[13] and published prior to 1940 followed children with epilepsy and examined IQ test data from one to greater than four years after the onset of seizures. Mild to moderate IQ reductions were reported in 10–40% of children with seizures. These early studies are useful to frame the current question, but since they could not diagnose the underlying etiology to the extent that can be done today, and the authors only reported standard IQ data, any nuances within the patient population or cognitive changes due to the underlying condition is unable to be determined. Since these early studies, a number of reports were published with varying results. Some of the more recent studies documented a definite cognitive decline [14–16] particularly in certain subgroups of patients, where as other studies showed no change in cognitive function over time [17].

One of the largest prospective cohort studies performed on children was the NIH Collaborative Perinatal project. Data were collected from this study to assess the effect of many factors on development [18]. Two of the areas of study in this project were to define the incidence of epilepsy in children and to monitor the effect of epilepsy on development [19, 20]. Patients with epilepsy, in this study, were found to have mean IQ scores that were similar to their sibling controls [17]. Although the mean IQ did not differ significantly between groups, more children with mental retardation were found in the epilepsy group. The low IQ in this subset was present before seizure onset, was not altered after the first seizures, and was most common in children with other neurological problems. The authors attributed the low IQ in this group to underlying brain abnormalities and not the seizures. In addition, approximately 4% of children in this study had recurrent non-symptomatic seizures [20], with no difference in IQ before and after the onset of seizures. These patients, however, had very few seizures on average. Therefore, the finding of this study, which shows that IQ is not altered by recurrent seizures, may be explained, in part, by the low seizure burden. The results of this study are somewhat discrepant with many of the other prospective studies where the patients had a higher seizure burden [20].

A seminal study on the cognitive effects of seizures was performed by Bourgeois et al. [15] Neuropsychological evaluations were performed on 72 new onset seizure patients between the age of 18 months and 16 years old within 2 weeks of treatment onset and then yearly for an average of four years. No overall IQ difference was found between the patients and 45 sibling controls. More importantly, there was no change in mean IQ over time in either group. However, over a quarter of the epilepsy patients had significant IQ changes, which were not seen in the control group. Over the duration of the 4 year study, 11% of the epileptic patients had a greater then 10 point decrease in IQ and 16.7% had an increase in IQ by more than 10 points. More frequent seizures, higher drug levels and earlier epilepsy onsets were found in the group with a decline in IQ. Consistent with the NIH Collaborative Perinatal project the authors also found that the symptomatic patients and patients with multiple seizure types had lower IQs at the initial epilepsy diagnosis regardless of age of seizure onset [15].

Similar findings were reported in two additional prospective longitudinal studies with comparable methodologies. First, a study from the Netherlands found no overall change in cognitive outcome of 45 patients followed for over 4.2 years with the Wechsler Intelligence Test for Children – Revised (WISC-R) performed at least twice over the study period [16]. Though no overall difference was found in the full cohort, a subgroup of children (24%) showed a greater than nine point decline in full scale IQ. These patients, typically those with frequent seizures, performed relatively poorly on the Information Coding, Digit Span, and Vocabulary subsections [16]. In addition, Bjornaes et. al. found no change in IQ during the 3.5 years mean follow-up when the entire cohort of children and adults with intractable epilepsy were analyzed together [14]. In this study, the children had a small decline of six points in both performance and full scale IQ, suggesting seizures have greater effect on the developing brain. In all three of these prospective studies, the patients had active seizure disorders so as only small changes were reported, infrequent seizures may not have significant effects on cognitive development.

In contrast to these previously described studies, a number of additional reports concluded that patients with epilepsy have a higher rate of cognitive impairment than controls. The design of these studies, however, makes it difficult to determine if the cognitive changes reported were from the seizures or from another confounding factor, such as underlying etiology or the detrimental effects of AEDs. For example, Schoenfeld and colleagues studied 57 children with complex partial seizures and sibling controls and reported that seizures affected both cognitive and behavioral measures. Age at onset of seizures was the strongest predictor of cognitive functioning in this study [21]. In another study showing similar results, 11 children with epilepsy were compared to age matched controls and there was a trend towards the epileptic children making smaller gains on the full scale IQ than the controls over the 1.5 years of follow-up [22]. Two other interesting findings were reported in this study. First, the biggest changes in IQ were in children followed for over 10 years with continued seizures. Second, children who were seizure free during the study period did not fare better then ones with continuing seizures [22]. All patients in this study were on continuous AED treatment.

These data suggest that seizures alone cannot be the cause of intellectual changes. AED treatment and/or underlying substrate are likely important contributors to the cognitive delay.

Although in the above studies, frequency of seizures had no predictive effect on cognition, other studies have shown a correlation between higher seizure frequency and poorer cognition [23]. In addition, these studies have found that early age at onset, symptomatic or cryptogenic etiology, and total lifetime seizure burden predict a detrimental cognitive change [23–26]. To attempt to remove the confounder of preexisting mental retardation, 74 idiopathic epileptic children, excluding children with mental retardation, symptomatic, and cryptogenic epilepsy were studied serially over three years. Impairments in intelligence, psychomotor speed, memory, academic achievement and behavior were found in the patients with epilepsy compared to sibling controls [27]. No correlation was found to seizure frequency or duration suggesting even children with normal intelligence and well-controlled seizures can experience learning problems. These studies suggest that changes in cognitive function may have some correlation to seizures, but the additional factors of the underlying substrate, AEDs, and learning differences can not be completely excluded from contributing to any cognitive difficulties.

As seizures do not clearly result in any detrimental effect on cognition, some authors concluded that the seizures are "concomitant rather then causal" [28] and the real problem is the underlying brain dysfunction, whether from malformations of cortical development or other remote insults. In support of this hypothesis are the studies that have reported that children with symptomatic epilepsy and underlying brain abnormalities are most adversely affected cognitively [29–31]. There is no doubt that the more severe the neurological problem, whether structural, physiological or molecular in etiology, the worse the patient's development. What remains not completely clear is the additive effect of any seizures. Does one seizure pose a problem for development or does the child need to be intractable before there is a serious adverse effect of seizures? Answers to these questions are particularly important because all drugs used to prevent seizures have potential developmental and clinical side effects. Specifically, as discussed in the next section of this chapter, there is evidence that seizure medications may affect the developing brain in the process of treatment.

5 Does Exposure to AEDs Affect the Developing Brain

There are two clinical situations where treatment of a patient with epilepsy (or other disease for which anti-epileptic drugs are used) with an AED can have adverse impact on brain development. The first situation, and the more thoroughly studied, is the exposure of the developing fetus during the pregnancy of a woman with epilepsy. The other clinical scenario is a child who develops epilepsy in the first few years of life. In this section, we will review the literature on in utero and early childhood exposure to AEDs focusing on the problems with these studies and the extent to which development may be impacted.

5.1 Pregnancy Related Studies

The clinical problem of fetal exposure to AEDs is not trivial. Approximately 24,000 children are born each year to women with epilepsy [32], and an untold number of other children were exposed to the same drugs while their mothers were being treated for psychiatric or pain related illnesses. The number of children exposed and at risk, therefore, for CNS malformations and long-term developmental effects makes this area of investigation an important public health problem. To place the developmental effects of in utero exposure in context, it is important to briefly review the literature of AEDs as a systemic teratogen. Large numbers of studies of phenobarbital, carbamazepine, phenytoin, and valproic acid (the older AEDs) have been performed over the last 40 years. These studies report, convincingly, that exposure during pregnancy increases the risk for minor or major malformation two to three times the risk in the general public [33–35]. Newer AEDs (many, but including lamotrigine, topiramate, zonisamide, and oxcarbezapine) have yet to be well studied. A number of on-going pregnancy registries are collecting the data on the teratogenicity of these drugs [36–38]. The published data, to-date, appears to be positive, in that there is less teratogenesis with newer medications such as lamotrigine [38, 39]. For more detailed information on the teratogenic effects of AEDs, see the informative reviews by Dolk, Holmes, Artama and Shorvon [36, 40–42].

With thorough investigations into the teratogenic effects of AEDs, what information has been reported on the cognitive impact of in utero exposure to AEDs? Unfortunately, most early studies did not test the developmental profile of the children exposed [32]. The available studies performed on the in utero exposure to AEDs have reported varying results. Some studies reported limited [43] or no [44] change in developmental quotients in children followed for up to one year after birth. The majority of studies, however, have found that children exposed to AEDs in utero have lower mean IQs as compared to children born to mothers with epilepsy not taking AEDs [45–48]. Using mothers with epilepsy, but who are not on AEDs is the necessary control as uncontrolled seizures could have an impact on the developing infant. Other potential confounders are mother's IQ, socioeconomic status, and educational level. Due to the multiple potential confounding variables that can impact developmental outcome, additional large well controlled studies are necessary to determine the long-term cognitive outcome of AED exposure during gestation. Although these multiple factors make studying the impact of AEDs on cognitive development difficult, the studies to date have generally shown that fetal exposure to AEDs is related to lower developmental and cognitive potential. The animal literature, which is discussed elsewhere in this volume, can alleviate many of these confounding factors and has shown that many AEDs affect the developing brain.

5.2 Early Childhood Exposure

The human and animal studies on in utero AED exposure suggest a negative impact on cognitive and school performance. What do studies on humans say about AED exposure during early childhood? As with studies of seizures and the in utero effect on development, there are many confounding factors in the human literature. A major ethical issue exists: child neurologists cannot study the effect of a drug on a normal child. One conceivable solution to this problem would be to study children who are on AEDs for conditions other than epilepsy (i.e., psychiatric or chronic pain). Unfortunately, both these patient populations have their own confounding factors. The ideal study would, of course, be testing the effects of the drugs on normal children but since this is not feasible, almost all cognitive and developmental studies of AED effects have been performed on children with epilepsy, raising all the related issues described above. The single exception was a febrile seizure study where the children were randomized to placebo or phenobarbital and underwent neuropsychological testing during treatment [49]. While on phenobarbital, the children's mean IQ was lowered [49]. Surprisingly after follow up for 3–5 years, even after phenobarbital treatment was discontinued, revealed a persistent difference in reading comprehension tests but the mean IQ had normalized [50]. Continued follow up of this cohort into teenage years would be necessary to see if the effects were truly permanent. Fortunately, febrile seizures currently are very rarely treated with a prophylactic medication and therefore child neurologists and pediatricians may never know if exposure to phenobarbital as a toddler really has mild lasting cognitive differences.

Many approaches have been attempted to control for a patient's underlying epilepsy and specifically look at the effects of AEDs on cognition and learning. For example, researchers have cognitively tested children with epilepsy before and after starting a medication. No difference in IQ, before, during, and after treatment was reported in a few studies [51, 52]. In these studies, measuring the latency of the evoked response to visual stimuli on the cortex (P300 latency) was lengthened on phenobarbital, but not other medications [51, 52]. Another approach has been testing medication(s) as add-on therapy[53] or as crossover studies between drugs [54]. The crossover study of Vining et. al. found that phenobarbital treatment reduced cognitive performance, decreased IQ, slowed reaction times and had negative effects on behavior compared to valproic acid [54]. Finally, many studies have looked at the effect of different drug levels on changes in IQ and other cognitive testing [55–60]. These studies have generally shown a mild to moderate effect on cognition depending on the drug and dosage taken by the patient. Overall, most studies have reported detrimental effects of phenobarbital on cognition and behavior[51, 52, 61, 62] while other drugs have more variable effects. Valproate has been shown to have variable effects, with the worst cognitive changes associated with poly therapy [45]. Importantly, none of the negative effects described in any of these studies has been reported to be permanent, though none of these researchers went back to study the cohort months to years after discontinuation of treatment.

Studies on the newer AEDs, those approved since 1990, are essentially non-existent except for data from younger patients in adult studies or the side effect data generated during clinical trials in children [60]. For example, the side effect profile of topiramate in pediatric trials appear to validate the adult data that topiramate can have an impact on cognition, particularly verbal performance [63, 64]. Of the newer medications, lamotrigine and gabapentin appear to have fewer cognitive side effects [63]. In general, most studies of AEDs in childhood, where cognitive testing was performed reported a trend toward an effect compared to placebo (see reviews by

Meador [57] and Drane [56]). In conclusion, the AED clinical studies show that most of the older drugs have a potential effect on cognition and behavior in some children. Though any cognitive change is likely temporary, there are suggestions of subtle but long term effects on brain development. More clinical research is needed on the newer medications to understand their influence on cognition and brain development. As with the literature on seizures and development, animal studies are able to address issues of the cognitive effects of AEDs in a more controlled fashion yielding insightful information (this subject is reviewed in the chapter by Moshe et. al in this volume). The information gathered by animal studies should be used to ask questions regarding drug effects in humans and begin to direct future human clinical studies.

6 Conclusions/Future Directions

In people, particularly children, seizures and anti-epileptic drugs affect the brain simultaneously as people are treated for epilepsy. The intertwined nature of the two potential insults makes deciphering the relative contribution of epilepsy treatment, any underlying neurological dysfunction, and seizures on brain development and cognitive outcome a complex issue. As discussed, there are many confounding factors in addition to the seizures and AEDs that limit the utility of human studies. However, the current literature suggests that both seizures and many AEDs, particularly the older generation of drugs, have the potential to affect cognitive development. What remains inconclusive is the extent of any impact and the clinical situation (i.e., symptomatic vs. idiopathic epilepsies) for which the seizures and/or AEDs have a detrimental effect.

These data point to the need for new studies to be performed that attempt to parse out what is the full extent of seizures or AEDs on development. Beginning to look at AED usage for non-epileptic conditions in childhood can possibly separate out AED effects from that of seizures. Studying children who, for whatever reason, did not get adequately treated (i.e., underprivileged countries, remote rural areas, or due to religious or social beliefs) can look at cognitive impact of seizures without the confounding factor of treatment. Finally, it is important to study the impact of seizures and AEDs in isolation on brain development and cognition using the available animal models. This basic science research can lead to new insights into these old clinical questions and can uncover mechanisms for the design of new therapies that would diminish the risks of both seizures and their treatment for our youngest and most vulnerable patients.

References

1. Wheless JW, Clarke DF, Carpenter D (2005) Treatment of pediatric epilepsy: expert opinion. J Child Neurol 20(Suppl 1):S1–S56 quiz S9–S60
2. Holmes GL (2004) Effects of early seizures on later behavior and epileptogenicity. Ment Retard Dev Disabil Res Rev 10(2):101–105

3. Marsh ED, Brooks-Kayal AR, Porter BE (2006) Seizures and anti-epileptic drugs: does exposure alter normal brain development? Epilepsia 47(12):1999–2010
4. Swann JW (2005) The impact of seizures on developing hippocampal networks. Prog Brain Res 147:347–354
5. Holmes GL, Khazipov R, Ben-Ari Y (2002) Seizure-induced damage in the developing human: relevance of experimental models. Prog Brain Res 135:321–334
6. Stafstrom CE (2002) Assessing the behavioral and cognitive effects of seizures on the developing brain. Prog Brain Res 135:377–390
7. Veliskova J, Claudio OI, Galanopoulou AS et al (2004) Seizures in the developing brain. Epilepsia 45(Suppl 8):6–12
8. Mikati MA, Holmes GL, Chronopoulos A et al (1994) Phenobarbital modifies seizure-related brain injury in the developing brain. Ann Neurol 36(3):425–433
9. Olney JW, Wozniak DF, Jevtovic-Todorovic V, Farber NB, Bittigau P, Ikonomidou C (2002) Drug-induced apoptotic neurodegeneration in the developing brain. Brain Pathol 12(4): 488–498
10. Brooks-Gunn J, McCormick MC, Klebanov PK, McCarton C (1998) Health care use of 3-year-old low birth weight premature children: effects of family and neighborhood poverty. J Pediatr 132(6):971–975
11. Klebanov PK, Brooks-Gunn J, McCormick MC (1994) School achievement and failure in very low birth weight children. J Dev Behav Pediatr 15(4):248–256
12. Winterkorn EB, Pulsifer MB, Thiele EA (2007) Cognitive prognosis of patients with tuberous sclerosis complex. Neurology 68:62–64
13. Dodrill CB (2004) Neuropsychological effects of seizures. Epilepsy Behav 5(Suppl 1):S21–S24
14. Bjornaes H, Stabell K, Henriksen O, Loyning Y (2001) The effects of refractory epilepsy on intellectual functioning in children and adults. A longitudinal study. Seizure 10(4):250–259
15. Bourgeois BF, Prensky AL, Palkes HS, Talent BK, Busch SG (1983) Intelligence in epilepsy: a prospective study in children. Ann Neurol 14(4):438–444
16. Aldenkamp AP, Alpherts WC, De Bruine-Seeder D, Dekker MJ (1990) Test-retest variability in children with epilepsy–a comparison of WISC-R profiles. Epilepsy Res 7(2):165–172
17. Ellenberg JH, Hirtz DG, Nelson KB (1986) Do seizures in children cause intellectual deterioration? N Engl J Med 314(17):1085–1088
18. Melnick M, Myrianthopoulos NC, Christian JC (1978) The effects of chorion type on variation in IQ in the NCPP twin population. Am J Hum Genet 30(4):425–433
19. Nelson KB, Ellenberg JH (1986) Antecedents of seizure disorders in early childhood. Am J Dis Child 140(10):1053–1061
20. Nelson KB, Ellenberg JH (1987) Predisposing and causative factors in childhood epilepsy. Epilepsia 28(Suppl 1):S16–S24
21. Schoenfeld J, Seidenberg M, Woodard A et al (1999) Neuropsychological and behavioral status of children with complex partial seizures. Dev Med Child Neurol 41(11):724–731
22. Neyens LG, Aldenkamp AP, Meinardi HM (1999) Prospective follow-up of intellectual development in children with a recent onset of epilepsy. Epilepsy Res 34(2–3):85–90
23. Bulteau C, Jambaque I, Viguier D, Kieffer V, Dellatolas G, Dulac O (2000) Epileptic syndromes, cognitive assessment and school placement: a study of 251 children. Dev Med Child Neurol 42(5):319–327
24. Freitag H, Tuxhorn I (2005) Cognitive function in preschool children after epilepsy surgery: rationale for early intervention. Epilepsia 46(4):561–567
25. Hoie B, Mykletun A, Sommerfelt K, Bjornaes H, Skeidsvoll H, Waaler PE (2005) Seizure-related factors and non-verbal intelligence in children with epilepsy. A population-based study from Western Norway. Seizure 14(4):223–231
26. Seidenberg M, Beck N, Geisser M et al (1986) Academic achievement of children with epilepsy. Epilepsia 27(6):753–759
27. Bailet LL, Turk WR (2000) The impact of childhood epilepsy on neurocognitive and behavioral performance: a prospective longitudinal study. Epilepsia 41(4):426–431

28. Lesser RP, Luders H, Wyllie E, Dinner DS, Morris HH 3rd (1986) Mental deterioration in epilepsy. Epilepsia 27(Suppl 2):S105–S123
29. Berg AT, Smith SN, Frobish D et al (2004) Longitudinal assessment of adaptive behavior in infants and young children with newly diagnosed epilepsy: influences of etiology, syndrome, and seizure control. Pediatrics 114(3):645–650
30. Northcott E, Connolly AM, Berroya A et al (2005) The neuropsychological and language profile of children with benign rolandic epilepsy. Epilepsia 46(6):924–930
31. Shinnar S, Hauser WA (2002) Do occasional brief seizures cause detectable clinical consequences? Prog Brain Res 135:221–235
32. Meador KJ, Zupanc ML (2004) Neurodevelopmental outcomes of children born to mothers with epilepsy. Cleve Clin J Med 71(Suppl 2):S38–S41
33. Waters CH, Belai Y, Gott PS, Shen P, De Giorgio CM (1994) Outcomes of pregnancy associated with anti-epileptic drugs. Arch Neurol 51(3):250–253
34. Holmes LB, Wyszynski DF, Lieberman E (2004) The AED (anti-epileptic drug) pregnancy registry: a 6-year experience. Arch Neurol 61(5):673–678
35. Yerby MS, Leavitt A, Erickson DM et al (1992) Anti-epileptics and the development of congenital anomalies. Neurology 42(4 Suppl 5):132–140
36. Artama M, Auvinen A, Raudaskoski T, Isojarvi I, Isojarvi J (2005) Anti-epileptic drug use of women with epilepsy and congenital malformations in offspring. Neurology 64(11):1874–1878
37. Holmes LB, Wyszynski DF (2004) North American anti-epileptic drug pregnancy registry. Epilepsia 45(11):1465
38. Cunnington MC (2004) The International Lamotrigine pregnancy registry update for the epilepsy foundation. Epilepsia 45(11):1468
39. Meador KJ, Baker GA, Finnell RH et al (2006) In utero anti-epileptic drug exposure: fetal death and malformations. Neurology 67(3):407–412
40. Dolk H, McElhatton P (2002) Assessing epidemiological evidence for the teratogenic effects of anticonvulsant medications. J Med Genet 39(4):243–244
41. Holmes LB, Harvey EA, Coull BA et al (2001) The teratogenicity of anticonvulsant drugs. N Engl J Med 344(15):1132–1138
42. Shorvon S (2002) Anti-epileptic drug therapy during pregnancy: the neurologist's perspective. J Med Genet 39(4):248–250
43. Leavitt AM, Yerby MS, Robinson N, Sells CJ, Erickson DM (1992) Epilepsy in pregnancy: developmental outcome of offspring at 12 months. Neurology 42(4 Suppl 5):141–143
44. Wide K, Winbladh B, Tomson T, Sars-Zimmer K, Berggren E (2000) Psychomotor development and minor anomalies in children exposed to anti-epileptic drugs in utero: a prospective population-based study. Dev Med Child Neurol 42(2):87–92
45. Adab N, Jacoby A, Smith D, Chadwick D (2001) Additional educational needs in children born to mothers with epilepsy. J Neurol Neurosurg Psychiatry 70(1):15–21
46. Adab N, Kini U, Vinten J et al (2004) The longer term outcome of children born to mothers with epilepsy. J Neurol Neurosurg Psychiatry 75(11):1575–1583
47. Granstrom ML, Gaily E (1992) Psychomotor development in children of mothers with epilepsy. Neurology 42(4 Suppl 5):144–148
48. Koch S, Titze K, Zimmermann RB, Schroder M, Lehmkuhl U, Rauh H (1999) Long-term neuropsychological consequences of maternal epilepsy and anticonvulsant treatment during pregnancy for school-age children and adolescents. Epilepsia 40(9):1237–1243
49. Farwell JR, Lee YJ, Hirtz DG, Sulzbacher SI, Ellenberg JH, Nelson KB (1990) Phenobarbital for febrile seizures – effects on intelligence and on seizure recurrence. N Engl J Med 322(6):364–369
50. Sulzbacher S, Farwell JR, Temkin N, Lu AS, Hirtz DG (1999) Late cognitive effects of early treatment with phenobarbital. Clin Pediatr (Phila) 38(7):387–394
51. Chen Y, Chi Chow J, Lee I (2001) Comparison the cognitive effect of anti-epileptic drugs in seizure-free children with epilepsy before and after drug withdrawal. Epilepsy Res 44(1):65–70

52. Chen YJ, Kang WM, So WC (1996) Comparison of anti-epileptic drugs on cognitive function in newly diagnosed epileptic children: a psychometric and neurophysiological study. Epilepsia 37(1):81–86
53. Meador KJ, Loring DW, Hulihan JF, Kamin M, Karim R (2003) Differential cognitive and behavioral effects of topiramate and valproate. Neurology 60(9):1483–1488
54. Vining EP, Mellitis ED, Dorsen MM et al (1987) Psychologic and behavioral effects of anti-epileptic drugs in children: a double-blind comparison between phenobarbital and valproic acid. Pediatrics 80(2):165–174
55. Trimble MR, Thompson PJ (1984) Sodium valproate and cognitive function. Epilepsia 25(Suppl 1): S60–S64
56. Drane DL, Meador KJ (2002) Cognitive and behavioral effects of anti-epileptic drugs. Epilepsy Behav 3(5S):49–53
57. Meador KJ, Gilliam FG, Kanner AM, Pellock JM (2001) Cognitive and behavioral effects of anti-epileptic drugs. Epilepsy Behav 2(4):SS1–SS17
58. Ortinski P, Meador KJ (2004) Cognitive side effects of anti-epileptic drugs. Epilepsy Behav 5(Suppl 1):S60–S65
59. Bourgeois BF (1998) Anti-epileptic drugs, learning, and behavior in childhood epilepsy. Epilepsia 39(9):913–921
60. Loring DW, Meador KJ (2004) Cognitive side effects of anti-epileptic drugs in children. Neurology 62(6):872–877
61. Calandre EP, Dominguez-Granados R, Gomez-Rubio M, Molina-Font JA (1990) Cognitive effects of long-term treatment with phenobarbital and valproic acid in school children. Acta Neurol Scand 81(6):504–506
62. Camfield CS, Chaplin S, Doyle AB, Shapiro SH, Cummings C, Camfield PR (1979) Side effects of phenobarbital in toddlers; behavioral and cognitive aspects. J Pediatr 95(3):361–365
63. Meador KJ, Loring DW, Vahle VJ et al (2005) Cognitive and behavioral effects of lamotrigine and topiramate in healthy volunteers. Neurology 64(12):2108–2114
64. Reith D, Burke C, Appleton DB, Wallace G, Pelekanos J (2003) Tolerability of topiramate in children and adolescents. J Paediatr Child Health 39(6):416–419

Prenatal Development of the Human Blood-Brain Barrier

Luca Cucullo

Abstract Mammalian, and more specifically human, brain development is the result of a long and complex evolutionary pathway. As we move from the simplest jellyfish nervous system, (which forms an undifferentiated network) to vertebrate animals (such as fish, amphibians, and reptiles), we observe the development of more complex and larger brains. The developmental process reaches its highest form of evolution and complexity in mammals with the addition of the neocerebellum and the neocortex where the highest cognitive functions take place. Among the mammals, humans show the highest form of brain development. The vasculature of the central nervous system (CNS) undergoes a parallel process of adaptation in order to provide a specialized structure such as the blood-brain barrier (BBB). The BBB is capable of maintaining the functional homeostasis of the brain, providing cellular nutritional support and allowing for finer tuning of neuronal activity. During embryonic brain development, this highly specialized vascular endothelium acquires brain-specific properties that set it apart from other vascular endothelial cells (EC). In this chapter, we will cover the initial stages of brain development and discuss in detail the vascular differentiation that leads to the establishment of the BBB. The data reported in this work have been obtained in experimental animal studies both in vivo and in vitro as well as human studies, therefore we included a dedicated session to discuss about cross-species comparability and the challenges inherent to the extrapolation of animal-derived data in humans. A dedicated session that summarize the functional characteristic of the BBB and its role in the development of neurological diseases is also provided.

Keywords Brain development • Blood-brain barrier • Tight junctions • Drug delivery • Pharmacodynamic • Drug resistance • Drug discovery • Neurons • Glial cells • Pregnancy

L. Cucullo
Cerebrovascular Research, Cleveland Clinic, 9500 Euclid Avenue, NB-20, Cleveland, OH44195, USA
e-mail: cuculll@ccf.org

D. Janigro (ed.), *Mammalian Brain Development*, Contemporary Neuroscience, DOI 10.1007/978-1-60761-287-2_4,

1 Brain Development

The most critical phase in the development of animal embryos is called gastrulation. During this phase the morphology of the embryo is dramatically restructured by cell migration. The initial blastula is reorganized into a trilaminar structured embryo from which all tissues and organs of the body will develop. The most internal germ layer, or *endoderm*, forms the lining of the gut and other internal organs. Connective tissues, myoblasts, blood, the lining of the pericardial, pleural, and peritoneal cavities, the circulatory system, the cardiovascular, lymphatic, and most of the urogenital system originate from the *mesoderm*. Skin and other internal organs are generated by the most exterior germ layer, the *ectoderm*. This layer undergoes an additional subdivision generating the *neuroectoderm* from which the nervous system is generated.

After gastrulation the mesoderm morphs into a rod-shaped cell formation called the notochord, which is a "patterning" embryonic structure that regulates the development of surrounding tissue and runs along the antero-posterior axis. During primary neurulation, the process that oversees at the formation of the dorsal nerve cord and the development of the central nervous system (CNS), the notochord sends signals to the overlying ectoderm. These signals induce it to become neuroectoderm (composed of neuronal precursor cells) and to form the neural plate.

After a few days, the neural plate forms the medial hinge point (MHP). Continuous growth of the epidermis causes the neural plate to fold thus generating a neural groove. The neural groove gradually deepens and ultimately the neural folds reunite and coalesce in the middle line through the expression of adhesion molecules (such as E-cadherin, N-cadherin and N-CAM) originating from the neural tube. Following the formation of the neural tube, the front portion will develop into the brain and the rest of the neural tube develops into the spinal cord. Neural crest cells become the peripheral nervous system (Fig. 1).

Primary neurulation occurs in response to soluble factors secreted by the notochord such as morphogen (a substance that regulates the pattern of tissue development and the spatial allocation of the various specialized cell types within a tissue), Sonic Hedgehog (Shh), and the neural inducers chordin [1], noggin [2], and follistatin. These factors regulate the organogenesis of the embryo, such as anterior to posterior patterning and specific tissue development (e.g. neuronal versus skeletal or muscular tissue). In particular, recent studies indicate that the expression pattern of neural-inducing factors noggin, follistatin, and chordin are consistent with the hypothesis that the nervous system is initially induced with an anterior character. These neural inducers act as inhibitory signals that antagonize the function of the bone morphogenetic protein (BMP)-4 (a member of the transforming growth factor-β (TGF-β) superfamily of proteins). This protein regulates the formation of teeth, limbs, and bone from mesoderm and shows strong mesoderm ventralizing and antineuralizing activity [3, 4].

Subsequent signals, such as fibroblast growth factor (FGF), retinoic acid [5] and wingless int (Wnt3) [6, 7], play a role by imparting posterior pattern. Recent studies

Fig. 1 Formation of the neural tube. Transverse sections that show the progression of the neural plate from the original neuroectoderm (*top*) followed by the formation and shaping of the neural plate, formation of the neural groove, and closure of the neural tube (*bottom*)

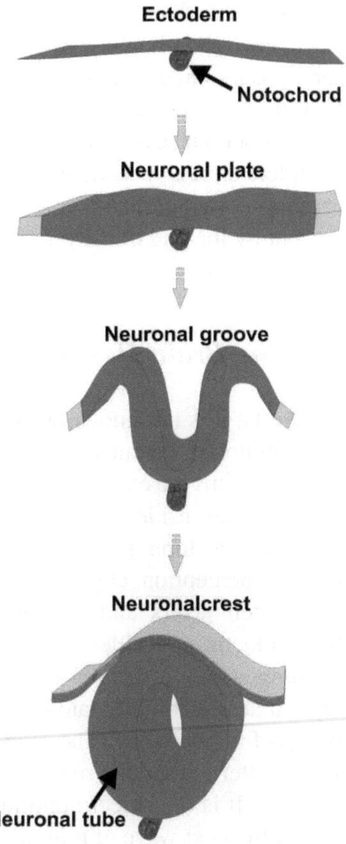

Table 1 The Developmental Sequence of the Brain Regions

Three-vesicle stage	Five-vesicle stage	Brain region
Proencephalon	Telencephalon	Cerebral hemisphere
	Diencephalon	Diencephalon
		Optic nerve and retina
Mesencephalon	Mesencephalon	Mesencephalon
Rhombencephalon	Metencephalon	Pons
		Cerebellum
	Myelencephalon	Medulla oblongata

have shown that other Wnt genes (e.g., Wnt 7B, and 8B) can play a variety of critical roles in early brain development [8, 9]. The anterior segment of the neural tube forms three main regions of the brain: (1) the forebrain or proencephalon, (2) the midbrain or mesencephalon, and (3) the hindbrain or rhombencephalon. Formation of these structures begins with a swelling of the neural tube in a pattern specified by homeobox (Hox) genes that are involved in the morphogenesis of tissues and organs. These brain regions further divide into sub-regions (see Table 1). The hindbrain divides into rhombomeres which originates the rhombencephalon. Neural crest cells form ganglia above each rhombomere. The neural tube becomes the germinal neuroepithelium and serves as a source of new neurons during brain development. By the 6 month of gestation, the cerebral hemispheres cover the whole top and sides of the brain including the cerebellum. Cerebellar development begins from this moment, but will not be complete until 2 years after birth. Six distinct layers are now differentiated within the cerebral cortex and almost all of the neurons within the central nervous system are present by the end of the 6 month of life (see Table 2).

1.1 The Brain Develops From the Inside-out

Currently, it is not known what initiates migration even though many theories have been proposed. Primitive neurons (or neuroblasts) migrate away from the ventricular border toward the external layer of thickening neural tube vesicle (the telencephalon) following an inside-out sequence of development. The telencephalon progressively thickens developing into the brain cortex where most higher level mental activity resides (perception, cognition, etc.). The cerebral cortex of higher forms is made up of six cell layers and the layer formation cycle is initiated by external signals (chemotactic molecules). Each successive migration ascends farther, progressively forming more superficial layers (fifth, fourth, third, second, and first) beyond the layer that was initially laid down and each layer has its distinct pattern of organization and connections. The later arriving cells appear to migrate out following the same pattern along the radial glial guide cells that were originally used by previous neurons. It is of critical importance that the earlier groups leave the glial guide cell before the next wave of migrating neurons start to ascend. The mechanism by which cells migrate in a timely manner to the appropriate position is not yet fully understood. Some studies suggest that astroglia cells play a role in the early establishment of the distribution pattern, growth, and maturation of the neural microvessels, which may also serve as guidance for further neuronal migration [10]. It is clear that if the mechanism is hindered, the ensuing traffic of cells pile-up producing clumps of neurons which determine the formation of heterotopic brain regions. These clumps of misallocated neurons are characteristic of cortical dysplasia which can facilitate

Table 2 Stages of brain development

Developmental stage	Main feature of developmental stage
Induction	Production of cells that will become nervous tissue
Proliferation	Mitosis
Migration	Cell movement toward the appropriate brain areas
Differentiation	Development of neurons into particular type
Synaptogenesis	Formation of appropriate synaptic connections
Selective cell death	Elimination of misallocated cells and cells that failed to form the proper synaptic connections
Functional validation	Strengthening of synapses in use and weakening of unused ones

or be a later cause of a variety of neurological diseases such as schizophrenia, severe seizure disorder, loss of motor skills, mental retardation, etc.

1.2 Maturation of Nerve Cells

Once the migrating nerve cells have reached their final position, the first dendritic protrusions start developing. Recent studies have shown that calcium signals plays a critical role in developing nerve cells by promoting the growth cone, which is the motile structure at the tip of an advancing axon or dendrite [11]. Dendritic branches (where the input to the neuron occurs) will come out from the cell body emerging from either the apical end of the pyramidal shaped cell growing toward the surface or from the basal side thus moving laterally and deeper into the cortex. Dendritic development provides a large connective area for the synaptic terminals which originate from other neurons and lead to the establishment of a neuronal network. The axon projection carries nerve signals away from the soma of the neuron at a distance tens of thousands of times the diameter of the soma in length. Many neurons have only one axon, but this axon may undergo extensive branching, which enables communication with multiple target cells. The initial tract of the emerging axon, or "axon hillock," presents the greatest density of voltage-dependent sodium channels and therefore having the most negative hyperpolarized action potential threshold. This characteristic makes the axon hillock the most easily-excited part of the neuron and the spike initiation zone for the axon. While the axon and axon hillock are generally involved in information outflow, this region can also receive input from other neurons. Recent studies have shown the protein PSD-95 seems to play an important role in building the physical scaffolding of the synapse [12].

Another critical aspect of brain development is the myelinization, or the deposit of electrically insulating phospholipid (myelin) sheaths around the axons. This process is performed by oligodendrocytes in the CNS and Schwann cells in the peripheral nervous system (PNS). These myelin sheaths have the primary function of increasing the resistance across the axonal membrane (reducing the leakage of electrical current). This allows the action potentials to travel much faster (over 100 m/s) than it normally would in unmyelinated axons of the same diameter. The myelinization has a rather well recognized time table during brain development. Fetal cerebral tissue (except basal ganglia) and cerebellar matter during pre-natal period are devoid of myelin sheaths. The myelinization process starts in the brain stem and basal ganglia after 29 gestational weeks. Oligodendrocytes derive from type 2 astrocytes also known as oligodendrocytes precursor cells (identified by the expression of specific antigens such as the ganglioside GD3 and the chondroitin sulfate proteoglycan NG2) [13, 14]. In rats, these oligodendrocyte precursor cells arise from the subventricular zone during late stage embryogenesis through early postnatal development [15, 16] and then migrate to populate the white and grey matter where they provide myelinization of the neuronal axons. Generally, fibers serving the primary sensory (touch, vision, audition etc.) and motor areas are myelinated shortly after birth while those involved with more complex cognitive and associative functions myelinate later.

2 Vascular Differentiation and BBB Development

2.1 Origin of the Vascular Endothelial Cells

Even before the onset of circulation, the neural tube of the developing embryo is vascularized by a capillary plexus (external vascularization of the brain) which later will differentiate into veins and leptomeningeal arteries [17]. From this external system, during the early embryonic neuroectoderm proliferation, blood vessels sprout radially from the perineural plexus into the neural tissue. This process gives rise to manifold branches and an astomose with other sprouts to form an undifferentiated capillary plexus. Interestingly, the onset of angiogenesis in the mammalian brain occurs at very reproducible stages (both temporally and spatially defined) of embryonic development [18, 19]. In more detail, blood vessel formation begins in the outer wall of the yolk sac. The yok sac endoderm causes the induction of numerous blood islands in the splanchnic mesoderm. The blood islands contain and are founded by hemangioblasts (bipotential stem cells) which can either give rise to angioblasts (endothelial cells progenitors or vasoformative cells) or blood-forming hematopoietic cells. The fusion of the hemangioblasts leads to the formation of blood islands and subsequently to the establishment of the primordial vascular plexus (vasculogenesis) from where the vasculature is further extended by sprouting of new capillaries (angiogenesis). Angioblasts can migrate and fuse with other angioblasts and capillaries, or form a vessel in situ. It is well known that the primary capillary plexus is remodeled several times, whereas the formation of new vessels occurs while others regress until a mature vascular system is formed. Even the direction of blood flow can change many times during this process.

The mechanisms underlying the differentiation of hemopoietic cells in blood islands are still poorly understood. Cell signaling, leading to differentiation of angioblast in function-specific endothelial cells, may occur through the release of soluble cytokines, cell-to-cell adhesion, and the synthesis of matrix proteins on which the endothelium adheres and grows. The involvement of brain derived factor, such as the vascular endothelial growth factor (VEGF) is determinant during embryonic brain angiogenesis. During the early stage of brain development the differential distribution of VEGF-A in the extracellular space plays a critical role for the regulation of vascular branching patterns during angiogenesis [20]. This is in agreement with the observation that the expression level of VEGF is high during the development of the embryonic brain but low in the adult brain (under normal physiological conditions). In animals, the expression level of VEGF receptors 1 and 2 (flt-1 and flk-1) as well as the tyrosine kinase angiopoietin receptors (tie-1 and tie-2) reflect the expression pattern observed for VEGF in the embryonic and adult brain [21–26]. Later during embryonic development, these molecules are restrictively expressed at the endothelial cell level where they promote endothelial cell growth and development of the vascular system which acquires organ-specific properties. Another important determinant for the differentiation of the vascular endothelium is the local environment. The interaction with surrounding cells (e.g., pericytes, astrocytes, smooth muscle, neurons, etc) modulates the acquisition and maintenance of tissue specific properties by which the endothelial cells can maintain the optimal homeostasis of different organs. Examples of these different barriers are represented by the arachnoid epithelium forming the middle layer of the meninges; the choroid plexus epithelium surrounding neurons specialized in the secretion of substances forming the cerebrospinal fluid (CSF) and the brain endothelium which at the capillary level forms the blood-brain barrier (BBB).

2.2 Induction and Formation of the Blood-Brain Barrier

The BBB represent a highly specialized dynamic and functional blood-brain interface that guarantees the brain homeostasis, greatly decreases the permeability of the paracellular pathways, regulates the passage of blood-borne substrates (ions, nutrients, amino acids, etc.) between the blood and the brain, provides a defense line against the passage of potentially harmful xenobiotic substances, and modulates the immune response at the brain parenchyma. Because brain vascular EC are derived from external endothelial cells by angiogenesis, there is an increasing amount of data suggesting the differentiation of endothelial cells (EC) into a BBB phenotype is modulated by the brain environment. This is demonstrated by the expression of early cell surface markers specific for the BBB endothelium [27]. Vascular endothelial cells are in fact adapted to the needs of the surrounding tissue and exhibit a remarkable heterogeneity at the functional and structural level.

One specific example is the BBB endothelium, which forms the vascular bed of the brain microcapillaries (Fig. 2). These particular vascular endothelial cells replaced the primordial glial cells as barrier constituent and dynamically adapted to minimize the interference between central and peripheral neurotransmitters [28, 29].

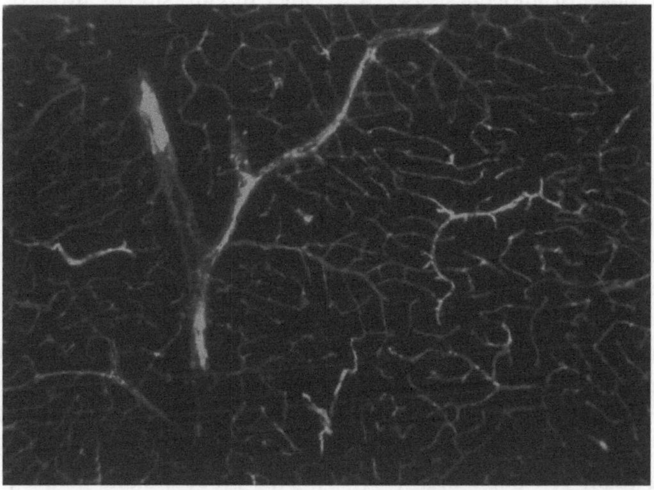

Fig. 2 Brain microvasculature. A mouse brain was perfused with FITC-albumin. Note how the presence of the BBB impede the extravasation of FITC albumin into the brain parenchyma

This evolutionary step allowed both glial and endothelial cells to acquire specialized functions to provide neuronal support (glial cells) [30] as well as barrier and regulatory functions (endothelial cells) [31]. This strategy allowed the establishment of an efficient system to precisely tune the ionic homeostasis of the CNS leading to the development of increasingly more complex brain functions. The induction of BBB properties on the brain vascular endothelium is strongly dependent on the association with perivascular glial cells [32, 33]. This is demonstrated by the fact that when the BBB endothelium is isolated and cultured in vitro, the cells undergo a partial dedifferentiating process reversing them to a phenotype which resembles "normal" vascular endothelial cells even though the genetic characteristic of the original phenotype are maintained [34]. Glial interaction with the cerebral endothelium regulates protein expression, modulates endothelium differentiation, and appears to be critical for the induction and maintenance of tight junctions and BBB properties [35]. This suggests that a close interaction (contacting and releasing factors such as glial-derived neurotrophic factor) with glial cells is necessary to induce and maintain the BBB phenotype. This has been demonstrated in vitro when endothelial cells cultured in presence of glial cells or glial conditioned media develop a higher trans-endothelial electrical resistance (TEER) [36].

The use of reverse transcription-polymerase chain reaction (RT-PCR) demonstrated that mRNAs for other representative BBB markers, including P-glycoprotein, gamma-GTP, transferrin receptor, and glucose transporter type 1 (GLUT-1) were also strongly upregulated in endothelial cells co-cultured with astrocytes [37]. A novel "BBB protective" role of astrocytes has also been described, including NO-mediated glial Interleukin-6 (IL-6) release which triggers the release of α2-macroglobulin, an MMPs inhibitor [38–40]. Despite the primary role played by glial cells in facilitating the differentiation of brain vascular endothelium into a BBB phenotype, several studies suggest that they are not the initiating trigger since BBB properties appear well ahead of the onset of gliogenesis [41, 42]. Other in vitro studies strongly support the hypothesis that the neural microenvironment plays a critical role for the early onset of BBB properties in microvascular endothelial cells during embryonic brain development [27, 28, 43]. For example, an increase in (-glutamyl transpeptidase activity (an enzyme responsible for amino acid transport across the BBB) [44] and an increased expression level of the tight junction protein occludin [45] has been shown in brain microvascular endothelial cells co-cultured with neurons. Another recent study demonstrated the development of enhanced barrier properties in the forms of increased TEER, lower permeability to the paracellular marker fluorescein, and positive modulation of the assembly of tight junction particles in rat brain microvascular endothelial cells co-cultured with embryonic neural progenitor cells (NPC) [42].

Other than glial cells and neurons, the association with microglia and pericytes also facilitate the BBB formation [46]. All these evidences together suggest that the development of unique properties of the brain microvasculature at the BBB level is a consequence of complex tissue-specific interactions between the endothelium and the surrounding brain cell population [47, 48]. This process starts taking place during neovascularization of the mammalian brain (when endothelial cells invade the neuroectoderm from pre-existing perineural plexus and establish the first vascular system) and

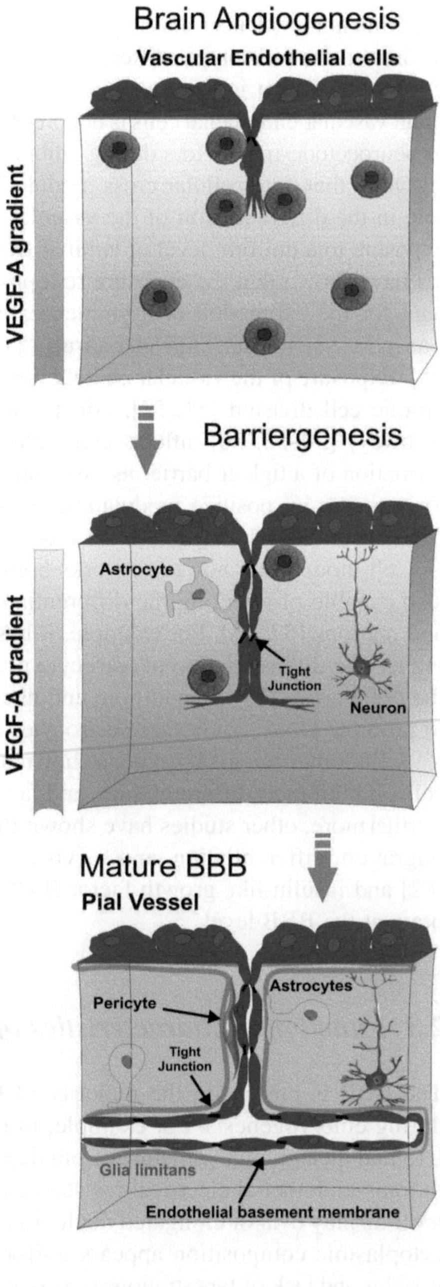

Fig. 3 Schematic representation of the steps of BBB differentiation. Starting from brain angiogenesis, vascular sprouts radially invade the embryonic neuroectoderm guided by a concentration gradient of VEGF-A. Following brain angiogenesis, we assist to the barriergenesis or the formation and maturation of the BBB. During the barriergenesis TJ complexes are formed and the endothelial cells develop polarized transport systems which endows the characteristic transport specificity of BBB endothelium. Maturation of the BBB endothelium is modulated by its close interaction with astrocytes, pericytes and neurons which also play a primary role for the maintenance of the BBB

continues until the establishment of a mature BBB (Fig. 3). The induction of distinctive BBB characteristics also includes the development of specialized transporter systems that evolved to support the changes in nutritional needs which parallel the brain development from the embryonic to postnatal stages [37, 49]. Despite the numerous experimental evidences and hypothesis that have been formulated so far, the process leading to the phenotypic differentiation of the BBB (barriergenesis) is not yet fully understood. What is most certain is that the development of BBB characteristic in brain vascular endothelial cells is not pre-determined but rather induced by the exposure to neuroectodermal factors during embryogenesis.

Other than intercellular cross modulation, another element which plays a crucial role in the differentiation of the vascular endothelium into a BBB phenotype is the exposure to a uniform level of laminar shear stress [36, 50–53]. Recent in vitro studies have shown that the exposure to a physiological level of shear stress is determinant for the expression of asymmetrically-localized enzymes and carrier-mediated transport systems that engender a truly "polarized" BBB endothelium phenotype [54, 55]. Exposure of the vascular endothelium to intraluminal shear stress also decreases mitotic cell division [51, 52], affects the cell metabolism (toward a more aerobic pathway) [50–52, 56], affects cytoskeletal gene expression [51], and leads to the formation of a tighter barrier as assessed by measurement of TEER [36, 52, 55, 56]. This occurs by positive modulation of expression, association, and localization of occludin and ZO-1 [53] which are crucial structure of tight junction particles.

Cell modulation however, works both ways. The brain vascular endothelium is also capable of affecting the differentiation of neighborhood cells such astrocytes and neurons [57, 58]. For example, it has been demonstrated that endothelial cells induce the differentiation of astrocyte precursor cells (APC) into astrocytes within the optic nerve [57]. In addition, endothelial cell conditioned medium was capable to promote DNA synthesis in astrocytes and pericytes but not in oligodendrocytes [59]. Endothelial cells have been shown to stimulate the self-renewal of neural stem cells, inhibit their differentiation, and facilitate neuroepithelial cell contact [60, 61]. Furthermore, other studies have shown that a variety of factors mediating neuronal migration, differentiation, and survival such as tissue plasminogen activator (tPA) [62] and insulin-like growth factor II (IGF-2) [63] are selectively expressed in the brain at the BBB level.

2.3 Functional Characteristics of the BBB

The BBB is present in the majority of brain capillaries and starts forming early during embryogenesis. For example, in rats the BBB starts forming at around day E13 and then gradually matures from day E18 as demonstrated by the rapid increase in trans-endothelial electrical resistance (TEER) [64]. Endothelial cells at the BBB level display oval or elongated nuclei in the direction of the capillary itself and their cytoplasmic composition appears uniform in thickness with very few pinocytotic vesicles and lack of fenestrations (i.e., openings). The outer leaflets toward the luminal side of adjacent cell membranes fuse together to form inter-endothelial connection structures called tight junctions (TJ) (Fig. 4). TJ form a diffusion barrier that selectively excludes most blood-borne and xenobiotic substances from entering the brain. In contrast to lipid soluble substances including alcohols, anesthetics and barbiturates, the BBB is highly impermeant to polar molecules or water soluble electrolytes.

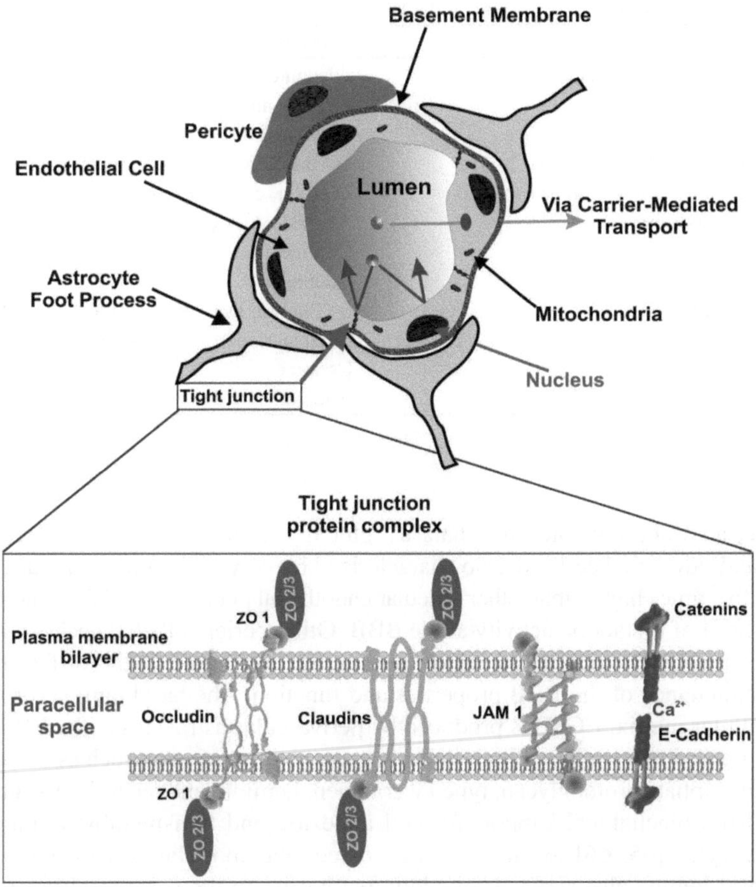

Fig. 4 Schematic representation of a brain microvasculature. The blood brain barrier is created by the tight apposition of contiguous endothelial cells. Note the endothelial cells lining blood vessels are in close contact with a variety of accessory cells such as astrocytes and pericytes which modulate the expression of BBB characteristics. Tight junctions between contiguous endothelial cells prevent the passage of large molecules and pathogens between the blood and the brain. Tight junctions consist of rows of transmembrane proteins (major types are claudins, occludins, and junctional adhesion molecules) anchored in the membranes of two adjacent endothelial cells while the intracellular portion is anchored to cytoskeletal proteins (e.g., actin) through scaffold protein such as ZO-1

However, the passage of certain water soluble, but biologically important substances, such as D-glucose or phenylamine are regulated by a variety of specific carrier-mediated transport systems (see Table 3) which allow the transit of nutrients and other important substances across the BBB. The membrane localization of these enzymes is indicative of the polarity of endothelial functions in the control of the blood-brain interface.

From a morphological point of view, the endothelial cytoplasm is highly enriched with a variety of dehydrogenases, adenosine triphosphatase, NADH, monoamine

Table 3 Specific transport systems at the BBB

Transport systems	Substrates
Large neutral amino acids	L-leucine, phenylalanine and tryptophan
Cationic amino acids	L-lysine and L-arginine
Anionic amino acids	L-glutamate and L-aspartate
Neurotransmitter transporters	GABA, serotonin, norepinephrine
Organic anion and cation transporters	Digoxin, carnitine, etc.
ABC transporters	ABCB1, ABCC1, ABCG2
Amine	Choline
Nucleoside	Adenosine
Energy transporters	D-Glucose, Dehydroascorbic acid
	L-Lactate, monocarboxylates
	Creatine
Purine	Adenine

oxidase, acid and alkaline phosphatases, glutamyl transpeptidase, and decarboxylase, etc. Endothelial cells are also characterized by very high mitochondrial density, almost five times higher than other vascular endothelial phenotypes. This demonstrates a high level of metabolic activity at the BBB. Other periendothelial accessory structures of the BBB, such as the basal membrane, contribute to the differentiation and maintenance of the BBB properties and function. The basal lamina (or mature extracellular matrix, ECM) is produced by perivascular astrocytes at the BBB level. It is composed of a variety of high-molecular-weight glycoprotein such as fibronectin, heparan sulphate proteoglycan, type IV collagen, laminin, and tenascin. Its functions range from mechanical support for cell adhesion and transmembrane migration of leukocytes [65, 66] to the regulation of cellular modulation across the BBB. The basal lamina also serve as an additional barrier to the passage of macromolecules between the cells and the vascular system [67] and might contribute to BBB maintenance [68] as well.

2.4 Modulation of the Blood-Brain Barrier Tight Junction

In vivo studies and freeze-fracture analysis of the vascular brain endothelium have shown that during brain development tight junction (TJ) particle association shifts from the outer membrane leaflet, or exocytoplasmic face (E-face), toward a more predominant inner membrane leaflet, the protoplasmic face (P-face) (post-natal). This indicates a progressive maturation of the BBB from the early stage of embryonic development and also is in agreement with the progressive increase of BBB tightness. In humans the TJ proteins ZO-1, claudin and occludin are present from the early phases of brain development, but BBB impermeability (due to the assembly of the TJ) increases over an extended gestational interval [69, 70] with an increasing amount of TJ formation at the inner membrane level. A number of proteins such as claudin-1, claudin-3, and claudin-5, occluding 1 and 5, cingulin, and the TJ associated protein 7H6 [64, 71–73] concur to the formation of TJ between juxtapose vascular endothelial cells. In addition, the cytoplasmic peripheral membrane proteins ZO1, 2, and 3 provide a critical linkage with the cytoskeletal protein actin [64, 74, 75], thus regulating the paracellular permeability. Recently data has shown that JAM-1 (a member of the junctional adhesion molecules family) plays a primary role in the organization of tight junctional structures [76, 77] and leukocyte extravasation observed in multiple sclerosis [78] and other neuroinflammatory diseases. The abundance of a particular type of protein such claudin-1 and claudin-3, versus claudin-5, is associated with a more or less stringent BBB permeability. This is determined by the predominant localization of the TJ between the P-face of two adjacent endothelial cells versus the E-face as shown by freeze fracture analysis [74, 79]. The signaling pathways involved in TJ regulation also comprise a complex regulation of a variety of factors such as TNF alpha, G-proteins, serine, threonine, and tyrosine kinases, extra- and intracellular calcium levels, cAMP levels, and others [80] which modulates the cytoskeletal regulation of barrier function [81] and the morphology of the BBB endothelium.

3 Role of BBB in the Development of Neurological Disease

Deterioration in BBB functions may play a major role in the pathogenesis of neurological disease since the BBB dynamically responds to many events associated with flow disturbances (e.g., focal ischemia), free radical release and cytokine generation. Many neurological disorders and lesions are associated with increased BBB permeability such as neoplasia, hypertension, dementia, epilepsy, infection, multiple sclerosis, Alzheimer's disease, and brain trauma [82–88]. Any disorder which affects BBB function will cause secondary effects on cerebral blood flow and vascular tone, further influencing transport across the BBB. Although the association between disease and BBB disruption is clear, the nature of this association is not always evident. An important question is whether impaired BBB function is a result of the pathological condition or whether the BBB impairment itself is the primary pathogenic factor.

Trans-endothelial leukocyte migration across an altered BBB is one of the most prominent features of many neuroimmune disorders. Leukocytes are found in large numbers in the brain following trauma and in certain neurodegenerative diseases such as multiple sclerosis. It is not clear whether the cells cross the endothelium through TJ, via a large pore or vacuole in the EC, or through some other mechanism [89]. Nevertheless, the passage of cells across the BBB happens only when several cell types (blood cells and endothelial and/or glia) are activated. The concept of EC activation was first proposed in the 1980s [90]. Locally secreted pro-inflammatory cytokines (e.g., interleukin-6 (IL-6), tumor necrosis factor-alpha (TNF-α), and interleukin-1 beta (IL-1β) mediate the interactions between leukocytes and microvascular EC which ultimately determines BBB failure by disorganizing cell-cell junctions, decreasing the brain solute barrier and enhancing leukocyte endothelial adhesion and migration. More specifically, vascular EC at the site of inflammation undergo a number of morphological and functional alterations, including increased permeability, hypertrophy, accumulation of intracellular organelles and the expression of a variety of adhesion molecules such as VCAM-1, E-selectin, P-selectin, ICAM-1. The interaction between these adhesion molecules with glycosylated ligands on circulating leukocytes allows for the initial contact and probing (tethering) of the BBB endothelial cells for the presence of chemokines immobilized by association with glycosaminoglycans on the luminal endothelial membrane surface [65]. The interaction between chemokines and G-protein-coupled chemokine receptors (GPCR) on the leukocyte membrane induces a shift from a low to high affinity state of the leukocyte integrins which facilitate their stable adhesion and their transmigration into the perivascular space [65, 91].

4 Pre-natal Pharmacological Treatment and Developmental Neurotoxicity

The outcome of exposure to potential teratogenic compounds yields different teratogenic effects depending on the developmental stage of the fetus (embryogenesis, organogenesis and fetal growth). For example, during embryogenesis, exposure to toxic substances can end up with the termination of the embryo or not have consequences at all. During organogenesis when the connective tissues, organs, and body systems are formed, a similar exposure can lead to major structural/developmental deficits (e.g., anencephaly, hydrocephaly, etc.) while during the final stage of fetal growth, the deficit will be more functional rather than structural. Toxicologic effects also largely depend on pharmacokinetic and pharmacodynamic parameters which include the nature of the compound, the delivered dose, the duration of the exposure, the presence of a placental barrier, and the metabolic ability of the fetus and the mother to neutralize the toxic agent [92]. In addition to the fetus' developmental stage, the vulnerability of the nervous system also depends on the presence of a functional BBB at the time of exposure.

In summary, all of these developmental variables yield a large number of different windows of vulnerability leading to different neurotoxic outcomes ranging from abnormal neural proliferation or irregular neuronal positioning during corticogenesis to dysfunctional receptor mediated neurotransmission. For example, the administration of nitrogen mustard alkylating agents used to treat various types of cancer and some autoimmune disorders such as cyclophosphamide has been shown to decrease the viability and growth of neurons, damage nuclear DNA, and induce early apoptotic morphological changes [93–95] thus demonstrating a toxic effect when proliferation is an active ongoing process in a specific region of the brain. This also determines the neurotoxic outcome on brain development. Similarly, to effects to nitrogen mustard alkylating agents will be shown by the exposure to ionizing radiation or the antimitotic and anti-angiogenic agent methylazoxymethanol (MAM) [96]. The administration of anticonvulsant and antiepileptic drugs such as phenobarbital and phenytoin have been shown to determine small head size as well cognitive impairments in new born children [97–99]. The exposure to organophosphorous insecticides diazinon (DZN) and chlorpyrifos (CPF) affect transmitter systems thought to support memory function [100]. They may also affect the development apoptosis, thus leading to an unwanted alteration of cell number in a specific brain region. Thalidomide, a drug prescribed during the late 1950s and early 1960s to pregnant women, as an antiemetic to combat morning sickness was found to be a potent teratogenic compound and more recently to act as an angiogenesis inhibitor which affect prenatal vasculogenesis [101]. This is a key step in brain development required to support normal neuronal migration and maturation.

More recently, a number of reports have shown potential developmental neurotoxicity of ethanol which affects neurotrophic factors, their signal transduction pathways and the involvement of muscarinic receptor-stimulated phosphoinositide hydrolysis [102–104]. Exposure to ethanol also seems to decrease and delay the myelinization of the neuronal axons [105, 106] which in humans progress through adolescence [107]. In addition, different industrial chemicals (e.g., toluene, lead, methylmercury, polychlorinated biphenyls [PCBs], arsenic, and toluene) have been recognized as primary causes of neurodevelopmental disorders and subclinical brain dysfunction [108, 109].

Cross-species comparability between experimental animals strongly supports the assumption that similar effects on brain development would be observed in humans taking into account the time and spatial differences in brain development. However, animal models may not be well representative of specifically human characteristics, thus posing a serious concern about their reliability. Given these concerns, the next question is "what drugs can be safely administered during pregnancy?" For now, what we have is a significant lack of adequate information on medications that sometimes must be used in pregnancy, such as antibiotics, drugs to treat seizure disorders, hypertension, and psychiatric conditions. In order to define a guideline for drugs to be used or not during pregnancy, the Food and Drug Administration established a Fetal Risk Summary and categorized the drugs (A, B, C, D, X) based on their safety of use during pregnancy (see Table 4).

For example, streptomycin, gentamicin, and tobramycin which are antibiotic drugs belonging to the class of aminoglycosides have been reported to cause cranial nerve damage in children whose mothers were prescribed these drugs as long-term therapy for tuberculosis [110]. In contrast, penicillin G, amoxicillin, levofloxacin and other antibiotics are considered to be safe for use during pregnancy [110]. Tricyclic drugs (imipramine derivatives) and monoamine oxidase inhibitors used as antidepressants have been shown to have teratogenic effects on the fetus. Other antidepressants such as fluvoxamine and paroxetine (serotonergic antidepressant drugs) have shown no adverse effects on animals, however, there are no sufficient clinical data to prove them safe in humans.

Given that the least understood factors involving potential fetal harm is teratogenicity, when taking into account all the maternal physiology, basic pharmacokinetic and pharmacodynamic factors, the drug-membranes physiochemical interactions, minimizing exposure to potentially toxic substances seems so far the only sustainable goal.

Table 4 Drug category based on safety during pregnancy

Category	Drug class	Drugs
Category A – drugs safe during pregnancy	Over the counter drugs	Folic acid, Vitamin B6, etc
Category B – drugs safe during pregnancy	Antibiotics	Penicillin G, Amoxicillin, Chloramphenicol, Ciprofloxacin, Doxycycline, Levofloxacin, Cephalosporins, Erythromycin, Clindamycin, Metronidazole
	Analgesic	Acetaminophen
	Synthetic corticosteroid	Prednisone
	Hormones	Insulin
	NSAID	Ibuprofen
	H_2-receptor antagonist	Famotidine
Category C – drugs likely to cause problems	Antibiotics	Streptomycin, Clindamycin, Gentamicin, Vancomycin, Tobramycin, Ciprofloxacin, etc.
	Antidepressant	Tricyclic drugs, monoamine oxidase inhibitors, serotonergic antidepressant drugs, etc
	Antipsychotic agent	Prochlorperazine
	Triazole antifungal drug	Fluconazole
Category D – drugs that have clear health risks for the fetus	Anticonvulsant	Phenytoin, Trimethadione, Valproic acid, Carbamazepine, Phenobarbital, etc
	Alcohol	
	Antineoplastics (Alkylating agents)	Mechlorethamine, Procarbazine, Cyclophosphamide, Chlorambucil, Melphalan, Lomustine, etc
	Antineoplastics (Taxanes)	Paclitaxel and Docetaxel
	Coumarin derivatives	Warfarin
	Antidepressant	Lithium
Category X – drugs that have been shown to cause birth defects	Sedative	Thalidomide
	Retinoids	Vitamin A
	Androgenic hormones	Testosterone
	Statins	Simvastatin

5 Brain Development and BBB: Extrapolating Animal Data to Human Studies

Cross-species comparability between experimental animals and humans presents a variety of challenges from developmental, pharmacodynamic and pharmacokinetic points of view. All these variables must be taken into consideration when extrapolating data from animal models to humans since animal models may not be well representative of specific human characteristics. In this chapter we will focus on the distinctive differences between human and rodent brain development and BBB characteristics.

One of the main differences between humans and rodents is the ontogeny of brain development. Even though the gradient of maturations of developing brain regions in humans and rodents follows the same general profile, the timeline in which this occurs is very different. This timeline is measured in months in humans and in weeks in rodents. In addition, the maturation of the nervous systems in rodents occurs mostly postnatally, while in humans it occurs primarily prenatally. However, in humans, brain growth continues for at least 2–3 years after birth [111] and synaptogenesis, according to the developmental pattern of different brain regions, continues through adolescence [112]. In terms of neurotoxicological outcome, this represents a critical difference that needs to be taken into account when extrapolating animal studies to humans. Because the process affected (cell migration, myelination, and synaptogenesis) is different, the windows of vulnerability and the neurotoxic effects on the CNS will also be different. In addition to the different timeframe of brain development, there is evidence that humans lack the ability to metabolize drugs efficiently at the prenatal stage (e.g., lacking the enzymes A-esterase and carboxylesterase) in comparison to common small laboratory animal [113] and therefore, animal data have poor relevance for the human fetus [114].

From a pharmacological point-of-view, the ability to predict the permeability of a drug into the CNS across the BBB is of primary importance for the development of novel pharmacological treatment for a variety of neurological diseases. The most common models used to assess the distribution of drugs across the BBB and into the CNS are rodent models. While the human and the rodent BBB are structurally very similar, there are however functional differences in the mechanisms that regulate the entry of drugs and xenobiotics into the CNS. In particular, differences in terms of substrate specificity between the various drug efflux transporters [115] which prevent or reduce the entry of these drugs (e.g., nerve growth factors (116, 117); anti-neoplastic agents [118]; anti-epileptic drugs (AEDs) [119–122], and anti-retrovirals [121, 123–125] into the CNS.

A prominent one among these efflux transport systems and probably the most studied is P-glycoprotein (P-gp), an ATP-binding cassette transporter encoded by the multidrug resistance 1 (*MDR1* or ABCB1) gene in humans and the mdr1a and b genes in rodents. Several studies now suggest that substrate recognition (or transport efficacy) by P-gp differs between human and rodents. This has been verified at least for certain AEDs [121]. Other transport systems, including the receptor–mediated transporters such as the IL-1 [126] also present a differential substrate specificity related to the species studied.

6 Conclusion

BBB development is a multi-step process starting from vasculogenesis followed by angiogenesis and barriergenesis which span the early phases of pre-natal brain development and continues into postnatal development until full maturation of the BBB is reached. This process is regulated by a complex pattern of cell signaling, soluble factors, and local environment. These factors prompt endothelial cells to acquire a distinct barrier phenotype at the microcapillary level in the central nervous system, to determine specific functions and peculiar characteristics of the BBB endothelial cells, and to affect their ability to respond to different environmental cues. This process can be truly defined as a cross modulation of the neurovascular unit which is a conceptual model that considers the BBB function from the perspective of the interaction of a variety of cells from different origins comprised in the BBB unit. The greatest challenges that have not yet been fully addressed are the need to increase our knowledge about the molecular and cellular biology, the gene and protein expression, and the regional differences of brain microvascular endothelial cells and pericytes and their interactions with adjacent brain cells. Understanding the basic biology of how the BBB develops and works under normal and disease conditions will also provide insight on the integrative function of the brain.

Acknowledgments This work was supported by *Philip Morris USA* and *Philip Morris International* external research awards to Dr. Luca Cucullo.

References

1. Mitchell T, Jones EA, Weeks DL, Sheets MD (2007) Chordin affects pronephros development in Xenopus embryos by anteriorizing presomitic mesoderm. Dev Dyn 236(1):251–261
2. Chiba S, Kurokawa MS, Yoshikawa H et al (2005) Noggin and basic FGF were implicated in forebrain fate and caudal fate, respectively, of the neural tube-like structures emerging in mouse ES cell culture. Exp Brain Res 163(1):86–99
3. Zhang YT, Lander AD, Nie Q (2007) Computational analysis of BMP gradients in dorsal-ventral patterning of the zebrafish embryo. J Theor Biol 248(4):579–589
4. Holm PC, Rodriguez FJ, Kele J, Castelo-Branco G, Kitajewski J, Arenas E (2006) BMPs, FGF8 and Wnts regulate the differentiation of locus coeruleus noradrenergic neuronal precursors. J Neurochem 99(1):343–352
5. Simeone A, Avantaggiato V, Moroni MC et al (1995) Retinoic acid induces stage-specific antero-posterior transformation of rostral central nervous system. Mech Dev 51(1):83–98
6. Sasai Y, De Robertis EM (1997) Ectodermal patterning in vertebrate embryos. Dev Biol 182(1):5–20
7. Shimogori T, Banuchi V, Ng HY, Strauss JB, Grove EA (2004) Embryonic signaling centers expressing BMP, WNT and FGF proteins interact to pattern the cerebral cortex. Development 131(22):5639–5647
8. Abu-Khalil A, Fu L, Grove EA, Zecevic N, Geschwind DH (2004) Wnt genes define distinct boundaries in the developing human brain: implications for human forebrain patterning. J Comp Neurol 474(2):276–288
9. Lako M, Lindsay S, Bullen P, Wilson DI, Robson SC, Strachan T (1998) A novel mammalian wnt gene, WNT8B, shows brain-restricted expression in early development, with sharply delimited expression boundaries in the developing forebrain. Hum Mol Genet 7(5):813–822
10. Virgintino D, Maiorano E, Errede M et al (1998) Astroglia-microvessel relationship in the developing human telencephalon. Int J Dev Biol 42(8):1165–1168
11. Bolsover SR (2005) Calcium signalling in growth cone migration. Cell Calcium 37(5):395–402
12. El Husseini AE, Schnell E, Chetkovich DM, Nicoll RA, Bredt DS (2000) PSD-95 involvement in maturation of excitatory synapses. Science 290(5495):1364–1368

13. Chekenya M, Enger PO, Thorsen F et al (2002) The glial precursor proteoglycan, NG2, is expressed on tumour neovasculature by vascular pericytes in human malignant brain tumours. Neuropathol Appl Neurobiol 28(5):367–380
14. Levine JM, Stincone F, Lee YS (1993) Development and differentiation of glial precursor cells in the rat cerebellum. Glia 7(4):307–321
15. Levison SW, Goldman JE (1993) Both oligodendrocytes and astrocytes develop from progenitors in the subventricular zone of postnatal rat forebrain. Neuron 10(2):201–212
16. Young GM, Levison SW (1996) Persistence of multipotential progenitors in the juvenile rat subventricular zone. Dev Neurosci 18(4):255–265
17. Risau W (1986) Developing brain produces an angiogenesis factor. Proc Natl Acad Sci USA 83(11):3855–3859
18. Robertson PL, Du BM, Bowman PD, Goldstein GW (1985) Angiogenesis in developing rat brain: an in vivo and in vitro study. Brain Res 355(2):219–223
19. Kuban KC, Gilles FH (1985) Human telencephalic angiogenesis. Ann Neurol 17(6):539–548
20. Ruhrberg C, Gerhardt H, Golding M et al (2002) Spatially restricted patterning cues provided by heparin-binding VEGF-A control blood vessel branching morphogenesis. Genes Dev 16(20):2684–2698
21. Ferrara N, Houck K, Jakeman L, Leung DW (1992) Molecular and biological properties of the vascular endothelial growth factor family of proteins. Endocr Rev 13(1):18–32
22. Risau W (1998) Development and differentiation of endothelium. Kidney Int Suppl 67:S3–S6
23. Kim KJ, Li B, Houck K, Winer J, Ferrara N (1992) The vascular endothelial growth factor proteins: identification of biologically relevant regions by neutralizing monoclonal antibodies. Growth Factors 7(1):53–64
24. Yamaguchi TP, Dumont DJ, Conlon RA, Breitman ML, Rossant J (1993) flk-1, an flt-related receptor tyrosine kinase is an early marker for endothelial cell precursors. Development 118(2):489–498
25. Millauer B, Wizigmann-Voos S, Schnurch H et al (1993) High affinity VEGF binding and developmental expression suggest Flk-1 as a major regulator of vasculogenesis and angiogenesis. Cell 72(6):835–846
26. Wulff C, Wilson H, Dickson SE, Wiegand SJ, Fraser HM (2002) Hemochorial placentation in the primate: expression of vascular endothelial growth factor, angiopoietins, and their receptors throughout pregnancy. Biol Reprod 66(3):802–812
27. Risau W, Hallmann R, Albrecht U, Henke-Fahle S (1986) Brain induces the expression of an early cell surface marker for blood-brain barrier-specific endothelium. EMBO J 5(12):3179–3183
28. Abbott NJ (2005) Dynamics of CNS barriers: evolution, differentiation, and modulation. Cell Mol Neurobiol 25(1):5–23
29. Konsman JP, Tridon V, Dantzer R (2000) Diffusion and action of intracerebroventricularly injected interleukin-1 in the CNS. Neuroscience 101(4):957–967
30. Sykova E (1998) HERLaNC. In: Laming PR, Syková E, Reichenbach A, Hatton GI, Bauer H (eds) Glial regulation of the neuronal microenvironment. Cambridge University Press, Cambridge, pp 130–163
31. McCarron RM, Chen Y, Tomori T et al (2006) Endothelial-mediated regulation of cerebral microcirculation. J Physiol Pharmacol 57(Suppl 11):133–144
32. Utsumi H, Chiba H, Kamimura Y et al (2000) Expression of GFRalpha-1, receptor for GDNF, in rat brain capillary during postnatal development of the BBB. Am J Physiol Cell Physiol 279(2):C361–C368
33. Prat A, Biernacki K, Wosik K, Antel JP (2001) Glial cell influence on the human blood-brain barrier. Glia 36(2):145–155
34. Reinhardt CA, Gloor SM (1997) Co-culture blood-brain barrier models and their use for pharmatoxicological screening. Toxicol In Vitro 11:513–518
35. Liberto CM, Albrecht PJ, Herx LM, Yong VW, Levison SW (2004) Pro-regenerative properties of cytokine-activated astrocytes. J Neurochem 89(5):1092–1100
36. Siddharthan V, Kim YV, Liu S, Kim KS (2007) Human astrocytes/astrocyte-conditioned medium and shear stress enhance the barrier properties of human brain microvascular endothelial cells. Brain Res 1147:39–50

37. Virgintino D, Robertson D, Monaghan P et al (1998) Glucose transporter GLUT1 localization in human foetus telencephalon. Neurosci Lett 256(3):147–150
38. Aschner M (1998) Immune and inflammatory responses in the CNS: modulation by astrocytes. Toxicol Lett 102–103:283–287
39. Aschner M (1998) Astrocytes as mediators of immune and inflammatory responses in the CNS. Neurotoxicology 19(2):269–281
40. Cucullo L, Marchi N, Marroni M, Fazio V, Namura S, Janigro D (2003) Blood-brain barrier damage induces release of {alpha}2-macroglobulin. Mol Cell Proteomics 2(4):234–241
41. LeVine SM, Goldman JE (1988) Embryonic divergence of oligodendrocyte and astrocyte lineages in developing rat cerebrum. J Neurosci 8(11):3992–4006
42. Weidenfeller C, Svendsen CN, Shusta EV (2007) Differentiating embryonic neural progenitor cells induce blood-brain barrier properties. J Neurochem 101(2):555–565
43. Bauer HC, Bauer H (2000) Neural induction of the blood-brain barrier: still an enigma. Cell Mol Neurobiol 20(1):13–28
44. Tontsch U, Bauer HC (1991) Glial cells and neurons induce blood-brain barrier related enzymes in cultured cerebral endothelial cells. Brain Res 539(2):247–253
45. Savettieri G, Di Liegro I, Catania C et al (2000) Neurons and ECM regulate occludin localization in brain endothelial cells. Neuroreport 11(5):1081–1084
46. Minakawa T, Bready J, Berliner J, Fisher M, Cancilla PA (1991) In vitro interaction of astrocytes and pericytes with capillary-like structures of brain microvessel endothelium. Lab Invest 65(1):32–40
47. Engelhardt B (2003) Development of the blood-brain barrier. Cell Tissue Res 314(1):119–129
48. Risau W, Hallmann R, Albrecht U (1986) Differentiation-dependent expression of proteins in brain endothelium during development of the blood-brain barrier. Dev Biol 117(2):537–545
49. Gerhart DZ, Enerson BE, Zhdankina OY, Leino RL, Drewes LR (1997) Expression of monocarboxylate transporter MCT1 by brain endothelium and glia in adult and suckling rats. Am J Physiol 273(1 Pt 1):E207–E213
50. Cucullo L, Couraud PO, Weksler B et al (2008) Immortalized human brain endothelial cells and flow-based vascular modeling: a marriage of convenience for rational neurovascular studies. J Cereb Blood Flow Metab 28(2):312–328
51. Desai SY, Marroni M, Cucullo L et al (2002) Mechanisms of endothelial survival under shear stress. Endothelium 9(2):89–102
52. Santaguida S, Janigro D, Hossain M, Oby E, Rapp E, Cucullo L (2006) Side by side comparison between dynamic versus static models of blood-brain barrier in vitro: a permeability study. Brain Res 1109(1):1–13
53. Colgan OC, Ferguson G, Collins NT et al (2007) Regulation of bovine brain microvascular endothelial tight junction assembly and barrier function by laminar shear stress. Am J Physiol Heart Circ Physiol 292(6):H3190–H3197
54. McAllister MS, Krizanac-Bengez L, Macchia F et al (2001) Mechanisms of glucose transport at the blood-brain barrier: an in vitro study. Brain Res 904(1):20–30
55. Stanness KA, Westrum LE, Fornaciari E et al (1997) Morphological and functional characterization of an in vitro blood-brain barrier model. Brain Res 771(2):329–342
56. Cucullo L, Hossain M, Rapp E, Manders T, Marchi N, Janigro D (2007) Development of a humanized in vitro blood-brain barrier model to screen for brain penetration of antiepileptic drugs. Epilepsia 48(3):505–516
57. Mi H, Haeberle H, Barres BA (2001) Induction of astrocyte differentiation by endothelial cells. J Neurosci 21(5):1538–1547
58. Banerjee S, Bhat MA (2007) Neuron-glial interactions in blood-brain barrier formation. Annu Rev Neurosci 30:235–258
59. Estrada C, Bready JV, Berliner JA, Pardridge WM, Cancilla PA (1990) Astrocyte growth stimulation by a soluble factor produced by cerebral endothelial cells in vitro. J Neuropathol Exp Neurol 49(6):539–549
60. Shen Q, Goderie SK, Jin L et al (2004) Endothelial cells stimulate self-renewal and expand neurogenesis of neural stem cells. Science 304(5675):1338–1340

61. Louissaint A Jr, Rao S, Leventhal C, Goldman SA (2002) Coordinated interaction of neurogenesis and angiogenesis in the adult songbird brain. Neuron 34(6):945–960
62. Seeds NW, Basham ME, Haffke SP (1999) Neuronal migration is retarded in mice lacking the tissue plasminogen activator gene. Proc Natl Acad Sci USA 96(24):14118–14123
63. Feldman EL, Sullivan KA, Kim B, Russell JW (1997) Insulin-like growth factors regulate neuronal differentiation and survival. Neurobiol Dis 4(3–4):201–214
64. Wolburg H, Lippoldt A (2002) Tight junctions of the blood-brain barrier: development, composition and regulation. Vascul Pharmacol 38(6):323–337
65. Man S, Ubogu EE, Ransohoff RM (2007) Inflammatory cell migration into the central nervous system: a few new twists on an old tale. Brain Pathol 17(2):243–250
66. Tilling T, Engelbertz C, Decker S, Korte D, Huwel S, Galla HJ (2002) Expression and adhesive properties of basement membrane proteins in cerebral capillary endothelial cell cultures. Cell Tissue Res 310(1):19–29
67. Rosenberg GA, Kornfeld M, Estrada E, Kelley RO, Liotta LA, Stetler-Stevenson WG (1992) TIMP-2 reduces proteolytic opening of blood-brain barrier by type IV collagenase. Brain Res 576(2):203–207
68. Barber AJ, Lieth E (1997) Agrin accumulates in the brain microvascular basal lamina during development of the blood-brain barrier. Dev Dyn 208(1):62–74
69. Virgintino D, Errede M, Robertson D et al (2004) Immunolocalization of tight junction proteins in the adult and developing human brain. Histochem Cell Biol 122(1):51–59
70. Anstrom JA, Thore CR, Moody DM, Brown WR (2007) Immunolocalization of tight junction proteins in blood vessels in human germinal matrix and cortex. Histochem Cell Biol 127(2):205–213
71. Hirase T, Staddon JM, Saitou M et al (1997) Occludin as a possible determinant of tight junction permeability in endothelial cells. J Cell Sci 110(Pt 14):1603–1613
72. Satoh H, Zhong Y, Isomura H et al (1996) Localization of 7H6 tight junction-associated antigen along the cell border of vascular endothelial cells correlates with paracellular barrier function against ions, large molecules, and cancer cells. Exp Cell Res 222(2):269–274
73. Zhong Y, Saitoh T, Minase T, Sawada N, Enomoto K, Mori M (1993) Monoclonal antibody 7H6 reacts with a novel tight junction-associated protein distinct from ZO-1, cingulin and ZO-2. J Cell Biol 120(2):477–483
74. Wolburg H, Neuhaus J, Kniesel U et al (1994) Modulation of tight junction structure in blood-brain barrier endothelial cells. Effects of tissue culture, second messengers and cocultured astrocytes. J Cell Sci 107(Pt 5):1347–1357
75. Hartsock A, Nelson WJ (2008) Adherens and tight junctions: structure, function and connections to the actin cytoskeleton. Biochim Biophys Acta 1778(3):660–669
76. Vorbrodt AW, Dobrogowska DH (2004) Molecular anatomy of interendothelial junctions in human blood-brain barrier microvessels. Folia Histochem Cytobiol 42(2):67–75
77. Aurrand-Lions M, Johnson-Leger C, Wong C, Du PL, Imhof BA (2001) Heterogeneity of endothelial junctions is reflected by differential expression and specific subcellular localization of the three JAM family members. Blood 98(13):3699–3707
78. Man S, Ubogu EE, Ransohoff RM (2007) Inflammatory cell migration into the central nervous system: a few new twists on an old tale. Brain Pathol 17(2):243–250
79. Liebner S, Kniesel U, Kalbacher H, Wolburg H (2000) Correlation of tight junction morphology with the expression of tight junction proteins in blood-brain barrier endothelial cells. Eur J Cell Biol 79(10):707–717
80. Kniesel U, Wolburg H (2000) Tight junctions of the blood-brain barrier. Cell Mol Neurobiol 20(1):57–76
81. Lai CH, Kuo KH, Leo JM (2005) Critical role of actin in modulating BBB permeability. Brain Res Brain Res Rev 50(1):7–13
82. Abbott NJ, Revest PA, Romero IA (1992) Astrocyte-endothelial interaction: physiology and pathology. Neuropathol Appl Neurobiol 18(5):424–433
83. Garcia-Nunez C, Saez J, Garcia-Nunez JM, Grau J, Molto-Jorda JM, Matias-Guiu J (2007) [Passive smoking as a cerebrovascular risk factor]. Rev Neurol 45(10):577–581
84. Hankey GJ (1999) Smoking and risk of stroke. J Cardiovasc Risk 6(4):207–211

85. Marchi N, Angelov L, Masaryk T et al (2007) Seizure-promoting effect of blood-brain barrier disruption. Epilepsia 48(4):732–742
86. Sundstrom P, Nystrom L, Hallmans G (2008) Smoke exposure increases the risk for multiple sclerosis. Eur J Neurol 15(6):579–583
87. Hawkes CH (2007) Smoking is a risk factor for multiple sclerosis: a metanalysis. Mult Scler 13(5):610–615
88. Almeida OP, Garrido GJ, Lautenschlager NT, Hulse GK, Jamrozik K, Flicker L (2008) Smoking is associated with reduced cortical regional gray matter density in brain regions associated with incipient Alzheimer disease. Am J Geriatr Psychiatry 16(1):92–98
89. Greenwood J, Howes R, Lightman S (1994) The blood-retinal barrier in experimental autoimmune uveoretinitis. Leukocyte interactions and functional damage. Lab Invest 70(1):39–52
90. Ogonowski AA, Kaesemeyer WH, Jin L, Ganapathy V, Leibach FH, Caldwell RW (2000) Effects of NO donors and synthase agonists on endothelial cell uptake of L-Arg and superoxide production. Am J Physiol Cell Physiol 278(1):C136–C143
91. Cavender DE, Edelbaum D, Ziff M (1989) Endothelial cell activation induced by tumor necrosis factor and lymphotoxin. Am J Pathol 134(3):551–560
92. Pacifici GM (2006) Placental transfer of antibiotics administered to the mother: a review. Int J Clin Pharmacol Ther 44(2):57–63
93. Xiao R, Yu HL, Zhao HF, Liang J, Feng JF, Wang W (2007) Developmental neurotoxicity role of cyclophosphamide on post-neural tube closure of rodents in vitro and in vivo. Int J Dev Neurosci 25(8):531–537
94. Moore IM (1995) Central nervous system toxicity of cancer therapy in children. J Pediatr Oncol Nurs 12(4):203–210
95. Autio K, Rassnick KM, Bedford-Guaus SJ (2007) Chemotherapy during pregnancy: a review of the literature. Vet Comp Oncol 5:61–75
96. Bassanini S, Hallene K, Battaglia G et al (2007) Early cerebrovascular and parenchymal events following prenatal exposure to the putative neurotoxin methylazoxymethanol. Neurobiol Dis 26(2):481–495
97. Dessens AB, Cohen-Kettenis PT, Mellenbergh GJ, Koppe JG, van De Poll NE, Boer K (2000) Association of prenatal phenobarbital and phenytoin exposure with small head size at birth and with learning problems. Acta Paediatr 89(5):533–541
98. Hatta T, Ohmori H, Murakami T, Takano M, Yamashita K, Yasuda M (1999) Neurotoxic effects of phenytoin on postnatal mouse brain development following neonatal administration. Neurotoxicol Teratol 21(1):21–28
99. Adams J, Vorhees CV, Middaugh LD (1990) Developmental neurotoxicity of anticonvulsants: human and animal evidence on phenytoin. Neurotoxicol Teratol 12(3):203–214
100. Timofeeva OA, Roegge CS, Seidler FJ, Slotkin TA, Levin ED (2008) Persistent cognitive alterations in rats after early postnatal exposure to low doses of the organophosphate pesticide, diazinon. Neurotoxicol Teratol 30(1):38–45
101. Hallene KL, Oby E, Lee BJ et al (2006) Prenatal exposure to thalidomide, altered vasculogenesis, and CNS malformations. Neuroscience 142(1):267–283
102. Fattori V, Abe SI, Kobayashi K, Costa LG, Tsuji R (2008) Effects of postnatal ethanol exposure on neurotrophic factors and signal transduction pathways in rat brain. J Appl Toxicol 28(3):370–376
103. Han JY, Joo Y, Kim YS et al (2005) Ethanol induces cell death by activating caspase-3 in the rat cerebral cortex. Mol Cells 20(2):189–195
104. Balduini W, Reno F, Costa LG, Cattabeni F (1994) Developmental neurotoxicity of ethanol: further evidence for an involvement of muscarinic receptor-stimulated phosphoinositide hydrolysis. Eur J Pharmacol 266(3):283–289
105. David P, Subramaniam K (2005) Prenatal alcohol exposure and early postnatal changes in the developing nerve-muscle system. Birth Defects Res A Clin Mol Teratol 73(11):897–903
106. Zoeller RT, Butnariu OV, Fletcher DL, Riley EP (1994) Limited postnatal ethanol exposure permanently alters the expression of mRNAS encoding myelin basic protein and myelin-associated glycoprotein in cerebellum. Alcohol Clin Exp Res 18(4):909–916

107. Paus T, Zijdenbos A, Worsley K et al (1999) Structural maturation of neural pathways in children and adolescents: in vivo study. Science 283(5409):1908–1911
108. Grandjean P, Landrigan PJ (2006) Developmental neurotoxicity of industrial chemicals. Lancet 368(9553):2167–2178
109. Burry M, Guizzetti M, Oberdoerster J, Costa LG (2003) Developmental neurotoxicity of toluene: in vivo and in vitro effects on astroglial cells. Dev Neurosci 25(1):14–19
110. Nahum GG, Uhl K, Kennedy DL (2006) Antibiotic use in pregnancy and lactation: what is and is not known about teratogenic and toxic risks. Obstet Gynecol 107(5):1120–1138
111. Rice D, Barone S Jr (2000) Critical periods of vulnerability for the developing nervous system: evidence from humans and animal models. Environ Health Perspect 108(Suppl 3):511–533
112. Uylings HB, van Eden CG (1990) Qualitative and quantitative comparison of the prefrontal cortex in rat and in primates, including humans. Prog Brain Res 85:31–62
113. Juchau MR, Chao ST, Omiecinski CJ (1980) Drug metabolism by the human fetus. Clin Pharmacokinet 5(4):320–339
114. Rane A, Tomson G (1980) Prenatal and neonatal drug metabolism in man. Eur J Clin Pharmacol 18(1):9–15
115. de Boer AG, van der Sandt IC, Gaillard PJ (2003) The role of drug transporters at the blood-brain barrier. Annu Rev Pharmacol Toxicol 43:629–656
116. Poduslo JF, Curran GL (1996) Permeability at the blood-brain and blood-nerve barriers of the neurotrophic factors: NGF, CNTF, NT-3, BDNF. Brain Res Mol Brain Res 36(2):280–286
117. Pardridge WM, Kang YS, Buciak JL (1994) Transport of human recombinant brain-derived neurotrophic factor (BDNF) through the rat blood-brain barrier in vivo using vector-mediated peptide drug delivery. Pharm Res 11(5):738–746
118. Wiranowska M (1992) Blood-brain barrier and treatment of central nervous system tumors. J Fla Med Assoc 79(10):707–710
119. Loscher W (2007) Drug transporters in the epileptic brain. Epilepsia 48(Suppl 1):8–13
120. Sills GJ, Kwan P, Butler E, de Lange EC, van den Berg DJ, Brodie MJ (2002) P-glycoprotein-mediated efflux of antiepileptic drugs: preliminary studies in mdr1a knockout mice. Epilepsy Behav 3(5):427–432
121. Baltes S, Gastens AM, Fedrowitz M, Potschka H, Kaever V, Loscher W (2007) Differences in the transport of the antiepileptic drugs phenytoin, levetiracetam and carbamazepine by human and mouse P-glycoprotein. Neuropharmacology 52(2):333–346
122. Kwan P, Brodie MJ (2005) Potential role of drug transporters in the pathogenesis of medically intractable epilepsy. Epilepsia 46(2):224–235
123. Yang Z, Brundage RC, Barbhaiya RH, Sawchuk RJ (1997) Microdialysis studies of the distribution of stavudine into the central nervous system in the freely-moving rat. Pharm Res 14(7):865–872
124. Kim S, Scheerer S, Geyer MA, Howell SB (1990) Direct cerebrospinal fluid delivery of an antiretroviral agent using multivesicular liposomes. J Infect Dis 162(3):750–752
125. Kim RB (2003) Drug transporters in HIV therapy. Top HIV Med 11(4):136–139
126. Plotkin SR, Banks WA, Maness LM, Kastin AJ (2000) Differential transport of rat and human interleukin-1alpha across the blood-brain barrier and blood-testis barrier in rats. Brain Res 881(1):57–61

Seizure Propensity and Brain Development: A Lesson from Animal Models

Jana Velíšková, Annamaria Vezzani, and Astrid Nehlig

Abstract Both clinical and animal studies suggest that seizure susceptibility is the highest during development. Data in immature rats using different seizure models show that ictal activity spreads faster in developing brain compared to adults. This may be related to immaturity of endogenous seizure controlling systems, multifocal seizure origin, or to a short refractory period following a seizure event in developing brain. Animal studies of cortical malformations as well as genetic models offer important clues for factors underlying neuronal hyperexcitability in the terrain of compromised brain.

Animal models also show that long-term seizure consequences have age-specific features. Seizure-induced changes in neurotransmitter receptors and ion channels, plasticity-related functional changes in neuronal networks, induction of neuromodulatory molecules such as inflammatory mediators, neurotrophins, neuropeptides are among the possible mechanisms, which define the immediate and long-term response of the developing brain to seizures. Further elucidation of these aspects and their role in ictogenesis and epileptogenesis using clinically relevant experimental models in developing rats is instrumental for the future translation of these findings into clinical practice.

Keywords Animal seizure model • Development • Seizure susceptibility • Epileptogenesis • Neuronal damage • Network plasticity

1 Introduction

Epidemiological studies indicate that children have the highest incidence of seizures and epilepsy [1, 2]. The developing brain has unusually high propensity to seizures, which may have special features such as multifocal origin or high incidence

A. Vezzani
Department Neuroscience, Mario Negri Institute for Pharmacological Research, Via La Masa 19, 20156, Milano, Italy
e-mail: Vezzani@marionegri.it

D. Janigro (ed.), *Mammalian Brain Development*, Contemporary Neuroscience,
DOI 10.1007/978-1-60761-287-2_5,

of recurrent status epilepticus (SE) [3–5]. Differences between children and adults in motor and EEG pattern, sensitivity to treatment, and outcomes have been observed. Many types of seizures and epilepsies are more typical or even exclusive in childhood.

Pioneering studies in developing animals by Vernadakis and Woodbury led to the recognition that immature brain is not a smaller version of an adult brain [6, 7]. It is now well established that the propensity of the brain to seizures is changing based on distinct developmental stages. Developmental studies in experimental animals parallel the influence of maturational age on seizures and epilepsy observed in humans. For example, bilateral, asynchronous, multifocal convulsions, as well as SE develop more easily in 2-week old than in adult rats [8–11]. Age-dependent patterns in behavioral expression of seizures and EEG ictal activity can be also demonstrated in rat seizure models [12, 13]. In this chapter, we will describe the differences in seizure thresholds, patterns, and consequences of ongoing seizure activity as a function of age depending on the seizure model used.

2 Are Immature Animals More Susceptible to Seizures?

2.1 Chemoconvulsant Drugs

2.1.1 Seizure Models Showing Higher Susceptibility in Immature Animals

In most chemical models of seizures, rat pups have a lower threshold for convulsions, seizures rapidly progress to more severe stages, as well as develop into SE faster compared to adult rats [8, 14–16]. Such increase in seizure susceptibility in developing rats is mostly observed with convulsant agents acting at $GABA_A$ receptors, ionotropic glutamate receptors, but also with metrazol or flurothyl, convulsants with multiple mechanisms of action.

$GABA_A$ receptor antagonists such as bicuculline or picrotoxin reliably produce generalized clonic and tonic-clonic seizures from the first postnatal week. Bicuculline has been used in two different forms, bicuculline hydrochloride or bicuculline methiodide. Systemic intraperitoneal administration of bicuculline hydrochloride during the first two postnatal weeks (as early as postnatal day – PN3) produces occasional myoclonic twitches, but generalized clonic seizures reliably occur in rats older than PN12 [16, 17]. Generalized tonic-clonic seizures during the first two postnatal weeks can be elicited by higher doses of bicuculline and consist of running, uncoordinated "swimming" movements. In some animals, the tonic phase is not well developed, lacking the full stretch and often affecting one limb only [17]. The mature pattern of behavioral expression of bicuculline-induced seizures occurs from the third postnatal week [16, 17]. At this age, clonic seizures consist usually of bilateral forelimb clonus with rearing and preserved righting reflex. Tonic-clonic seizures usually begin with a wild run, followed by loss of righting reflex, tonic flexion or extension of the limbs, and subsequent long-lasting clonic phase [13].

The sensitivity to bicuculline hydrochloride decreases with age [16, 17]. CD50 for bicuculline is similar during the first three postnatal weeks and sharply increases from the fourth postnatal week to the adult values. The increased sensitivity of immature rats to bicuculline seems to be related to different pharmacokinetics of the drug during development. The significant difference in DC50 during the first three postnatal weeks from adulthood disappears by using a different administration route [17].

Bicuculline methiodide, in contrast to the hydrochloric form, does not cross the mature blood-brain barrier [18]. Thus, the seizure phenomena in PN18 and older rats are poor and consist mostly of automatisms such as sniffing, wet dog shakes, or scratching, and the generalized clonic seizures occur very rarely [19]. The highest susceptibility to bicuculline methiodide is during the first two postnatal weeks, when the blood-brain barrier is undeveloped and immature [20] and thus, both clonic and tonic-clonic seizures can be demonstrated during this developmental period [19].

Agonists to ionotropic glutamate receptors are commonly used to produce seizures during development [21]. Developing animals have also increased sensitivity to seizures induced by excitatory amino acids (EAA). The behavioral expression of seizures induced by glutamate receptor agonists is age- and receptor type-specific. Different EAA-related agents have been used, including glutamate, kainic acid, quisqualic acid, domoic acid, N-methyl-D-aspartic acid (NMDA), homocystein, and homocysteic acid [22–24]. The most popular EAA agent is kainic acid, an agonist to kainate receptors, which can be used to create a model of temporal lobe epilepsy in adult rats [25]. Lower doses produce self-sustaining SE in rats as young as PN1 [26, 27]. Automatisms are the first sign of kainic acid-induced seizures. The character of automatisms is changing as a function of age. During first two postnatal weeks, automatisms consist of scratching [15, 27, 28]. Scratching behavior is then gradually replaced by wet dog shakes from about PN18 and lasting up to adulthood [15, 27]. The origin of wet dog shakes seems to be in the dentate gyrus of the hippocampus [29, 30]. Interestingly, EEG recordings during kainic acid-induced seizures do not show hippocampal involvement up to the age of PN18, the time of wet dog shakes occurrence [27]. The clonic seizures are not well developed till PN14 and rather consist of uncoordinated swimming-like movements, tremor, and occasional hyperextension of the limbs [15, 27]. Since the end of second postnatal week, clonic seizures involve rearing with forelimb clonus. Following high doses of kainic acid, generalized tonic-clonic seizures may develop, which are characterized by running, tonic flexion/extension of all limbs and long-lasting clonus with loss of righting reflex. Occurrence of tonic-clonic seizures is usually associated with death of the animal [15].

Lately, seizures induced by an EAA receptor agonist NMDA attracted attention because of the relevance of the model to infantile spams, especially in combination with prenatal exposure to corticosteroids [31]. Using high doses of NMDA, seizures can be elicited in adult rats, however immature brain is extremely sensitive to this EAA agent, especially during first three postnatal weeks [32]. The interesting feature of the NMDA-induced seizures during the first three postnatal weeks (but not in older rats) is the presence of flexion (emprosthotonic) seizures, which resemble the flexion spasms in children with West syndrome.

Higher seizure susceptibility of immature rats compared to adults can be also demonstrated in models using flurothyl or metrazol (pentylenetetrazole, PTZ). Besides other effects, both agents act as antagonists at $GABA_A$ receptor sites [33, 34], so seizure phenotype is similar to that induced by other $GABA_A$ receptor antagonists. In flurothyl model, seizures during the first 2 weeks start as swimming-like movements resembling clonic seizures and rapidly progress into tonic-clonic seizures. This is different in older age groups (Fig. 1). At the end of third postnatal week, multiple clonic seizures precede the tonic-clonic seizure [13, 35]. In the PTZ model, clonic seizures very rarely occur early in development up to the third postnatal week due to rapid occurrence of tonic-clonic seizures [14]. After the third postnatal week, clonic seizures regularly precede tonic-clonic seizures in the PTZ model [14].

In adult rats, forebrain structures are most sensitive to the majority of chemical convulsants. Since forebrain is the origin of forelimb clonic seizures [36], this type of seizures occurs first, while tonic-clonic seizures represent a progression and spread of ictal activity to brainstem structures [37]. On the other hand, tonic-clonic seizures are often the first sign of seizures in immature animals [14, 16, 17]. This would suggest that early in development, brainstem might be the leading and the most sensitive structure to chemoconvulsants. Another possibility is a fast spread of seizures in immature animals, which could be supported by observations from flurothyl model, in which clonic seizures occur as the initial seizure type even during the first postnatal week but immediately progress into tonic-clonic seizure (Fig. 1). Interestingly, a study investigating the spread of cortical ictal activity from a focus shows that developing cortex may not be involved in the fast seizure propagation. EEG ictal activity induced by focal bicuculline application on frontal cortex propagated to contralateral frontal and to occipital cortices much slower in developing rats compared to mature animals (Fig. 2) [12].

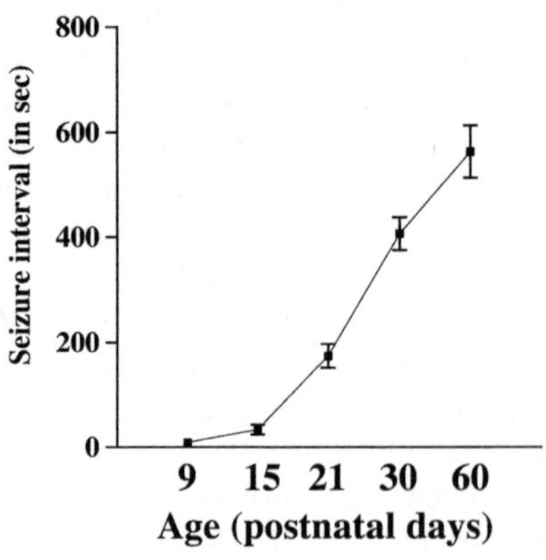

Fig. 1 Seizure progression pattern in flurothyl seizure model. Flurothyl first induces clonic seizures and later tonic-clonic seizures develop suggesting that the seizures originate in the forebrain structures and later spread to the brainstem. The interval between the first clonic and the tonic-clonic seizure is very short during development and increases with age demonstrating that the seizure spread is faster in immature compared to mature brain

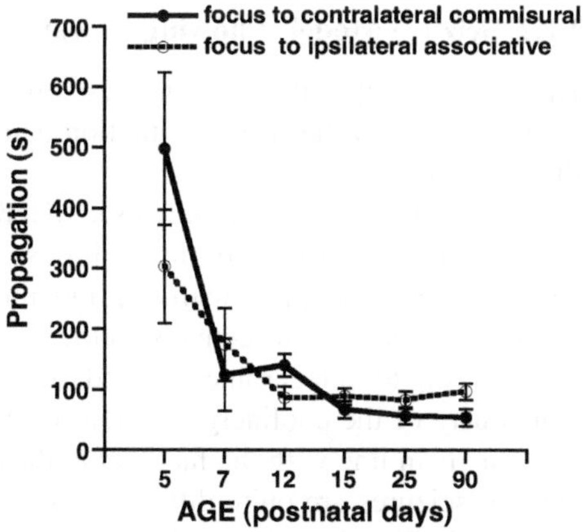

Fig. 2 Spread of focal bicuculline-induced ictal activity within the cerebral cortex. Focal bicuculline-induced activity propagates slower in developing compared to mature brain. Thus, the cerebral cortex does not seem to be involved in the rapid seizure spread during development

[^{14}C]2-deoxyglucose (2-DG) autoradiographic studies have been used to identify involvement of structures in initiation and termination of seizures. Changes (an increase or a decrease) in glucose uptake mark the regions involved during a seizure. The 2-DG studies clearly show differences in the pattern of glucose uptake in developing rats compared to adults. For example, one of the structures identified as part of the endogenous seizure-controlling network is the substantia nigra pars reticulata (SNR) [38–40]. In adult rats, the 2-DG studies identified the SNR as a structure regularly involved during seizures in different seizure models [41] and the changes in glucose uptake are especially striking during seizure spread and generalization [42–45]. On the contrary, in 2 week-old rats, the SNR shows no changes in glucose uptake in different seizure models and regardless of seizure severity [28, 46, 47]. A developmental study using PTZ-induced seizures demonstrated that the pattern of 2-DG metabolic changes in the SNR varies as a function of age [48]. These data suggest that involvement of the SNR in seizures is different in developing animals compared to established seizure-controlling function in adults. This may account for early generalization and spread of seizures, as well as increased seizure severity in developing brain due to the immaturity of endogenous seizure controlling systems.

Finally, multifocal seizure onset may be responsible for fast occurrence of more severe types of seizures in developing animals. Administration of NMDA receptor antagonists such as MK-801, ketamine, or AP7 leads to suppression of tonic-clonic seizures but does not affect the incidence or onset of clonic seizures [49–51]. Even very low doses of MK-801 can completely block PTZ tonic-clonic seizures. Such tonic-clonic seizure suppression reveals the forebrain originating PTZ clonic seizures in the youngest age groups [51]. Interestingly, the unmasked onset of the clonic seizures by MK-801 treatment corresponds approximately to the onset of tonic-clonic seizures in age-matched animals exposed to the convulsant alone, suggesting concurrent onset of both forebrain and brainstem seizure types in developing rats following PTZ administration. These data support the idea of multifocal origin of seizures in developing animals.

2.1.2 Seizure Models Showing Lower Susceptibility in Immature Animals

The lower susceptibility of immature rodents to chemical convulsants was reported more rarely than the reverse situation detailed before. Only two convulsive drugs showed this effect.

The first one very commonly used to induce seizures and SE is pilocarpine alone or associated to lithium. In the high dose pilocarpine model, the susceptibility of rats to seizures increases with age, rats under 2 weeks being relatively resistant to seizures induced by this convulsant [8, 52]. The resistance of the very immature rats (3–12 days) to pilocarpine-induced seizures has been considered to reflect the immaturity of the cholinergic system [53] rather than the inability of rats to seize [8]. This is in line with the fact that at these very early ages, behavioral manifestations of seizures are quite different and less severe than in older rats. Beyond the age PN18–20, when the cholinergic neurons attain their functional maturity the doses of pilocarpine necessary to induce SE remain relatively stable [8, 52]. Besides the direct action of pilocarpine on the CNS muscarinic receptors, some peripheral factors such as increase in blood-brain barrier permeability for pro-epileptogenic blood borne substances may be involved in pilocarpine-induced neuronal excitability in mature animals [54, 55]. Only one group reported an age-dependent decrease in susceptibility to pilocarpine alone after the age of 100 days suggesting the effect of aging [56].

When pilocarpine is associated with lithium, a similar increase in susceptibility with age to the chemoconvulsant was reported. Only in one study, the authors failed to induce electrographic seizures with lithium-pilocarpine between PN3–8. In PN7–10 rats, 60 mg/kg pilocarpine are usually needed to induce SE and this was reproduced by other groups [57–61]. Thereafter, the susceptibility to the chemoconvulsant decreases and most often dose of 30 mg/kg is used at about 3 weeks [57–60].

Another example of lower susceptibility of immature rats compared to older animals relates to inhibitors of glutamate decarboxylase (GAD), such as isonicotinehydrazide [a noncompetitive GAD inhibitor [62, 63]] or 3-mercaptopropionic acid [a specific GAD inhibitor[64]]. The latency to tonic-clonic seizures following administration of either of these GAD inhibitors is significantly longer in the youngest age-groups compared to older rats [65, 66]. The CD50 for the induction of tonic-clonic seizures decreases dramatically with age and becomes similar to the adult value by PN18 [65, 66]. The authors speculate that slower GABA turnover, GABA uptake, distinct expression of the two GAD isoforms in the immature animals may be involved in this phenomena [67, 68].

2.2 *Electrically-Induced Seizures*

Electrical kindling represents the best model of age-specific patterns of seizure propagation. Kindling phenomenon has been studied in rats from the end of first postnatal week [69]. Studies in the kindling model indicate that in the immature brain focal afterdischarges are not easily confined to the stimulated focus with fast development of kindled seizures recurring at short time intervals [9, 69–71]. Most importantly, seizure propagation from the stimulation site occurs faster in developing rats during the first 3 weeks of life than in mature animals [70]. This is illustrated by the fact that immature rats spend less time in kindling stages marking the focal seizure origin (stages 1–2) compared to older rats. Also, more severe seizures such as explosive jumping and tonic seizures (stages 6–7) occur rapidly in immature rats requiring less than 30 stimulations [72, 73] compared to around 100 stimulations in adult rats [74]. Alternating stimulations of two sites in immature rats lead to the development of seizures from both sites [72, 73]. This is not the case in adult rats as the kindling antagonism develops [75]. These kindling data suggest that in the immature brain, the refractory period that follows a seizure is very short compared to the long refractory period observed in adulthood. The decreased refractory period may underlie the propensity of the immature brain for faster progression of seizures and development of SE.

2.3 Febrile Seizures

Febrile seizures are commonly affecting about 3–5% of infants and young children [76, 77]. This type of seizures is mostly prevalent in the first years of life and their occurrence dramatically decreases with age [78]. As a model of human febrile seizures in the rat, induction of hyperthermia that is associated with seizures is commonly used. The susceptibility to hyperthermia-induced seizures is age-dependent. The highest sensitivity is recorded between PN10 and PN13 and susceptibility decreases between PN15 and PN17 [79–82]. This stage of cerebral maturity in the rat corresponds to the period of high sensitivity to fever-induced seizures in human infants [83–85]. The various animal models for febrile seizures have been reviewed in details recently [86, 87].

The first group that induced hyperthermic seizures used a copper sheet heated by an infrared source. They used PN5–6 rats that are younger than a full-term newborn human baby, probably because their paradigm was lethal by PN10 [88]. Their model reproduced the pharmacological reactivity of human febrile seizures, i.e., the anticonvulsant potency of phenobarbital or valproic acid and no effect of phenytoin [82]. Other groups used very similar devices to induce hyperthermic seizures, i.e., a heated metal chamber with an infrared light source [89, 90]. Several groups have used microwaves[80, 91], circulation of warm air by means of a commercial fan[81, 92] or a hair dryer [79]. Other groups have put rat pups in a tank of water heated at 45°C [93–95], in a container floating in warm water [96] or directly on a heated surface [97]. Finally, using in vitro hippocampal slice preparation also shows that increasing the temperature of the recording solution to 38.5°C leads to development of epileptiform activity in the CA1 region following a single orthodromic stimulation, which elicits a single population spike when slices are incubated at the baseline temperature of 35°C [98].

In general, hyperthermic seizures were induced at core temperature (measured rectally as well as in the brain) reaching 40–42°C in PN5–12 rats [88, 99] while in

PN14 or PN22–29 rats, seizures occurred only with 42–44°C [81, 93, 96]. With microwaves, threshold temperatures were lower, 37–39°C [80]. However, while the temperature is an important factor, the most critical factor is also the rate at which temperature increases. This rate underlies both the generation of seizures and their consequences [81, 86].

The model in which the characteristics and consequences of febrile seizures have been best characterized is the one developed by Baram et al. [79]. In this model, the core and brain temperature is slowly increased by a stream of heated hair and seizures occur at a threshold temperature of 40.9°C. Hyperthermia is maintained for 30 min and the mean seizure duration is 22 min and hence corresponds to febrile SE [100]. The seizures present behaviorally a major tonic expression and depth recordings showed that they involve primarily the hippocampus but not the neocortex [79, 99]. In this model, hyperthermia-induced seizures do not lead to any neuronal death [101], although there is transient neuronal injury detectable over 2 weeks in hippocampus, amygdala and perirhinal cortex [102]. Spontaneous seizures, develop in adulthood in about in 35% of the rats with early febrile seizures and interictal spiking was recorded in about 88% of these rats [87].

Although the hyperthermia-induced seizures very closely model the human febrile seizures, the additional factor, the inflammatory response associated with the infectious disease in the human condition, also contributes to the neuronal excitability and seizures. Intraventricular administration of a pro-inflammatory cytokine interleukin-1β (IL-1β) in rat pups prior to the hyperthermia-induced seizures significantly decreased the temperature necessary to induce the seizure [103] supporting the role of inflammatory cytokines in febrile seizures. Recently, a new model for febrile seizures has been introduced by Heida et al. [104, 105], involving combined injection of a bacterial endotoxin lipopolysaccharide (LPS) with a subthreshold dose of kainic acid in rat pups. Systemic LPS injection induced fever within 1.5 h following administration. Subsequent injection of a subthreshold dose of kainic acid during the fever period then resulted in seizures in about 50% of pups. In adulthood, rats suffering febrile seizures during development had lower afterdischarge threshold, longer afterdischarge duration, and more severe seizure-induced hippocampal damage in amygdala kindling compared to animals which previously did not develop febrile seizures [105].

2.4 Cortical Malformations

Cortical malformations in rodents are usually induced by the injection of toxic compounds crossing the placental barrier, the irradiation of the pregnant dam during the gestational period, a freeze, excitotoxic or undercut lesion in the neonatal rat [106].

The toxic compounds interfere with DNA synthesis and the most widely used model is induced by the administration of methylazoxymethanol acetate (MAM). In comparison to other teratogenic drugs, MAM does not affect gestational parameters and is not teratogenic outside of the CNS. It kills neuroepithelial cells only in the state of active S phase state of division during a narrow time window, sparing postmitotic neurons or neuroblasts in the G0 phase during the 2–24 h following its administration; hence it is possible to target specific brain cell populations. A recent study reported that MAM exposure also leads to disturbances in angiogenesis and hence contributes to general cerebrovascular dysfunction besides its direct neurotoxic effects [107].

The earlier studies used a single MAM injection on either embryonic day (E) 14–16. The injection on E14 induced a reduction in thickness of all cortical layers while treatments on the other 2 days selectively suppressed layers II-IV and spared layers V–VI [108]. Later depending on the groups, either one or two injections, 12 h apart were given on E15. This leads to severe cortical hypoplasia and layering abnormalities with a clear rostro-caudal and mediolateral gradient of the abnormalities [109–116]. This supports the concept that the heterotopias were formed by neurons originally migrating to the cortical layers and ending in an abnormal cortico-subcortical circuit involving hippocampus and cortex. MAM-treated rats display anatomical and functional features reminiscent of those found in human with cortical malformations. These are abnormally located vessels, fusiform neurons at the margin of the heterotopias, dense GABA immunoreactivity in the neuropil. Furthermore, a large fraction of heterotopic neurons are prone to abnormal burst firing [113, 114, 116] and the heterotopic neurons are hyperexcitable [for review see [110]]. All these characteristics mimic quite closely those reported in human periventricular nodular heterotopia [117, 118].

Adult offspring of MAM-treated dams have a lower threshold in kainate- [96], kindling- [113, 119], pentylenetetrazol- [113], or flurothyl-induced seizures than controls [109]. Furthermore, these seizures induce neuronal damage never seen in normal animals of the same age [96, 115, 120]. Increased seizure susceptibility was also reported in immature offspring of MAM-treated dams. Indeed, these rats are more susceptible to seizures induced by hyperthermia at PN14 [120], and bicuculline or kainate at PN15 [121]. Likewise, immature offspring (PN21) exposed to MAM needed a lower dose of pilocarpine than control rats to seize but were unable to survive lithium-pilocarpine SE (Dubé and Nehlig, unpublished data).

In utero gamma irradiation is also quite currently used as a model of cortical dysplasia. Rats are most often irradiated on E17, which leads to microcephaly, diffuse cortical dysplasia, heterotopic neurons in the hippocampus and agenesis or hypoplasia of the corpus callosum. This model, which presumably interferes with DNA replication in proliferating cells, most closely mimics "acquired cortical dysplasia" in humans [122]. The thickness of the affected cortex is reduced about twofold compared to control rats [123]. The most severely affected part is the dorsomedial cortex while normal lamination of the lateral cortex near the rhinal sulcus is relatively preserved. The severely affected cortex has lost all recognizable lamination [124]. The GABAergic system has a reduced capacity to recover after in utero irradiation [125]. At moderate intensity of irradiation but not at low or high intensity, the rats develop spontaneous seizures [126, 127]. Thus, the treatment window for producing seizures is quite narrow and too extended damage impairs the occurrence of seizures [for review see [124]]. In this model, sedating agents like acepromazine and xylazine paradoxically increased propensity for seizures [128] possibly related to impaired inhibitory synaptic transmission [129].

The neocortical freeze lesion in the newborn rodent induces a three- or four-layered cortex (microgyrus), focal heterotopia in the layer I, or a cortical cleft (schizencephaly). The epileptogenic area is not located at the exact site of the lesion but rather in a small region surrounding the microgyrus [130–133]. The epileptiform activity propagates over large distances in areas with normal histology but with alterations in the GABAergic and glutamatergic receptors [134, 135]. This is in accordance with the clinical features of human cortical dysplasia [136]. The age of the animal at the time of the freeze lesion is critical. Cortical migration disorders can be generated only during the period of cortical migration, up to PN4. The best reproducible data are obtained when the lesion is performed over the first 24 h of life [for review, see [137]]. In this model no spontaneous epileptiform activity can be recorded, but the susceptibility to hyperthermia-induced seizures is higher [133]. Brain metabolism in adult rats subjected to neonatal freeze lesions is decreased up to 1 mm from the microgyrus but normal in remote cortical areas [138]. The binding to glutamate receptor agonists is increased and the one to GABA receptor agonists is decreased in the dysplastic cortex. This imbalance was also found in remote areas [135].

The injection of ibotenate in the cortex of the neonatal rat was also used to induce neural depopulation, mainly in cortical layers V and VI, ectopic cells in superficial layers, and aberrant sulcus formation (microgyria) [139, 140]. Other neurons are normally organized. As in the freeze lesion model, there is widespread hyperexcitability in response to stimulation both in the dysplastic region and in widespread cortical regions around the dysplastic zone. However, glucose metabolism was only increased in layer I of the area of structural abnormality. No hypometabolism, typical of human dysplastic lesions was observed [141]. However these rats do not develop spontaneous seizures.

None of these four models of cortical dysplasia replicates all features of the human disease. In particular, they do not produce balloon cells characteristic of tuberous sclerosis and Taylor's type cortical dysplasia. In addition, in most models, there are no spontaneous seizures.

2.5 Genetic Models

There are numerous genetic models of epilepsy but not that many have been studied as models of developmental epilepsies. The procedure of knocking out (KO) genes has very often led to the spontaneous occurrence of seizures and some of these genetic manipulations do not allow the long survival of animals because of the frequency and severity of seizures. This part will be limited to a few genetic models of developmental epilepsies.

Several spontaneous mutations induce cortical malformations and seizures in rats. Genetic manipulations, knocking out specific genes, have also led to similar pathologies in mice. In rats, the *flathead*, the *Eker* and the *Tish* mutant rat have been well studied. In mice, the KO types studied are the *Otx−/−* mouse, the *p35−/−* mouse and KO animals for the *TSC1* and *TSC2* genes, which results in abnormal cellular differentiation, migration, and proliferation leading to cortical tubers.

The *flathead* rat is a spontaneous mutant characterized by marked neurologic impairment, frequent seizures and premature death [97]. It originates in the autosomal recessive mutation in the Citron kinase gene (CitK) [142]. These rats suffer from microcephaly, abnormal cell death, neuronal and glial cytomegaly, GABAergic interneuron loss, and neurogenic cytokinesis failure [143, 144]. Generalized tonic-clonic seizures start during the second postnatal week and the phenotype is lethal by 3 weeks.

The *Tish* mutant rat is also an autosomal recessive mutation, which is not lethal. The gene mutation responsible for the disease is not yet identified. This rat is characterized by the presence of well-developed bilateral subcortical band non-laminated heterotopias of variable size; their typical extension is from frontal to occipital cortices. The homozygous mutants develop seizures from 1 to at least 6 months of age [145].

The *Eker* rat is a model for human tuberous sclerosis caused by a mutation in the *Tsc2* gene. Carriers of the mutations develop subcortical and subependymal hamartomas [146], cortical tubers and anaplastic gangliomas [147]. However these rats do not develop tubers and do not exhibit spontaneous seizures [148], but display increased responses to chemical kindling [149] and increased paired-pulse inhibition at hippocampal synapses [150].

The *OTX1−/−* mouse displays microencephaly, characterized by a 25% reduction of the neocortex thickness. The reduction in neuronal density ranges from 4 to 35%. Subsets of cortical layer V neurons are missing thus disturbing the appropriate cortical output to subcortical areas [151, 152]. There is also a disproportionate decrease in the number of interneurons [153]. The pattern is similar to, but less severe than in the *flathead* rat. All the homozygous KO mice display generalized tonic-clonic seizures and 30% die over the first month of life. There is no corresponding human disorder associated with mutations of the *OTX1* gene [145].

The p35−/− mouse suffers from migration disorders in the neocortex, dysplasia in hippocampus accompanied by spontaneous seizures in some animals and reduced threshold to convulsive drugs [154, 155]. About 75% of mice develop seizures by 3–5 months with 25% of the tonic-clonic type. In these mice, sprouting of the dentate granule cells is present and the granule cell layer is more diffuse than in the wild type. This model is a mixture of migration disorder and temporal lobe features and the axonal sprouting results in recurrent feedback excitation that may create hyperexcitable pathways in hippocampus [156].

The mutation of the *TSC1* and *TSC2* genes is responsible in humans for tuberous sclerosis characterized by cortical tubers and seizures. Recently, a model of mice losing most of their *Tsc1* expression during embryogenesis was developed. This loss leads to the enlargement or dysplasia of neurons that are ectopic in many cortical and hippocampal locations. There is a marked delay in myelination caused by an inductive neuronal defect. These mice display neurological abnormalities starting on PN5 and survive for about a month. Seizures occur both spontaneously and after physical stimulation and may end with a lethal tonic phase. This new model replicates several features of human disorders after TSC gene mutations [157].

The other well-known models of developmental epilepsies are the genetic models of childhood absence epilepsy. Genetic models with spontaneous mutations have been developed in rats and mice and some groups have used the gene knocking out procedure to engineer absence epilepsy models in mice.

In rats, two groups reported the occurrence of spontaneous spike-and-wave discharges (SWDs) on the EEG that could evoke absence epilepsy in rats [158, 159]. The two models were developed, one in France, the GAERS (Genetic Absence Epilepsy Rat from Strasbourg) and one in the Netherlands, the WAG/Rij rat. In these models, SWDs (7–11 cps, 300–1,000 μV, 0.5–7.5 s) start and end abruptly on normal, low amplitude desynchronized EEG background and are accompanied by behavioral arrest, staring and sometimes twitching of the vibrissae. SWDs are recorded bilaterally in a thalamo-cortical pathway. These seizures share many common features with human childhood absence epilepsy. They are suppressed by antiepileptic medication that is effective in humans and are aggravated by drugs that aggravate the human symptoms [160, 161]. They also represent excellent pharmacologically predictive models of the response of the disease to new drugs in humans. Most neurotransmitters are involved in the control of SWDs but GABA and gamma-hydroxybutyrate seem to play a critical role. The main difference between the two models and the human features is the ontogenetic development. In GAERS, the first SWDs are detected by about PN30, at PN40 about 30% of the rats are affected and 100% by 3 months [162]. In WAG/Rij, the first SWDs occur by PN60–80 and only 50% of the rats display SWDs by 3 months, all rats being epileptic by 6 months [163]. SWDs in GAERS and WAG/Rij are genetically determined with an autosomal dominant inheritance [164, 165] of polygenic origin [166, 167].

In mice as in rats, spontaneous mutants were found that display SWDs and absence epilepsy features. Some of those are the tottering (tg) mouse with a mutation on a P/Q type calcium channel alpha subunit [168], the lethargic (lh) mouse with a mutation on the calcium channel beta 4 subunit [169], the ducky (du) mouse with a mutation on an alpha2-delta2 calcium channel subunit [170], the stargazer (stg) mouse with a mutation on the gamma2 calcium channel subunit [171], the SWE (swe) mouse with a mutation on the sodium hydrogen exchanger [172], the Mocha2j (mh) mouse with a mutation on the delta subunit AP3 adaptor protein [173] and the Coloboma (Cm) mouse with a microdeletion including SNA25 and phospholipase C isoform b1 [174]. In these models, multiple abnormalities in network synchronization and excitability were found [for review see [175]]. At least, two transgenic models with mice exhibiting SWDs have been engineered, one with the deletion of the pacemaker channel HCN2 [176] and the other one with the deletion alpha 1G T-type calcium channel [177].

In humans, a genetic component in the etiology of absence epilepsy is well established but the mechanisms and genes underlying the inheritance of the disease are still not clear. The complex pattern of inheritance suggests the involvement of a large number of susceptibility genes [178]. A polymorphism in the promoter of the $GABA_A$ and $GABA_B$ receptor subunit genes GABRB3, GABRG3 and GABRA5 [179, 180] and variants in the T-type calcium channel gene CACNA1H [181–184] have been associated with childhood absence epilepsy. Variants in CACNA1H are not sufficient by themselves to cause epilepsy [185]. The genes coding for voltage-gated calcium channels, including CACNG3 [179, 186] and the chloride channel gene CLCN2 [187] may also be a susceptibility loci in some cases of absence epilepsy. For detailed review of genes responsible for human epilepsies and comparisons with rodent mutants see [188].

3 Long-Term Consequences

3.1 Cell Damage and Network Plasticity

Seizure activity in adult brain leads to neuronal damage, while the immature brain is relatively resistant to SE-induced morphological damage despite the fact that seizures appear more severe and seizure susceptibility is highest during development [189–191]. Even long-lasting SE regardless of seizure origin (non-limbic or limbic) induced by flurothyl [192, 193], PTZ [194], NMDA [195], or KA [28, 196–198] and pilocarpine [8], produces little or no neuronal damage in the hippocampus or extrahippocampal regions in rats younger than 3 weeks, although the hippocampus and other cortical structures exhibit prolonged electrographic discharges [28, 193, 199]. Synaptic reorganization in the hippocampus also does not occur following SE up to the third postnatal week [61, 197, 200, 201].

Many factors may contribute to the relative resistance of immature hippocampus to SE-induced damage. High levels of neurotrophic factors such as BDNF seem to play the protective role since administration of antisense to BDNF during KA seizures in P19 rats, in which the damage normally does not occur, results in neuronal loss in the CA1 and CA3 hippocampal regions [202]. Interestingly, reactive oxygen species are not formed following KA-induced SE in immature brain [203], which may be related to the protective role of maternal milk containing mitochondria uncoupling protein [204]. Moreover, glia activation and production of inflammatory cytokines following KA-induced SE is limited up to age PN21 and specifically corresponds with occurrence of neuronal injury [26]. GABA synthesis during SE is better maintained in early stages of development compared to older rats [205]. The resistance of immature brain may be related to differential regulation of $GABA_A$ receptor $\alpha1$ subunit expression in the hippocampus; it is increased in developing but decreased in adult rats [206, 207].

Although majority of animal studies indicate that the rat immature brain is relatively resistant to SE-induced damage, there are few reports suggesting that a degree of hippocampal injury may occur during early development. Occasional hippocampal damage can be detected in rats from the second postnatal week. SE induced by lithium/pilocarpine produced acute damage in the mediodorsal thalamic nucleus [208] but hippocampal damage does not occur until PN15 [61]. In PN15 rats, the damage is restricted to the CA1 region. This is in contrast to 3-week old and older rats, which show a special vulnerability of the hilar region to SE-induced damage [61].

Continuous stimulation of perforant pathway for 16 h is another seizure model, in which hippocampal damage can be detected in PN15 rats [209]. In these rats, acute neuronal damage was also observed in CA1, CA3 regions and hilus. This is interesting because the behavioral expression of seizures in this model consisted only of wet dog shakes and scratching, which, early on is similar to the seizure expression following KA at the same age [209], and less severe than the seizure expression in lithium/pilocarpine model [61]. In the KA model, this type of SE does not produce any hippocampal injury [28, 196–198]. However, this long-lasting continuous stimulation paradigm during the pre-weaning age may be associated with enormous stress not only from the seizure burden but also because of the prolonged separation from the mother. In fact, activation of the stress-response network by corticotropin-releasing hormone (CRH) injection in rats during the second postnatal week, leads to seizures and transient neuronal damage in the hippocampus [210–212]. The sensitivity to excitatory effects of CRH is inversely related to age [213]. At PN21, perforant path stimulation for 8 h produces only minimal damage in CA3 and hilar region, although more severe behavioral pattern occurred in these rats than at PN15 rats (including occasional bilateral forelimb clonus) [214]. This unexpected decrease in the extent of neuronal injury may be related to shorter duration of the stimulation paradigm or to decreased sensitivity to stressful event [212].

Interestingly, several studies showed normal neuronal cell counts in adult rats previously subjected to SE at PN15, although acute neuronal injury has been present. SE induced by CRH, hyperthermia, or PTZ during the second postnatal week causes acute neuronal damage in several regions including the hippocampus, amygdala, piriform cortex, thalamus, and hypothalamus. However, cell counts in animals surviving up to adulthood show no significant difference from controls [102, 194, 210, 211].

Thus, developmental studies demonstrate that SE-induced damage may occur in immature brain, but the damage may be only transient and far less prominent compared to SE-induced changes in adulthood. Although SE early in life may not produce detectable morphological changes persisting till adulthood, it still may constitute a priming condition for some type of brain insults such as seizure-induced damage later in life [215] or ischemic injury [216].

3.2 Spontaneous Seizures

After an initial precipitating injury such as SE induced by pilocarpine, lithium-pilocarpine or kainate, almost 100% of adult rats develop spontaneous recurrent seizures (SRS) after a mean latency phase of about 2 weeks [217, 218]. When this type of injury is applied to younger rats, the percentage of animal developing SRS is lower.

For example, in the pilocarpine model, no motor SRS were recorded when SE was induced up to PN17. Between PN18 and PN24, 22% of the rats developed SRS after a mean latency of 37 days. In the age ranges of PN25–35 and PN36–45, 54% and 93% of the rats developed SRS after a mean latency of 23 and 19 days, respectively. From PN50, the vast majority of rats undergoing SE became epileptic after a latency of 14–19 days [52]. In the lithium-pilocarpine model, the data were rather similar: no animal exposed to SE before 2 weeks became epileptic, 27% of rats undergoing SE at PN14 developed SRS while 72–75% of rats subjected to SE at 3 and 4 weeks developed SRS [57, 58, 61]. However, more recent data showed that young animals are able to develop SRS but mostly with no behavioral expression, which explains why they were missed in most studies not using careful video-EEG monitoring. Indeed, 3 months after lithium-pilocarpine SE in PN12–15 rats, 25 and 75% developed non-convulsive spontaneous seizures, respectively [219, 220]. These seizures had no motor expression and were rather characterized by freezing and/or automatisms like chewing, licking, vibrissae twitching. Likewise the repetition of pilocarpine SE on three consecutive days, PN7–9, led to electrographic seizures accompanied by behavioral arrest followed sometimes by masticatory and orofacial automatisms. However, the severity of the seizures increased with age and in 10% of the animals, spontaneous clonic seizures could be recorded in rats over PN60 [221].

Finally, prolonged febrile seizures induced at PN11 also lead to spontaneous electro-behavioral seizures in 35% of the rats. These seizures consisted in sudden freezing followed by limbic automatisms, as recorded in rats subjected to SE at a young age [87].

Thus, it appears that there is an age-dependent evolution in the percentage of rats able to develop spontaneous seizures after an early initial insult. The severity of the recurrent seizures varies also with age. The youngest rats usually develop seizures with no motor component while motor SRS occur most often after an initial insult induced in rats aged PN21 and older. The exact mechanisms (extent and nature of neuronal loss) and the nature and properties of the new circuits underlying the expression of the different types of seizures at various ages need to be clarified.

4 Mechanisms Underlying Changes in Seizure Propensity

4.1 Ictogenesis

Increased seizure susceptibility of developing brain is multifactorial and can be accounted for by factors such as ongoing changes in neurotransmitter systems, ion channels, or synaptogenesis during the postnatal period as the higher neuronal excitability is critical for normal neuronal development [222–225]. The first 2 weeks of life mark a period of dramatic increase in synaptogenesis, axonal and dendritic outgrowth [226, 227]. Maturational pruning follows this initial early developmental synaptic density overshoot [228]. The higher number of spine density early in development seems to correspond to higher binding density of glutamate receptors compared to adults [229, 230]. Functional properties of glutamate receptors also seem to contribute to increased excitability of developing brain. Immature NMDA receptors exhibit slower decay times and longer synaptic currents due to higher NR2B/NR2A ratio during development [231–233]. AMPA receptors during development have higher permeability to Ca^{2+} due to lower expression of GluR2 subunit [234]. GABA (as well as glycine), the main inhibitory transmitter in the mature brain, undergoes a shift from depolarizing to hyperpolarizing action at $GABA_A$ receptors during development [235] [for a recent extensive review see [225]]. This shift has age- and brain region-specific pattern. For example, the hyperpolarizing GABA action in the hippocampus, one of structures involved in seizure initiation, is present at PN6 in the CA1 region [236] and by PN13 in the CA3 region [237], while in the substantia nigra pars reticulata (SNR), one of structures involved in seizure termination, the depolarizing to hyperpolarizing shift of GABA action occurs later around PN17 [238]. The disproportional maturation of GABA effects in regions responsible for seizure generation or suppression may participate in the increased excitability.

Several lines of evidence indicate that the IL-1β signaling may contribute critically to fever-induced hyperexcitability underlying seizures in immature rodents. Dubé et al. [103] showed that the threshold temperature for the onset of hyperthermia-induced seizures is reduced in PN14–15 mice by the intraventricular injection of IL-1β while it is increased by the application of IL-1 receptor antagonist, a naturally occurring competitive antagonist of IL-1 type 1 receptor, the type of receptor that transduces the biological actions of IL-1β. Moreover, mice with an impaired IL-1β signaling due to a genetic deletion of the IL-1 receptor type 1 gene, have a higher threshold temperature for seizure induction [103]. Heida and Pittman [239]. showed that lipopolysaccharide injection in PN14 rats, mimicking bacterial infection and inducing fever, enhanced the susceptibility of rats to kainate-induced seizures and this effect was associated with an increase in hippocampal levels of IL-1β at the onset of seizures. Intraventricular administration of IL-1β increased the core body temperature as well as the number of animals that experienced convulsions in a dose-dependant manner. Whereas increasing doses of IL 1 receptor antagonist given to separate groups of animals were anticonvulsant.

These data indicate that a raise in IL-1β in the forebrain, and more specifically in the hippocampus, can contribute to set the threshold for seizure induction in models of fever and infection [240]. Recent evidence in adult rats exposed to chemical- or electrical-induced SE has shown that pro-inflammatory cytokines produced by glia during seizure activity or following an initial brain-damaging event, can increase neuronal excitability contributing to seizure onset and their maintenance [241]. The mechanisms underlying these novel actions of cytokines are mostly unknown but may involve functional and molecular interactions between cytokines and ionotropic glutamate [242], AMPA and $GABA_A$ receptors [243, 244], and interactions with glutamate re-uptake mechanisms by astrocytes leading to increased extracellular glutamate levels [245]. Recent evidence points to the critical role of astrocytic glutamate release involvement in genesis of epileptic activity [246]. The astrocytic glutamate release can regulate the strength of seizure-like events [247] and cytokines can promote glutamate release from glie in pathophysiological conditions [248].

4.2 Epileptogenesis

Mechanisms of epileptogenesis in developing rats have been mostly investigated so far in SE and febrile seizure models [for review see [87, 249]]. Prolonged seizures and SE in developing animals demonstrate age- and model-dependent propensity for brain injury. Even in the absence of overt brain injury, plasticity phenomena including synaptic reorganization and neurogenesis, changes in neurotransmitter receptors or voltage-gated ion channels have been described although it is still unknown which are the critical mediators of epileptogenicity [249, 250]. Importantly, long-term behavioral and cognitive effects have been demonstrated in animal models of early-life seizures therefore supporting that prolonged seizures and SE can induce negative consequences. A whole cascade of molecular events has been described in febrile seizure models [for review see [87, 249, 250]]. The most remarkable change is the concurrent decrease of the gene and protein level of expression of the hyperpolarizing-activated, cyclic nucleotide-gated channel 1 (HCN) and the increase of the expression of HCN2 in specific populations of hippocampal neurons [251, 252]. Similar changes have been observed also after kainic acid – induced seizures although in this model these changes are transient as opposed to their endurance after febrile-like seizures. In the long-term, adult animals with early febrile seizures become more susceptible than controls to kainate-induced seizures and undergo changes in GABAergic neurotransmission in hippocampal interneurons [57]. Changes in the excitability of the hippocampal circuit can be recorded from 1 week after the initial insult up to adulthood [253].

5 Conclusions

The information from various models of seizures in developing rats indicates that susceptibility of immature brain to seizures and the long-term consequences of seizures are unique age-dependent features, which differ in many respects from adulthood. One typical aspect is the dissociation of epileptogenesis from excitoxicity as demonstrated in developmental febrile seizure models as opposed to SE models in adult rats, which is invariably associated with induction and progression of cell loss. Transcription-dependent and post-translational changes in neurotransmitter receptors and ion channels, plasticity-related functional changes in neuronal networks, the induction of neuromodulatory molecules such as inflammatory mediators, neurotrophins, neuropeptides are among the possible mechanisms which define the immediate and long-term response of the developing brain to seizures. Further elucidation of these aspects and their role in ictogenesis and epileptogensis using clinical relevant experimental model in developing rats as well as the study of genetic influences on the process of epileptogenesis in immature brain is instrumental for the future translation of these findings into clinical practice.

Acknowledgments Supported by grants from Fondazione Monzino (A.V.), EPICURE LSH-CT-2006-037315 (A.V.), Negri Weizmann Programme (A.V.), NIH/NINDS NS056093 (J.V.), and INSERM U666 (A.N.).

References

1. Hauser WA (1997) Incidence and prevalence. In: Engel J Jr, Pedley TA (eds) Epilepsy: a comprehensive textbook. Lippincott-Raven Publishers, Philadelphia, pp 47–57

2. Shinnar S, Pellock JM (2002) Update on the epidemiology and prognosis of pediatric epilepsy. J Child Neurol 17(Suppl 1):S4–S17
3. Hauser WA, Kurland LT (1975) The epidemiology of epilepsy in Rochester, Minnesota, 1935 through 1967. Epilepsia 16:1–66
4. Hauser WA (1990) Status epilepticus: epidemiologic considerations. Neurology 40:9–13
5. Moshé SL, Shinnar S, Swann JW (1995) Partial (focal) seizures in developing brain. In: Schwarzkroin PA, Moshe SL, Noebels JL, Swann JW (eds) Brain development and epilepsy. Oxford University Press, New York, pp 34–65
6. Vernadakis A, Woodbury DM (1969) The developing animal as a model. Epilepsia 10:163–178
7. Vernadakis A, Woodbury D (1969) Maturational factors in development of seizures. In: Jasper H, Ward A, Pope A (eds) Mechanisms of the epilepsies. Little, Brown and Co., Boston, pp 535–541
8. Cavalheiro EA, Silva DF, Turski WA, Calderazzo-Filho LS, Bortolotto ZA, Turski L (1987) The susceptibility of rats to pilocarpine-induced seizures is age-dependent. Brain Res 465:43–58
9. Moshé SL, Albala BJ, Ackermann RF, Engel JJ (1983) Increased seizure susceptibility of the immature brain. Dev Brain Res 7:81–85
10. Moshé SL (1989) Ontogeny of seizures and substantia nigra modulation. In: Kellaway P, Noebels JF (eds) Problems and concepts in developmental neurophysiology. Johns Hopkins, Baltimore, pp 247–262
11. Schwartzkroin PA (1984) Epileptogenesis in the immature CNS. In: Schwartzkroin PA, Wheal HV (eds) Electrophysiology of epilepsy. Academic, London, pp 389–412
12. Velíšková J, Velíšek L, Mareš P, Rokyta R, Budko KP (1991) Bicuculline-induced neocortical epileptiform foci and the effects of 6-hydroxydopamine in developing rats. Brain Res Bull 26:693–698
13. Velíšková J (2006) Behavioral characterization of seizures in rats. In: Pitkanen A, Schwartzkroin PA, Moshé SL (eds) Models of seizures and epilepsy. Elsevier, San Diego, pp 601–611
14. Mareš P, Schickerová R (1980) Seizures elicited by subcutaneous injection of metrazol during ontogenesis in rats. Act Nerv Super (Praha) 22:264–268
15. Velíšková J, Velíšek L, Mareš P (1988) Epileptic phenomena produced by kainic acid in laboratory rats during ontogenesis. Physiol Bohemoslov 37:395–405
16. Baram TZ, Snead OC III (1990) Bicuculline induced seizures in infant rats: ontogeny of behavioral and electrocortical phenomena. Dev Brain Res 57:291–295
17. Zouhar A, Mareš P, Lišková-Bernášková K, Mudrochová M (1989) Motor and electrocorticographic epileptic activity induced by bicuculline in developing rats. Epilepsia 30:501–510
18. Remler MP, Marcussen WH (1985) Bicuculline methiodide in the blood-brain barrier-epileptogen model of epilepsy. Epilepsia 26:69–73
19. Mareš P, Chino M, Kubová H, Mathern P, Veliký M (2000) Convulsant action of systemically administered glutamate and bicuculline methiodide in immature rats. Epilepsy Res 42:183–189
20. Joo F (1987) Current aspects of the development of the blood-brain barrier. Int J Dev Neurosci 5:369–372
21. Velíšek L (2006) Model of chemically-induced acute seizures. In: Pitkanen A, Schwartzkroin PA, Moshé SL (eds) Models of seizures and epilepsy. Elsevier, San Diego, pp 127–152
22. Schoepp DD, Gamble AY, Salhoff CR, Johnson BG, Ornstein PL (1990) Excitatory amino acid-induced convulsions in neonatal rats mediated by distinct receptor subtypes. Eur J Pharmacol 182:421–427
23. Mareš P, Folbergrová J, Langmeier M, Haugvicová R, Kubová H (1997) Convulsant action of D, L-homocysteic acid and its stereoisomers in immature rats. Epilepsia 38:767–776
24. Mareš P, Folbergrová J, Kubová H (2004) Excitatory aminoacids and epileptic seizures in immature brain. Physiol Res 53(Suppl 1):S115–S124
25. Nadler JV, Perry BW, Cotman CW (1978) Intraventricular kainic acid preferentially destroys hippocampal pyramidal cells. Nature 271:676–677

26. Rizzi M, Perego C, Aliprandi M et al (2003) Glia activation and cytokine increase in rat hippocampus by kainic acid-induced status epilepticus during postnatal development. Neurobiol Dis 14:494–503
27. Cherubini E, DeFeo MR, Mecarelli O, Ricci GF (1983) Behavioral and electrographic patterns induced by systemic administration of kainic acid in developing rats. Dev Brain Res 9:69–77
28. Albala BJ, Moshé SL, Okada R (1984) Kainic-acid-induced seizures: a developmental study. Dev Brain Res 13:139–148
29. Frush DP, Giacchino JL, McNamara JO (1986) Evidence implicating dentate granule cells in development of entorhinal kindling. Exp Neurol 92:92–101
30. Grimes LM, Earnhardt TS, Mitchell CL, Tilson HA, Hong JS (1990) Granule cells in the ventral, but not dorsal, dentate gyrus are essential for kainic acid-induced wet dog shakes. Brain Res 514:167–170
31. Velíšek L, Jehle K, Asche S, Velíšková J (2007) Model of infantile spasms induced by N-methyl-D-aspartic acid in prenatally impaired brain. Ann Neurol 61:109–119
32. Mareš P, Velíšek L (1992) N-methyl-D-aspartate (NMDA)-induced seizures in developing rats. Brain Res Dev Brain Res 65:185–189
33. Olsen RW (1981) The GABA postsynaptic membrane receptor-ionophore complex. Site of action of convulsant and anticonvulsant drugs. Mol Cell Biochem 39:261–279
34. Wakamori M, Ikemoto Y, Akaike N (1991) Effects of two volatile anesthetics and a volatile convulsant on the excitatory and inhibitory amino acid responses in dissociated CNS neurons of the rat. J Neurophysiol 66:2014–2021
35. Velíšková J (1999) Age related mechanisms involved in the control of seizures. In: Nehlig A, Motte J, Moshé SL, Plouin P (eds) Childhood epilepsies and brain development. John Libbey, London, pp 39–51
36. Browning RA, Nelson DK (1986) Modification of electroshock and pentylenetetrazol seizure patterns in rats after precollicular transections. Exp Neurol 93:546–556
37. Browning RA (1985) Role of the brain-stem reticular formation in tonic-clonic seizures: lesion and pharmacological studies. Fed Proc 44:2425–2431
38. Velíšková J, Moshé SL (2006) Update on the role of substantia nigra pars reticulata in the regulation of seizures. Epilepsy Curr 6:83–87
39. Iadarola MJ, Gale K (1981) Cellular compartments of GABA in brain and their relationship to anticonvulsant activity. Mol Cell Biochem 39:305–330
40. Depaulis A, Vergnes M, Marescaux C (1994) Endogenous control of epilepsy: the nigral inhibitory system. Prog Neurobiol 42:33–52
41. Pazdernik TL, Cross RS, Giesler M, Samson FE, Nelson SR (1985) Changes in local cerebral glucose utilization induced by convulsants. Neuroscience 14:823–835
42. Engel J Jr, Wolfson L, Brown L (1978) Anatomical correlates of electrical and behavioral events related to amygdaloid kindling. Ann Neurol 3:538–544
43. Ben-Ari Y, Riche D, Tremblay E, Charton G (1981) Alterations in local glucose consumption following systemic administration of kainic acid, bicuculline or metrazol. Eur Neurol 20:173–175
44. Lothman EW, Collins RC (1981) Kainic acid induced limbic seizures: metabolic, behavioral, electroencephalographic and neuropathological correlates. Brain Res 218:299–318
45. Velíšková J, Miller AM, Nunes ML, Brown LL (2005) Regional neural activity within the substantia nigra during peri-ictal flurothyl generalized seizure stages. Neurobiol Dis 20:752–759
46. Ackermann RF, Moshé SL, Albala BJ (1989) Restriction of enhanced 14C–2-deoxyglucose utilization to rhinencephalic structures in immature amygdala-kindled rats. Exp Neurol 104:73–81
47. Sperber EF, Brown LL, Moshé SL (1992) Functional mapping of different seizure states in the immature rat using 14C–2-deoxyglucose. Epilepsia 33:44
48. Pereira de Vasconcelos A, el Hamdi G, Vert P, Nehlig A (1992) An experimental model of generalized seizures for the measurement of local cerebral glucose utilization in the immature rat. II mapping of the brain metabolism using quantitative [14C]-2-deoxyglucose technique. Dev Brain Res 69:243–259

49. Velíšek L, Mikolášová R, Mareš P (1989) Effects of ketamine on metrazol-induced seizures during ontogenesis in rats. Pharmacol Biochem Behav 32:405–410
50. Velíšek L, Kusá R, Kulovaná M, Mareš P (1990) Excitatory amino acid antagonists and pentylenetetrazol-induced seizures during ontogenesis. I. The effects of 2-amino-7-phosphonoheptanoate. Life Sci 46:1349–1357
51. Velíšek L, Verešová S, Pobišová H, Mareš P (1991) Excitatory amino acid antagonists and pentylenetetrazol-induced seizures during ontogenesis. II. The effects of MK-801. Psychopharmacology (Berl) 104:510–514
52. Priel MR, dos Santos NF, Cavalheiro EA (1996) Developmental aspects of the pilocarpine model of epilepsy. Epilepsy Res 26:115–121
53. Coyle JT, Enna SJ (1976) Neurochemical aspects of the ontogenesis of GABAnergic neurons in the rat brain. Brain Res 111:119–133
54. Marchi N, Oby E, Batra A et al (2007) In vivo and in vitro effects of pilocarpine: relevance to ictogenesis. Epilepsia 48:1934–1946
55. Uva L, Librizzi L, Marchi N et al (2008) Acute induction of epileptiform discharges by pilocarpine in the in vitro isolated guinea-pig brain requires enhancement of blood-brain barrier permeability. Neuroscience 151:303–312
56. Patel S, Meldrum BS, Fine A (1988) Susceptibility to pilocarpine-induced seizures in rats increases with age. Behav Brain Res 31:165–167
57. Dubé C, Boyet S, Marescaux C, Nehlig A (2000) Progressive metabolic changes underlying the chronic reorganization of brain circuits during the silent phase of the lithium-pilocarpine model of epilepsy in the immature and adult rat. Exp Neurol 162:146–157
58. Dubé C, Boyet S, Marescaux C, Nehlig A (2001) Relationship between neuronal loss and interictal glucose metabolism during the chronic phase of the lithium-pilocarpine model of epilepsy in the immature and adult rat. Exp Neurol 167:227–241
59. Fernandes MJ, Dube C, Boyet S, Marescaux C, Nehlig A (1999) Correlation between hypermetabolism and neuronal damage during status epilepticus induced by lithium and pilocarpine in immature and adult rats. J Cereb Blood Flow Metab 19:195–209
60. Hirsch E, Baram TZ, Snead OC III (1992) Ontogenic study of lithium-pilocarpine-induced status epilepticus in rats. Brain Res 583:120–126
61. Sankar R, Shin DH, Liu H, Mazarati A, Pereira de Vasconcelos A, Wasterlain CG (1998) Patterns of status epilepticus-induced neuronal injury during development and long-term consequences. J Neurosci 18:8382–8393
62. Wood JD, Peesker SJ (1971) Sequential lowering and raising of brain gamma-aminobutyric acid levels by isonicotinic acid hydrazide. Can J Physiol Pharmacol 49:780–781
63. Wood JD, Abrahams DE (1971) The comparative effects of various hydrazides on -aminobutyric acid and its metabolism. J Neurochem 18:1017–1025
64. Wu JY, Roberts E (1974) Properties of brain L-glutamate decarboxylase: inhibition studies. J Neurochem 23:759–767
65. Mareš P, Trojan S (1991) Ontogenetic development of isonicotinehydrazide-induced seizures in rats. Brain Dev 13:121–125
66. Mareš P, Kubová H, Zouhar A, Folbergrová J, Koryntová H, Sta ková L (1993) Motor and electrocorticographic epileptic activity induced by 3-mercaptopropionic acid in immature rats. Epilepsy Res 16:11–18
67. Roberts E (1974) Gamma-aminobutyric acid and nervous system function – a perspective. Biochem Pharmacol 23:2637–2649
68. Erlander MG, Tobin AJ (1991) The structural and functional heterogeneity of glutamic acid decarboxylase: a review. Neurochem Res 16:215–226
69. Baram TZ, Hirsch E, Schultz L (1993) Short-interval amygdala kindling in neonatal rats. Brain Res Dev Brain Res 73:79–83
70. Moshé SL (1981) The effects of age on the kindling phenomenon. Dev Psychobiol 14:75–81
71. Moshé SL, Albala BJ (1983) Maturational changes in postictal refractoriness and seizure susceptibility in developing rats. Ann Neurol 13:552–557
72. Haas K, Sperber EF, Moshé SL (1990) Kindling in developing animals: expression of severe seizures and enhanced development of bilateral foci. Dev Brain Res 56:275–280

73. Haas KZ, Sperber EF, Moshé SL (1992) Kindling in developing animals: interactions between ipsilateral foci. Dev Brain Res 68:140–143
74. Pinel JPJ, Rovner LI (1978) Experimental epileptogenesis: kindling induced epilepsy in rats. Exp Neurol 58:190–202
75. Burchfiel JL, Applegate CD, Konkol RJ (1986) Kindling antagonism: a role for norepinephrine in seizure suppression. In: Wada JA (ed) Kindling 3. Raven, New York, pp 213–229
76. Shinnar S, Berg AT, Moshe SL et al (1990) Risk of seizure recurrence following a first unprovoked seizure in childhood: a prospective study. Pediatrics 85:1076–1085
77. Verity CM, Butler NR, Golding J (1985) Febrile convulsions in a national cohort followed up from birth. I – Prevalence and recurrence in the first five years of life. Br Med J (Clin Res Ed) 290:1307–1310
78. Fishman MA (1979) Febrile seizures: the treatment controversy. J Pediatr 94:177–184
79. Baram TZ, Gerth A, Schultz L (1997) Febrile seizures: an appropriate-aged model suitable for long-term studies. Brain Res Dev Brain Res 98:265–270
80. Hjeresen DL, Diaz J (1988) Ontogeny of susceptibility to experimental febrile seizures in rats. Dev Psychobiol 21:261–275
81. Morimoto T, Nagao H, Sano N, Takahashi M, Matsuda H (1990) Hyperthermia-induced seizures with a servo system: neurophysiological roles of age, temperature elevation rate and regional GABA content in the rat. Brain Dev 12:279–283
82. Olson JE, Scher MS, Holtzman D (1984) Effects of anticonvulsants on hyperthermia-induced seizures in the rat pup. Epilepsia 25:96–99
83. Alling C (1985) Biochemical maturation of the brain and the concept of vulnerable periods. In: Rydberg UEA (ed) Alcohol and the developing brain. Raven, New York, pp 5–10
84. Dobbing J, Sands J (1979) Comparative aspects of the brain growth spurt. Early Hum Dev 3:79–83
85. Romijn HJ, Hofman MA, Gramsbergen A (1991) At what age is the developing cerebral cortex of the rat comparable to that of the full-term newborn human baby? Early Hum Dev 26:61–67
86. Baram T (2002) Animal models for febrile seizures. In: Baram T (ed) Febrile seizures. Academic, San Diego, pp 189–201
87. Dubé CM, Baram TZ (2006) Complex febrile seizures – An animal model in rodents. In: Pitkanen A, Schwartzkroin PA, Moshé SL (eds) Models of seizures and epilepsy. Elsevier, San Diego, pp 333–340
88. Holtzman D, Obana K, Olson J (1981) Hyperthermia-induced seizures in the rat pup: a model for febrile convulsions in children. Science 213:1034–1036
89. Chisholm J, Kellogg C, Franck JE (1985) Developmental hyperthermic seizures alter adult hippocampal benzodiazepine binding and morphology. Epilepsia 26:151–157
90. Zhao DY, Wu XR, Pei YQ, Zuo QH (1985) Long-term effects of febrile convulsion on seizure susceptibility in P77PMC rat – resistant to acoustic stimuli but susceptible to kainate-induced seizures. Exp Neurol 88:688–695
91. Carratala F, Moya M (1991) Febrile convulsions induced by microwaves and the alteration in behavior of albino mouse OF1. Biol Neonate 60:62–68
92. Gilbert M, Cain D (1985) A single neonatal pentylenetetrazol or hyperthermia convulsion increases kindling susceptibility in the adult rat. Dev Brain Res 22:169–180
93. Jiang W, Duong TM, de Lanerolle NC (1999) The neuropathology of hyperthermic seizures in the rat. Epilepsia 40:5–19
94. Klauenberg BJ, Sparber SB (1984) A kindling-like effect induced by repeated exposure to heated water in rats. Epilepsia 25:292–301
95. Palmer GC, Borrelli AR, Hudzik TJ, Sparber S (1998) Acute heat stress model of seizures in weanling rats: influence of prototypic anti-seizure compounds. Epilepsy Res 30:203–217
96. Germano IM, Sperber EF, Moshé SL (1996) Molecular and experimental aspects of neuronal migration disorders. In: Guerrini R, Andermann F, Canapichi R (eds) Dysplasias of cerebral cortex and epilepsy. Lipponcott-Raven, New York, pp 473–488
97. Sarkisian MR, Rattan S, D'Mello SR, LoTurco JJ (1999) Characterization of seizures in the flathead rat: a new genetic model of epilepsy in early postnatal development. Epilepsia 40:394–400

98. Tancredi V, D'Arcangelo G, Zona C, Siniscalchi A, Avoli M (1992) Induction of epileptiform activity by temperature elevation in hippocampal slices from young rats: an in vitro model for febrile seizures? Epilepsia 33:228–234
99. Dubé C, Chen K, Eghbal-Ahmadi M, Brunson K, Soltesz I, Baram TZ (2000) Prolonged febrile seizures in the immature rat model enhance hippocampal excitability long term. Ann Neurol 47:336–344
100. Jensen FE, Baram TZ (2000) Developmental seizures induced by common early-life insults: short- and long-term effects on seizure susceptibility. Ment Retard Dev Disabil Res Rev 6:253–257
101. Dubé C, Yu H, Nalcioglu O, Baram TZ (2004) Serial MRI after experimental febrile seizures: altered T2 signal without neuronal death. Ann Neurol 56:709–714
102. Toth Z, Yan XX, Haftoglou S, Ribak CE, Baram TZ (1998) Seizure-induced neuronal injury: vulnerability to febrile seizures in an immature rat model. J Neurosci 18:4285–4294
103. Dubé C, Vezzani A, Behrens M, Bartfai T, Baram TZ (2005) Interleukin-1beta contributes to the generation of experimental febrile seizures. Ann Neurol 57:152–155
104. Heida JG, Boisse L, Pittman QJ (2004) Lipopolysaccharide-induced febrile convulsions in the rat: short-term sequelae. Epilepsia 45:1317–1329
105. Heida JG, Teskey GC, Pittman QJ (2005) Febrile convulsions induced by the combination of lipopolysaccharide and low-dose kainic acid enhance seizure susceptibility, not epileptogenesis, in rats. Epilepsia 46:1898–1905
106. Schwartzkroin PA, Roper SN, Wenzel HJ (2004) Cortical dysplasia and epilepsy: animal models. Adv Exp Med Biol 548:145–174
107. Bassanini S, Hallene K, Battaglia G et al (2007) Early cerebrovascular and parenchymal events following prenatal exposure to the putative neurotoxin methylazoxymethanol. Neurobiol Dis 26:481–495
108. Ashwell K (1987) Direct and indirect effects on the lateral geniculate nucleus neurons of prenatal exposure to methylazoxymethanol acetate. Brain Res 432:199–214
109. Baraban SC, Schwartzkroin PA (1996) Flurothyl seizure susceptibility in rats following prenatal methylazoxymethanol treatment. Epilepsy Res 23:189–194
110. Battaglia G, Bassanini S (2006) MAM and other "lesion" models of developmental epilepsy. In: Pitkanen A, Schwartzkroin PA, Moshé SL (eds) Models of seizures and epilepsy. Elsevier, San Diego, pp 305–313
111. Castro PA, Cooper EC, Lowenstein DH, Baraban SC (2001) Hippocampal heterotopia lack functional Kv4.2 potassium channels in the methylazoxymethanol model of cortical malformations and epilepsy. J Neurosci 21:6626–6634
112. Chevassus-Au-Louis N, Congar P, Represa A, Ben-Ari Y, Gaiarsa JL (1998) Neuronal migration disorders: heterotopic neocortical neurons in CA1 provide a bridge between the hippocampus and the neocortex. Proc Natl Acad Sci USA 95:10263–10268
113. Chevassus-Au-Louis N, Rafiki A, Jorquera I, Ben-Ari Y, Represa A (1998) Neocortex in the hippocampus: an anatomical and functional study of CA1 heterotopias after prenatal treatment with methylazoxymethanol in rats. J Comp Neurol 394:520–536
114. Colacitti C, Sancini G, DeBiasi S et al (1999) Prenatal methylazoxymethanol treatment in rats produces brain abnormalities with morphological similarities to human developmental brain dysgeneses. J Neuropathol Exp Neurol 58:92–106
115. Germano IM, Sperber EF (1998) Transplacentally induced neuronal migration disorders: an animal model for the study of the epilepsies. J Neurosci Res 51:473–488
116. Sancini G, Franceschetti S, Battaglia G et al (1998) Dysplastic neocortex and subcortical heterotopias in methylazoxymethanol-treated rats: an intracellular study of identified pyramidal neurones. Neurosci Lett 246:181–185
117. Battaglia G, Arcelli P, Granata T et al (1996) Neuronal migration disorders and epilepsy: a morphological analysis of three surgically treated patients. Epilepsy Res 26:49–58
118. Battaglia G, Granata T, Farina L, D'Incerti L, Franceschetti S, Avanzini G (1997) Periventricular nodular heterotopia: epileptogenic findings. Epilepsia 38:1173–1182
119. Germano IM, Sperber EF, Ahuja S, Moshé SL (1998) Evidence of enhanced kindling and hippocampal neuronal injury in immature rats with neuronal migration disorders. Epilepsia 39:1253–1260

120. Germano IM, Zhang YF, Sperber EF, Moshé SL (1996) Neuronal migration disorders increase susceptibility to hyperthermia-induced seizures in developing rats. Epilepsia 37:902–910
121. de Feo MR, Mecarelli O, Ricci GF (1995) Seizure susceptibility in immature rats with microencephaly induced by prenatal exposure to methylazoxymethanol acetate. Pharmacol Res 31:109–114
122. Marin-Padilla M (1999) Developmental neuropathology and impact of perinatal brain damage. III: gray matter lesions of the neocortex. J Neuropathol Exp Neurol 58:407–429
123. Marin-Padilla M, Tsai RJ, King MA, Roper SN (2003) Altered corticogenesis and neuronal morphology in irradiation-induced cortical dysplasia: a Golgi-Cox study. J Neuropathol Exp Neurol 62:1129–1143
124. Lin D, Roper SN (2006) In utero irradiation as a model of cortical dysplasia. In: Pitkanen A, Schwartzkroin PA, Moshé SL (eds) Models of seizures and epilepsy. Elsevier, San Diego, pp 315–322
125. Deukmedjian AJ, King MA, Cuda C, Roper SN (2004) The GABAergic system of the developing neocortex has a reduced capacity to recover from in utero injury in experimental cortical dysplasia. J Neuropathol Exp Neurol 63:1265–1273
126. Kellinghaus C, Kunieda T, Ying Z, Pan A, Luders HO, Najm IM (2004) Severity of histopathologic abnormalities and in vivo epileptogenicity in the in utero radiation model of rats is dose dependent. Epilepsia 45:583–591
127. Kondo S, Najm I, Kunieda T, Perryman S, Yacubova K, Luders HO (2001) Electroencephalographic characterization of an adult rat model of radiation-induced cortical dysplasia. Epilepsia 42:1221–1227
128. Roper SN, Gilmore RL, Houser CR (1995) Experimentally induced disorders of neuronal migration produce an increased propensity for electrographic seizures in rats. Epilepsy Res 21:205–219
129. Zhu WJ, Roper SN (2000) Reduced inhibition in an animal model of cortical dysplasia. J Neurosci 20:8925–8931
130. Jacobs KM, Gutnick MJ, Prince DA (1996) Hyperexcitability in a model of cortical maldevelopment. Cereb Cortex 6:514–523
131. Jacobs KM, Kharazia VN, Prince DA (1999) Mechanisms underlying epileptogenesis in cortical malformations. Epilepsy Res 36:165–188
132. Luhmann HJ, Raabe K (1996) Characterization of neuronal migration disorders in neocortical structures: I. Expression of epileptiform activity in an animal model. Epilepsy Res 26:67–74
133. Scantlebury MH, Ouellet PL, Psarropoulou C, Carmant L (2004) Freeze lesion-induced focal cortical dysplasia predisposes to atypical hyperthermic seizures in the immature rat. Epilepsia 45:592–600
134. Redecker C, Luhmann HJ, Hagemann G, Fritschy JM, Witte OW (2000) Differential downregulation of GABAA receptor subunits in widespread brain regions in the freeze-lesion model of focal cortical malformations. J Neurosci 20:5045–5053
135. Zilles K, Qu M, Schleicher A, Luhmann HJ (1998) Characterization of neuronal migration disorders in neocortical structures: quantitative receptor autoradiography of ionotropic glutamate, GABA(A) and GABA(B) receptors. Eur J Neurosci 10:3095–3106
136. Palmini A, Gambardella A, Andermann F et al (1995) Intrinsic epileptogenicity of human dysplastic cortex as suggested by corticography and surgical results. Ann Neurol 37:476–487
137. Luhman HJ (2006) The cortical freeze lesion model. In: Pitkanen A, Schwartzkroin PA, Moshé SL (eds) Models of seizures and epilepsy. Elsevier, San Diego, pp 295–303
138. Kraemer M, Roth-Haerer A, Bruehl C, Luhmann HJ, Witte OW (2001) Metabolic and electrophysiological alterations in an animal model of neocortical neuronal migration disorder. Neuroreport 12:2001–2006
139. Marret S, Mukendi R, Gadisseux JF, Gressens P, Evrard P (1995) Effect of ibotenate on brain development: an excitotoxic mouse model of microgyria and posthypoxic-like lesions. J Neuropathol Exp Neurol 54:358–370
140. Redecker C, Hagemann G, Witte OW, Marret S, Evrard P, Gressens P (1998) Long-term evolution of excitotoxic cortical dysgenesis induced in the developing rat brain. Brain Res Dev Brain Res 109:109–113

141. Redecker C, Lutzenburg M, Gressens P, Evrard P, Witte OW, Hagemann G (1998) Excitability changes and glucose metabolism in experimentally induced focal cortical dysplasias. Cereb Cortex 8:623–634
142. Sarkisian MR, Li W, Di Cunto F, D'Mello SR, LoTurco JJ (2002) Citron-kinase, a protein essential to cytokinesis in neuronal progenitors, is deleted in the flathead mutant rat. J Neurosci 22:RC217
143. LoTurco JJ, Sarkisian MR, Cosker L, Bai J (2003) Citron kinase is a regulator of mitosis and neurogenic cytokinesis in the neocortical ventricular zone. Cereb Cortex 13:588–591
144. Roberts MR, Bittman K, Li WW et al (2000) The flathead mutation causes CNS-specific developmental abnormalities and apoptosis. J Neurosci 20:2295–2306
145. Ramos R (2006) Models with spontaneous seizures and developmental disruption of genetic etiology. In: Pitkanen A, Schwartzkroin PA, Moshé SL (eds) Models of seizures and epilepsy. Elsevier, San Diego, pp 249–259
146. Yeung RS, Katsetos CD, Klein-Szanto A (1997) Subependymal astrocytic hamartomas in the Eker rat model of tuberous sclerosis. Am J Pathol 151:1477–1486
147. Mizuguchi M, Takashima S, Yamanouchi H, Nakazato Y, Mitani H, Hino O (2000) Novel cerebral lesions in the Eker rat model of tuberous sclerosis: cortical tuber and anaplastic ganglioglioma. J Neuropathol Exp Neurol 59:188–196
148. Tschuluun N, Wenzel HJ, Schwartzkroin PA (2007) Irradiation exacerbates cortical cytopathology in the Eker rat model of tuberous sclerosis complex, but does not induce hyperexcitability. Epilepsy Res 73:53–64
149. Waltereit R, Welzl H, Dichgans J, Lipp HP, Schmidt WJ, Weller M (2006) Enhanced episodic-like memory and kindling epilepsy in a rat model of tuberous sclerosis. J Neurochem 96:407–413
150. von der Brelie C, Waltereit R, Zhang L, Beck H, Kirschstein T (2006) Impaired synaptic plasticity in a rat model of tuberous sclerosis. Eur J Neurosci 23:686–692
151. Acampora D, Mazan S, Avantaggiato V et al (1996) Epilepsy and brain abnormalities in mice lacking the Otx1 gene. Nat Genet 14:218–222
152. Weimann JM, Zhang YA, Levin ME, Devine WP, Brulet P, McConnell SK (1999) Cortical neurons require Otx1 for the refinement of exuberant axonal projections to subcortical targets. Neuron 24:819–831
153. Cipelletti B, Avanzini G, Vitellaro-Zuccarello L et al (2002) Morphological organization of somatosensory cortex in Otx1(-/-) mice. Neuroscience 115:657–667
154. Pringle AK, Iannotti F, Wilde GJ, Chad JE, Seeley PJ, Sundstrom LE (1997) Neuroprotection by both NMDA and non-NMDA receptor antagonists in in vitro ischemia. Brain Res 755:36–46
155. Wenzel HJ, Robbins CA, Tsai LH, Schwartzkroin PA (2001) Abnormal morphological and functional organization of the hippocampus in a p35 mutant model of cortical dysplasia associated with spontaneous seizures. J Neurosci 21:983–998
156. Patel LS, Wenzel HJ, Schwartzkroin PA (2004) Physiological and morphological characterization of dentate granule cells in the p35 knock-out mouse hippocampus: evidence for an epileptic circuit. J Neurosci 24:9005–9014
157. Meikle L, Talos DM, Onda H et al (2007) A mouse model of tuberous sclerosis: neuronal loss of Tsc1 causes dysplastic and ectopic neurons, reduced myelination, seizure activity, and limited survival. J Neurosci 27:5546–5558
158. van Luijtelaar EL, Coenen AM (1986) Two types of electrocortical paroxysms in an inbred strain of rats. Neurosci Lett 70:393–397
159. Vergnes M, Marescaux C, Micheletti G et al (1982) Spontaneous paroxysmal electroclinical patterns in rat: a model of generalized non-convulsive epilepsy. Neurosci Lett 33:97–101
160. Danober L, Deransart C, Depaulis A, Vergnes M, Marescaux C (1998) Pathophysiological mechanisms of genetic absence epilepsy in the rat. Prog Neurobiol 55:27–57
161. Depaulis A (2006) Genetic models of absence epilepsy in the rats. In: Pitkanen A, Schwartzkroin PA, Moshé SL (eds) Models of seizures and epilepsy. Elsevier, San Diego, pp 233–248
162. Vergnes M, Marescaux C, Depaulis A, Micheletti G, Warter JM (1986) Ontogeny of spontaneous petit mal-like seizures in Wistar rats. Dev Brain Res 30:85–87

163. Coenen AM, Van Luijtelaar EL (1987) The WAG/Rij rat model for absence epilepsy: age and sex factors. Epilepsy Res 1:297–301
164. Marescaux C, Vergnes M, Depaulis A (1992) Genetic absence epilepsy in rats from Strasbourg – A review. J Neural Transm [Suppl] 35:37–69
165. Peeters BW, Kerbusch JM, Coenen AM, Vossen JM, van Luijtelaar EL (1992) Genetics of spike-wave discharges in the electroencephalogram (EEG) of the WAG/Rij inbred rat strain: a classical mendelian crossbreeding study. Behav Genet 22:361–368
166. Gauguier D, van Luijtelaar G, Bihoreau MT et al (2004) Chromosomal mapping of genetic loci controlling absence epilepsy phenotypes in the WAG/Rij rat. Epilepsia 45:908–915
167. Rudolf G, Therese Bihoreau M, Godfrey FR et al (2004) Polygenic control of idiopathic generalized epilepsy phenotypes in the genetic absence rats from Strasbourg (GAERS). Epilepsia 45:301–308
168. Fletcher CF, Lutz CM, O'Sullivan TN et al (1996) Absence epilepsy in tottering mutant mice is associated with calcium channel defects. Cell 87:607–617
169. Burgess DL, Jones JM, Meisler MH, Noebels JL (1997) Mutation of the Ca2+ channel beta subunit gene Cchb4 is associated with ataxia and seizures in the lethargic (lh) mouse. Cell 88:385–392
170. Brodbeck J, Davies A, Courtney JM et al (2002) The ducky mutation in Cacna2d2 results in altered Purkinje cell morphology and is associated with the expression of a truncated alpha 2 delta-2 protein with abnormal function. J Biol Chem 277:7684–7693
171. Letts VA, Felix R, Biddlecome GH et al (1998) The mouse stargazer gene encodes a neuronal Ca2+-channel gamma subunit. Nat Genet 19:340–347
172. Cox GA, Lutz CM, Yang CL et al (1997) Sodium/hydrogen exchanger gene defect in slow-wave epilepsy mutant mice. Cell 91:139–148
173. Kantheti P, Diaz ME, Peden AE et al (2003) Genetic and phenotypic analysis of the mouse mutant mh2J, an Ap3d allele caused by IAP element insertion. Mamm Genome 14:157–167
174. Hess EJ, Collins KA, Copeland NG, Jenkins NA, Wilson MC (1994) Deletion map of the coloboma (Cm) locus on mouse chromosome 2. Genomics 21:257–261
175. Noebels JL (2006) Spontaneous epileptic mutations in the mouse. In: Pitkanen A, Schwartzkroin PA, Moshé SL (eds) Models of seizures and epilepsy. Academic, San Diego, pp 223–232
176. Ludwig A, Budde T, Stieber J et al (2003) Absence epilepsy and sinus dysrhythmia in mice lacking the pacemaker channel HCN2. EMBO J 22:216–224
177. Song I, Kim D, Choi S, Sun M, Kim Y, Shin HS (2004) Role of the alpha1G T-type calcium channel in spontaneous absence seizures in mutant mice. J Neurosci 24:5249–5257
178. Hempelmann A, Taylor KP, Heils A et al (2006) Exploration of the genetic architecture of idiopathic generalized epilepsies. Epilepsia 47:1682–1690
179. Robinson R, Taske N, Sander T et al (2002) Linkage analysis between childhood absence epilepsy and genes encoding GABAA and GABAB receptors, voltage-dependent calcium channels, and the ECA1 region on chromosome 8q. Epilepsy Res 48:169–179
180. Urak L, Feucht M, Fathi N, Hornik K, Fuchs K (2006) A GABRB3 promoter haplotype associated with childhood absence epilepsy impairs transcriptional activity. Hum Mol Genet 15:2533–2541
181. Chen Y, Lu J, Pan H et al (2003) Association between genetic variation of CACNA1H and childhood absence epilepsy. Ann Neurol 54:239–243
182. Chioza B, Everett K, Aschauer H et al (2006) Evaluation of CACNA1H in European patients with childhood absence epilepsy. Epilepsy Res 69:177–181
183. Liang J, Zhang Y, Wang J et al (2006) New variants in the CACNA1H gene identified in childhood absence epilepsy. Neurosci Lett 406:27–32
184. Liang J, Zhang Y, Chen Y et al (2007) Common polymorphisms in the CACNA1H gene associated with childhood absence epilepsy in Chinese Han population. Ann Hum Genet 71:325–335
185. Heron SE, Khosravani H, Varela D et al (2007) Extended spectrum of idiopathic generalized epilepsies associated with CACNA1H functional variants. Ann Neurol 62:560–568

186. Everett KV, Chioza B, Aicardi J et al (2007) Linkage and association analysis of CACNG3 in childhood absence epilepsy. Eur J Hum Genet 15:463–472
187. Everett K, Chioza B, Aicardi J et al (2007) Linkage and mutational analysis of CLCN2 in childhood absence epilepsy. Epilepsy Res 75:145–153
188. Meisler MH, Kearney J, Ottman R, Escayg A (2001) Identification of epilepsy genes in human and mouse. Annu Rev Genet 35:567–588
189. Haut SR, Velíšková J, Moshé SL (2004) Susceptibility of immature and adult brains to seizure effects. Lancet Neurol 3:608–617
190. Pitkanen A, Sutula TP (2002) Is epilepsy a progressive disorder? Prospects for new therapeutic approaches in temporal-lobe epilepsy. Lancet Neurol 1:173–181
191. Swann JW, Moshé SL (1997) Developmental issues in animal models. In: Engel J Jr, Pedley TA (eds) Epilepsy: a comprehensive textbook. Lippincott-Raven Publishers, Philadelphia, pp 467–480
192. Wasterlain CG (1978) Neonatal seizures and brain growth. Neuropadiatrie 9:213–228
193. Sperber EF, Haas KZ, Romero MT, Stanton PK (1999) Flurothyl status epilepticus in developing rats: behavioral, electrographic histological and electrophysiological studies. Brain Res Dev Brain Res 116:59–68
194. Nehlig A, de Vasconselos AP (1996) The model of pentylenetetrazol-induced status epilepticus in the immature rat: short- and long-term effects. Epilepsy Res 26:93–103
195. Stafstrom CE, Sasaki-Adams DM (2003) NMDA-induced seizures in developing rats cause long-term learning impairment and increased seizure susceptibility. Epilepsy Res 53:129–137
196. Ben-Ari Y, Tremblay E, Berger M, Nitecka L (1984) Kainic acid seizure syndrome and binding sites in developing rats. Brain Res 316:284–288
197. Haas KZ, Sperber EF, Opanashuk LA, Stanton PK, Moshé SL (2001) Resistance of immature hippocampus to morphologic and physiologic alterations following status epilepticus or kindling. Hippocampus 11:615–625
198. Stafstrom CE, Thompson JL, Holmes GL (1992) Kainic acid seizures in the developing brain: status epilepticus and spontaneous recurrent seizures. Dev Brain Res 65:227–236
199. Tremblay E, Nitecka L, Berger ML, Ben-Ari Y (1984) Maturation of kainic acid seizure-brain damage syndrome in the rat. I. Clinical, electrographic and metabolic observations. Neuroscience 13:1051–1072
200. Sperber EF, Haas KZ, Stanton PK, Moshe SL (1991) Resistance of the immature hippocampus to seizure-induced synaptic reorganization. Dev Brain Res 60:88–93
201. Yang Y, Tandon P, Liu Z, Sarkisian MR, Stafstrom CE, Holmes GL (1998) Synaptic reorganization following kainic acid-induced seizures during development. Dev Brain Res 107:169–177
202. Tandon P, Yang Y, Das K, Holmes GL, Stafstrom CE (1999) Neuroprotective effects of brain-derived neurotrophic factor in seizures during development. Neuroscience 91:293–303
203. Patel M, Li QY (2003) Age dependence of seizure-induced oxidative stress. Neuroscience 118:431–437
204. Sullivan PG, Dube C, Dorenbos K, Steward O, Baram TZ (2003) Mitochondrial uncoupling protein-2 protects the immature brain from excitotoxic neuronal death. Ann Neurol 53:711–717
205. Sankar R, Shin DH, Wasterlain CG (1997) GABA metabolism during status epilepticus in the developing rat brain. Dev Brain Res 98:60–64
206. Brooks-Kayal AR, Shumate MD, Jin H, Rikhter TY, Coulter DA (1998) Selective changes in single cell GABA(A) receptor subunit expression and function in temporal lobe epilepsy. Nat Med 4:1166–1172
207. Zhang G, Raol Y, Hsu F-C, Coulter D, Brooks-Kayal A (2004) Effects of status epilepticus on hippocampal GABAA receptors are age-dependent. Neuroscience 125:299–303
208. Kubová H, Druga R, Lukasiuk K et al (2001) Status epilepticus causes necrotic damage in the mediodorsal nucleus of the thalamus in immature rats. J Neurosci 21:3593–3599
209. Thompson K, Holm AM, Schousboe A, Popper P, Micevych P, Wasterlain C (1998) Hippocampal stimulation produces neuronal death in the immature brain. Neuroscience 82:337–348

210. Ribak CE, Baram TZ (1996) Selective death of hippocampal CA3 pyramidal cells with mossy fiber afferents after CRH-induced status epilepticus in infant rats. Brain Res Dev Brain Res 91:245–251
211. Baram TZ, Ribak CE (1995) Peptide-induced infant status epilepticus causes neuronal death and synaptic reorganization. Neuroreport 6:277–280
212. Avishai-Eliner S, Brunson KL, Sandman CA, Baram TZ (2002) Stressed-out, or in (utero)? Trends Neurosci 25:518–524
213. Baram TZ, Schultz L (1991) Corticotropin-releasing hormone is a rapid and potent convulsant in the infant rat. Brain Res Dev Brain Res 61:97–101
214. Sankar R, Shin D, Mazarati AM, Liu H, Wasterlain CG (1999) Ontogeny of self-sustaining status epilepticus. Dev Neurosci 21:345–351
215. Koh S, Storey TW, Santos TC, Mian AY, Cole AJ (1999) Early-life seizures in rats increase susceptibility to seizure-induced brain injury in adulthood. Neurology 53:915–921
216. Giorgi FS, Malhotra S, Hasson H, Velíšková J, Rosenbaum DM, Moshé SL (2005) Effects of status epilepticus early in life on susceptibility to ischemic injury in adulthood. Epilepsia 46:490–498
217. Sperk G (1994) Kainic acid seizures in the rat. Prog Neurobiol 42:1–32
218. Turski L, Ikonomidou C, Turski WA, Bortolotto ZA, Cavalheiro EA (1989) Cholinergic mechanisms and epileptogenesis. The seizures induced by pilocarpine: a novel experimental model of intractable epilepsy. Synapse 3:154–171
219. Kubová H, Mareš P, Suchomelová L, Brožek G, Druga R, Pitkanen A (2004) Status epilepticus in immature rats leads to behavioural and cognitive impairment and epileptogenesis. Eur J Neurosci 19:3255–3265
220. Suchomelová L, Baldwin RA, Kubová H, Thompson KW, Sankar R, Wasterlain CG (2006) Treatment of experimental status epilepticus in immature rats: dissociation between anticonvulsant and antiepileptogenic effects. Pediatr Res 59:237–243
221. dos Santos NF, Arida RM, Filho EM, Priel MR, Cavalheiro EA (2000) Epileptogenesis in immature rats following recurrent status epilepticus. Brain Res Brain Res Rev 32:269–276
222. Holmes GL (1997) Epilepsy in the developing brain: lessons from the laboratory and clinic. Epilepsia 38:12–30
223. Moshé SL, Velísek L, Holmes GL (1994) Developmental aspects of experimental seizures. In: Genton P, Hirsch E, Malafosse A, Marescaux C (eds) International workshop on idiopathic generalized epilepsies: clinical, experimental and genetic aspects. John Libbey, London, pp 51–64
224. Swann JW, Lee CL, Smith KL, Hrachovy RA (2000) Developmental neuroplasticity and epilepsy. Epilepsia 41:1078–1079
225. Ben-Ari Y, Gaiarsa JL, Tyzio R, Khazipov R (2007) GABA: a pioneer transmitter that excites immature neurons and generates primitive oscillations. Physiol Rev 87:1215–1284
226. Harris KM, Jensen FE, Tsao B (1992) Three-dimensional structure of dendritic spines and synapses in rat hippocampus (CA1) at postnatal day 15 and adult ages: implications for the maturation of synaptic physiology and long-term potentiation. J Neurosci 12:2685–2705
227. Rakic P, Bourgeois J, Eckenhoff M, Zecevic N, Goldman-Rakic P (1986) Concurrent overproduction of synapses in diverse regions of the primate cerebral cortex. Science 232:232–235
228. Swann JW, Pierson MG, Smith KL, Lee CL (1999) Developmental neuroplasticity: roles in early life seizures and chronic epilepsy. Adv Neurol 79:203–216
229. Insel TR, Miller RP, Gelhard RE (1990) The ontogeny of excitatory amino acid receptors in rat forebrain – I. N-methyl-D-aspartate receptors and quisqualate receptors. Neuroscience 35:31–43
230. Miller LP, Johnson AE, Gelhard RE, Insel TR (1990) The ontogeny of excitatory amino acid receptors in the rat forebrain-II. Kainic acid receptors. Neuroscience 35:45–51
231. Monyer H, Burnashev N, Laurie DJ, Sakmann B, Seeburg PH (1994) Developmental and regional expression in the rat brain and functional properties of four NMDA receptors. Neuron 12:529–540
232. Flint AC, Maisch US, Weishaupt JH, Kriegstein AR, Monyer H (1997) NR2A subunit expression shortens NMDA receptor synaptic currents in developing neocortex. J Neurosci 17:2469–2476

233. Williams K, Russell SL, Shen YM, Molinoff PB (1993) Developmental switch in the expression of NMDA receptors occurs in vivo and in vitro. Neuron 10:267–278
234. Pellegrini-Giampietro D, Bennett M, Zukin R (1991) Differential expression of three glutamate receptor genes in developing rat brain: an in situ hybridization study. Proc Natl Acad Sci USA 88:4157–4161
235. Mueller AL, Taube JS, Schwartzkroin PA (1984) Development of hyperpolarizing inhibitory postsynaptic potentials and hyperpolarizing response to gamma-aminobutyric acid in rabbit hippocampus studied in vitro. J Neurosci 4:860–867
236. Ben-Ari Y, Cherubini E, Corradetti R, Gaiarsa JL (1989) Giant synaptic potentials in immature rat CA3 hippocampal neurones. J Physiol (Lond) 416:303–325
237. Tyzio R, Holmes GL, Ben-Ari Y, Khazipov R (2007) Timing of the developmental switch in GABA(A) mediated signaling from excitation to inhibition in CA3 rat hippocampus using gramicidin perforated patch and extracellular recordings. Epilepsia 48(Suppl 5):96–105
238. Kyrozis A, Chudomel O, Moshé SL, Galanopoulou AS (2006) Sex-dependent maturation of GABA(A) receptor-mediated synaptic events in rat substantia nigra reticulata. Neurosci Lett 398:1–5
239. Heida JG, Pittman QJ (2005) Causal links between brain cytokines and experimental febrile convulsions in the rat. Epilepsia 46:1906–1913
240. Vezzani A, Baram TZ (2007) New roles for interleukin-1 Beta in the mechanisms of epilepsy. Epilepsy Curr 7:45–50
241. Vezzani A, Granata T (2005) Brain inflammation in epilepsy: experimental and clinical evidence. Epilepsia 46:1724–1743
242. Viviani B, Bartesaghi S, Gardoni F et al (2003) Interleukin-1beta enhances NMDA receptor-mediated intracellular calcium increase through activation of the Src family of kinases. J Neurosci 23:8692–8700
243. Beattie EC, Stellwagen D, Morishita W et al (2002) Control of synaptic strength by glial TNFalpha. Science 295:2282–2285
244. Viviani B, Gardoni F, Marinovich M (2007) Cytokines and neuronal ion channels in health and disease. Int Rev Neurobiol 82:247–263
245. Ye ZC, Sontheimer H (1996) Cytokine modulation of glial glutamate uptake: a possible involvement of nitric oxide. Neuroreport 7:2181–2185
246. Tian GF, Azmi H, Takano T et al (2005) An astrocytic basis of epilepsy. Nat Med 11:973–981
247. Fellin T, Gomez-Gonzalo M, Gobbo S, Carmignoto G, Haydon PG (2006) Astrocytic glutamate is not necessary for the generation of epileptiform neuronal activity in hippocampal slices. J Neurosci 26:9312–9322
248. Bezzi P, Domercq M, Brambilla L et al (2001) CXCR4-activated astrocyte glutamate release via TNFalpha: amplification by microglia triggers neurotoxicity. Nat Neurosci 4:702–710
249. Sankar R, Rho JM (2007) Do seizures affect the developing brain? Lessons from the laboratory. J Child Neurol 22:21S–29S
250. Bender RA, Baram TZ (2007) Epileptogenesis in the developing brain: what can we learn from animal models? Epilepsia 48(Suppl 5):2–6
251. Brewster A, Bender RA, Chen Y, Dubé C, Eghbal-Ahmadi M, Baram TZ (2002) Developmental febrile seizures modulate hippocampal gene expression of hyperpolarization-activated channels in an isoform- and cell-specific manner. J Neurosci 22:4591–4599
252. Brewster AL, Bernard JA, Gall CM, Baram TZ (2005) Formation of heteromeric hyperpolarization-activated cyclic nucleotide-gated (HCN) channels in the hippocampus is regulated by developmental seizures. Neurobiol Dis 19:200–207
253. Chen K, Baram TZ, Soltesz I (1999) Febrile seizures in the developing brain result in persistent modification of neuronal excitability in limbic circuits. Nat Med 5:888–894

Seizures and Antiepileptic Drugs: Does Exposure Alter Normal Brain Development in Animal Models?

Aristea S. Galanopoulou, Libor Velíšek, and Solomon L. Moshé

Abstract Clinical studies indicate that recurrent or prolonged seizure early in development may alter learning and behavior. Similarly, antiepileptic drugs administered during the early postnatal age may interfere with healthy brain development. Experimental studies indicate that there are indeed effects of frequent or prolonged seizures (status epilepticus) on the brain development. However these effects vary as a function of the seizure/epilepsy model used, age of implementation, and associated conditions (such as inflammation). Further, there are differences between developmental effects of seizures in naïve or already impaired brain (with the confounding effects of the underlying impairment, which may not be easy to separate). Antiepileptic drugs may also have profound effects on brain development, through a variety of actions. Particular focus has been given to drugs interfering with $GABA_A$ and NMDA receptor neurotransmission as both these systems are essential for normal patterns of neuronal activity and plasticity-driven development, early in life. In contrast, the existing studies support that antiepileptic drugs may not have as many adverse effects in the brain that has experienced seizures, although further investigation is required. The effects of seizures and antiepileptic drugs are further modified by genetic and epigenetic factors, biological and metabolic underlying conditions, or environmental influences. Sex of subjects may significantly influence not only normal brain development but also the effects of seizures and drugs on brain development.

A.S. Galanopoulou (✉)
Albert Einstein College of Medicine, Saul R. Korey Department of Neurology,
Dominick P. Purpura Department of Neuroscience, Einstein/Montefiore Comprehensive
Epilepsy Management Center, Kennedy Center Rm 306, 1410 Pelham Parkway South,
Bronx, NY 10461, USA
e-mail: agalan@aecom.yu.edu

L. Velíšek (✉)
Saul R. Korey Department of Neurology, Dominick P. Purpura Department of Neuroscience,
Einstein/Montefiore Comprehensive Epilepsy Management Center, Albert Einstein College
of Medicine, Kennedy Center Rm 314, 1410 Pelham Parkway South, Bronx, NY 10461, USA
e-mail: velisek@aecom.yu.edu

D. Janigro (ed.), *Mammalian Brain Development*, Contemporary Neuroscience,
DOI 10.1007/978-1-60761-287-2_6,

Keywords Development • Prenatal • Postnatal • Seizure • Epilepsy • Antiepileptic drugs • Genetic defects • Impaired brain

1 Introduction

Neonates and infants are highly susceptible to seizures [1, 2]. Clinical studies suggest, that recurrent and prolonged seizures during development can induce long-term deficits, such as learning and behavioral, particularly if they occur during the first year of life [3, 4]. Similarly, antiepileptic drugs may interfere with brain function. As a result, a key question is what happens if the antiepileptic drugs or the condition they are used for (i.e., seizures) are presented in the developing brain. This "million dollar question" has been raised repeatedly in the past two decades and even before [5–7]. The outcome of the interaction of brain development, seizures, and antiepileptic therapy cannot be easily resolved in children with epilepsy. This is partially due to the variability in pre-existing abnormalities in brain development (i.e., symptomatic epilepsies), the heterogeneity of the clinical manifestations, and the ethical limitations in withholding treatment to produce appropriate drug-free controls. Thus, the aims of long-term developmental side effects of seizures and/or antiepileptic therapy may be addressed with advantage in developing animals. Indeed, excellent reviews on this topic should be mentioned here [7–17]. In this chapter, we will briefly overview various aspects of seizures and antiepileptic therapy and their influence on the developing brain morphology and function.

2 Effects of Seizures on Brain Development

Seizures may indeed affect brain development in humans yet the magnitude of the impact depends on the following critical factors [18–20]. Specific epilepsy syndrome is the first. Infants with infantile spasms are highly likely to develop mental retardation or cognitive deficits; though in symptomatic cases, the underlying condition may contribute to the cognitive decline. On the other hand, typical absence or myoclonic seizures, in the context of an idiopathic epilepsy syndrome, pose only a minor threat for further cognitive development unless of course, they are so frequent that interfere with the process of acquiring information. The second factor is the precise developmental stage. Seizures may have serious influence on brain development only during certain windows of maturation [21, 22]. The importance of the role of developmental stage cannot be emphasized enough. The correlation between children and developing animals in terms of brain development is therefore of paramount significance. Several excellent papers and reviews have been published on this topic [23–26]. Our interpretation of comparative developmental stages in rats and humans is presented in the Table 1. Thirdly, the number and duration of seizures needs to be considered. Frequent or prolonged seizures are more

Table 1 Comparative (brain) developmental stages in humans and rats

Milestone	Human	Rat
Brain development at birth	Full term newborn	PN10–PN12
Puberty onset	9–10 years	PN32–PN36
End of puberty	14–17 years	Around PN55
Life expectancy	74–79 years	2 years

In this table, the estimated equivalency periods of "brain development at birth" are based on studies comparing brain growth spurts, DNA, cholesterol or water content, as well as vulnerability patterns between humans and rats, around the embryonic and perinatal periods. Each specific developmental process, i.e., synaptogenesis, neurogenesis, neuronal differentiation, may follow their own species-, region-, and sex-specific ontogenetic patterns
Modified from [20, 23–25, 187]

likely to result in detectable long-term changes in brain development as compared to a single or infrequent brief seizures. Other modifiers include the genetic background, any underlying pre-existing metabolic or organic abnormalities, sex, and epigenetic factors which may deteriorate (i.e., stress) or improve (enriched environment) the functional outcome [7, 27–29].

3 Effects of Seizures

The way seizures are produced may affect brain development. In naïve animals, seizures can be induced by chemoconvulsant drugs or physical means, such as electrical stimulation or temperature changes. Additionally, chemical or physical methods to predispose animals to develop seizures in response to otherwise subconvulsant stimuli have been devised, to model symptomatic types of epilepsies (Sect. 3.2). Finally, methods of genetic selection or manipulations have been implemented to study how the genetic substrate may contribute to abnormal brain development. Each model has its advantages and disadvantages as will be further discussed.

3.1 Triggered Seizures in Naive Brain

3.1.1 Chemical Models

Flurothyl

Flurothyl is a liquid convulsant ether, which easily evaporates under normal conditions. The advantage is that flurothyl can be introduced to the hermetically sealed environment, in which the animal breaths vapors until seizures occur. Thus, no restrictive handling and potentially painful injection are involved [30]. Additionally, flurothyl may be precisely dosed for the desired effect, simulating either status epilepticus (SE) or clusters of repetitive brief seizures [31]. An advantage of flurothyl-induced seizures is that seizures can be elicited in a controlled manner, as they occur only during flurothyl inhalation. As a result, this model has been utilized to address whether few or many seizures have different outcomes. Flurothyl-induced seizures have primarily generalized character, first consisting of face and forelimb clonus and later of tonic-clonic seizures with loss of righting. In immature rats, clonic seizures may not occur or may be masked by a fast succession of tonic-clonic seizures. In 1976, Wasterlain studied effects of moderate and severe flurothyl-induced SE in 4-day-old rats [6]. While the SE at this age did not produce any gross morphological lesions in the brain, DNA synthesis was inhibited and therefore, decreased brain DNA content long-term (for almost 30 days after the status). Behavioral milestones were impaired and rats displayed decreased seizure thresholds. Flurothyl has been later employed by Holmes and colleagues, to induce multiple brief seizures [31]. Neonatal rats subjected to 25 or 50 brief flurothyl seizures over 5 or 10 days respectively had impaired learning and decreased activity. In adulthood, these rats had decreased seizure thresholds consistent with the Wasterlain data. Further these rats also displayed morphological changes in the hippocampus, specifically mossy fiber sprouting and increased neurogenesis in the dentate granular layer compared to controls [7, 32]. Similar experiments inducing 25 or 50 brief flurothyl seizures between ages PN11–16 or PN11–23, respectively, also showed mossy fiber sprouting in adulthood. However, unlike the neonatal data, spatial learning as assessed by water maze testing at PN30, did not show any deficits in the rats that had experienced 25 flurothyl seizures in infancy [33], indicating that older ages may have decreased vulnerability to certain sequelae of frequent seizures. These differential findings indicate that the timing of seizure impact is as important as the character of the seizures. Studies of neuronal excitation (using c-fos immunoreactivity, which is associated with voltage-gated influx of calcium and therefore acts as a transsynaptic marker of neuronal activation after seizures [34]) showed moderate excitation in the neocortex, entorhinal cortex and amygdala in the rats after 25 seizures. However in the rats with 50 seizures there was a widespread c-fos staining in the neocortex, entorhinal cortex, amygdala, hypothalamus, and dentate gyrus of the hippocampus. Despite this intense excitation, cresyl violet and silver staining did not reveal any cell injury or neuronal loss [31, 35].

An interesting recent study investigated a possible involvement of cyclooxygenase-2 (COX-2) in recurrent seizures. Flurothyl seizures repeatedly delivered to PN7–10 rats indeed were associated with increased COX-2 expression and COX-2 inhibitors were able to increase the seizure threshold in this model suggesting possible involvement of COX-2 pathways on the long-term effects of seizures on brain development [36]. Swann et al. demonstrated that 15–25, but not 5, brief (less than 2 min each) flurothyl seizures, elicited in PN9–13 C57BL/6 mice, were necessary to decrease the expression of NR2A subunit of NMDA receptors or its associated scaffolding protein PSD95 in the hippocampus and neocortex, as assayed 10 days later. Furthermore, these effects were age-specific, as they were not seen in adult mice exposed to flurothyl seizures [37]. However, a single, prolonged (up to 30 min) episode or flurothyl-induced SE in PN6 rats increased NR2C-immunoreactivity

(NR2C-ir), decreased the expression of $GABA_A$ receptor subunit α1 in the hippocampus at PN46, increased susceptibility to pentylenetetrazole (PTZ) seizures, without inducing cell loss [38]. As the seizure load however increased, the rats became more vulnerable to seizure-induced damage in adolescence or adulthood [38, 39]. Although it is difficult to compare different endpoints and experimental conditions, these results raise the possibility that a single flurothyl seizure or infrequent brief seizures may not always lead to measurable side effects. As the age of subjects, seizure duration or frequency increases, some histological or functional changes may become apparent.

Kainic Acid

Kainic acid is a neurotoxin activating the kainic acid (KA) subtypes of glutamate receptors. After systemic or local intracranial administration it induces prolonged seizures, which often culminate in SE. These seizures are believed to arise from limbic structures. Early studies have indicated [22, 40] that a single SE in immature rats was relatively without long-term consequences compared to adult rats [41, 42]. The idea has been expanded [43] to test the effects and consequences of repeated KA-induced status epilepticus (KA-SE) in immature versus adult rats. Immature rats between P20–26 were given 4 KA injections. The rats were tested for spatial learning prior to KA seizures and the retention was tested more than 1 month later. A similar schedule was used for adult rats. While the immature rats suffering recurrent SE displayed no deficits in spatial learning retention, there was a significant impairment in the adult group. No lesions were found in the rats suffering SE in young age while the adult rats had profound cell loss in the CA4, CA3 and CA1 of the hippocampus. On the other hand, rats subjected to a KA-SE at PN10 or PN25 had impaired acquisition of avoidance behavior at PN45 [44], though the effect may have occurred because of the lesion. KA-SE at different developmental ages (PN1, 7, 14, 24 or 75) had impaired learning in adulthood, as assessed by the radial arm maze test, and this was not necessarily attributed to ensuing neuronal injury [45], and indeed, the effects increased with increasing age. These results would indicate that early life KA-SE may cause some cognitive deficits, but in certain aspects these appear to be milder than in adults.

One candidate mechanism through which early life KA-SE may alter normal brain development is the impairment of calcium-sensitive signaling processes that are controlled by the depolarizing effects of $GABA_A$ receptors [46, 47]. The depolarizing effects mediated in rats by $GABA_A$ receptors early postnatally are important for normal neuronal development, as they activate a number of calcium-sensitive signaling cascades [47]. Each neuronal cell type switches from depolarizing to hyperpolarizing $GABA_A$ signaling at its own pace, but usually during the first three postnatal weeks [46, 48–50]. However, the timing of the switch is significantly disrupted in brain regions that have been exposed to three episodes of KA-SE at PN4–6, due to changes in the activity of chloride cotransporters, and these effects are sex-specific [46]. At the molecular level, KA-SE in immature rats has also been shown to alter the expression of GABA and glutamate receptor subunits. A single KA-SE at PN9 results in region and subunit-specific changes in expression at the different hippocampal regions, measured a week later [51]. These include decrease in $\alpha1$, $\alpha4$, increase in $\alpha2$, $\alpha3$, $\beta3$, $\gamma2$, and occasionally $\alpha5$ and $\beta2$. Following a protocol of 3 KA-SE elicited in PN4, 5, and 6 rats, we have also found a decrease in $\alpha1$ and $\gamma1$ subunit expression in the substantia nigra pars reticulata at PN30 [52]. Three episodes of KA-SE in PN6–13 rats result in upregulation of the mGluR1alpha subunit in the CA1 hippocampal interneurons, in the amygdala, and piriform cortex [53]. It is difficult to interpret how this morphological finding reflects functional changes. These subunits play a significant role in seizure-induced impairments of synaptic plasticity [54]. Antagonists of mGluR1 have anticonvulsant and analgesic features [55, 56] and in higher doses also produce memory impairments [57].

Pilocarpine

Pilocarpine (or lithium/pilocarpine) is a cholinergic muscarinic agonist producing after systemic administration prolonged seizures and SE initiated in the limbic structures. Lithium/pilocarpine SE increases significantly the expression of BDNF in PN7–12 rats [58]. This is an interesting finding since BDNF in immature rats has been associated with neuroprotective effects [59] rendering the immature brain more resistant to the SE-induced neuronal injury and keeping its functional integrity compared to the adult brain, in which BDNF may be proconvulsant [60]. Though BDNF may play a role in synaptic plasticity in the developing brain working together with NMDA receptors in stabilization and potentiation of immature synapses [61], no such effects have been demonstrated after seizures in immature animals. Recurrent pilocarpine-induced SE on PN1, PN4 and PN7 result in biphasic changes in neoneurogenesis. Early after seizures, there was a decrease in dentate gyrus neoneurogenesis whereas 40 days later, there was an increase in neoneurogenesis associated with mossy fiber sprouting [62]. The data indicate that the arrest in neurogenesis may be later compensated by overproduction of new neurons resulting in aberrant connections of mossy fibers in the dentate gyrus. While these aberrant connections may have an association with impaired performance in the radial maze [63], such an effect has not yet been shown after early life pilocarpine SE. The report of Xiu-Yu indicates that no spontaneous seizures occurred (at least during the light period of the day). If however lithium-pilocarpine SE was induced at PN20, neoneurogenesis increased at the dentate occurs, but eventually declined with age [64]. In long-term follow up, rats that continued to manifest spontaneous seizures were found to sustain increased rates of neoneurogenesis in the dentate 24 h later. A different study used older rats and followed BrdU incorporation after lithium/pilocarpine SE on PN20. Their data indicate that at this age only the rats showing spontaneous seizures have increased BrdU incorporation compared to the rats with no spontaneous seizures and no seizures [65]. These differences indicate differential age-specific sensitivity of dividing dentate gyrus granule cells to the impact of SE. While early in life, SE is capable in decreasing production of granule cells, later in development SE or spontaneous seizures increase neoneurogenesis.

Several studies have reported late cognitive deficits as a result of early life pilocarpine SE, whether single [66, 67] or repetitive [68]. However, the degree of subsequent impairment, at least in Morris water maze testing, increased with age (none if single SE was induced at PN12, worse if induced at PN20 rather than PN16) [67]. However, in younger ages, multiple SE episodes can result in cognitive impairment. Three episodes of status epilepticus were induced by pilocarpine on PN7–9 [68]. Rats were followed from PN30–90. All the rats displayed severe cognitive deficits. In vitro studies also revealed hyperexcitability, a finding consistent with increased seizure susceptibility after multiple seizures in infancy. In a subsequent study, a similar protocol resulted in acquired disruption of neocortical development, with decreased rate of apoptosis, increased glutamate decarboxylase, earlier rise in parvalbumin staining, as well as altered intracortical circuitry [69]. These may contribute to the abnormal behavioral scores.

3.1.2 Models Induced by Physical Means

Electrical Kindling

Electrical kindling is delivery of repeated electrical subthreshold stimulations to elicit motor seizures [70] to prone brain structures (such as amygdala, hippocampus, entorhinal cortex, etc.) [71]. With repetitions, motor seizures develop and become permanent [70]. Interestingly, immature rats feature several important differences in kindled seizures from adults: First, effective kindling in immature rats can be achieved with very short interstimulus intervals (as short as 5 min). Second, kindled immature rats easily express severe seizures, which in adult rats require over 200 stimulations. Third, kindling effect persists till adulthood [72–76]. Long-term behavioral effects were also evaluated; rats fully kindled at PN20 with 15 additional stimulations over stage five seizures did not display any differences in activity levels or in Morris water maze learning. The only difference versus controls was an increase in emotionality [77]. Thus, the available data may indicate that despite the ease of kindling induction in infant rats, the long-term consequences of kindling are less pronounced than in the chemical models. This again evokes a question of the effect of seizures on long-term sequelae versus the effect of the precipitating factor. It seems that additional mechanisms are involved in long-term effects of convulsogenic chemicals. Alternatively, these results are reminiscent of the lack of cognitive impairment after 25 flurothyl seizures elicited at similar developmental ages [33], supporting the idea that cognitive dysfunction occurs after a certain age-specific threshold of seizure load is surpassed.

Continuous Hippocampal Stimulation

Continuous hippocampal stimulation is an electrically-induced analogue of the SE. Forty five minutes of continuous stimulation in PN20 rats did not elicit any gross morphological lesions [78]. Similarly, there was no effect on the activity of the rats in the open field or on the learning in the Morris Water Maze [79].

Hyperthermic Seizures

Hyperthermic seizures use the finding that increases of ambient temperature may sufficiently raise core temperature in immature animals to trigger seizures similar to the febrile seizures in humans. Some models do not use physical induction of hyperthermia but an injection of LPS followed by an increase of body temperature. This model will be also overviewed here. Hyperthermia-induced seizures have very narrow window of opportunity at PN11–12 [80], during which almost 94% of rats develop seizures. These seizures are associated with neuronal damage persisting for at least 2 weeks, but no long-term decreases in neuronal numbers were found [81]. In long-term consequences, rats experiencing hyperthermic seizures develop spontaneous limbic seizures in 35% of cases and interictal epileptiform activity in 88% of cases [82]. If the rats are subjected to multiple (9) repeated episodes of hyperthermic seizures between PN10–12, they exhibit long-term memory deficits in the Morris Water Maze. This is not the case in the rats after one or three episodes of hyperthermic seizures [83]. A different approach has been used by Heida et al., who elicited febrile seizures as a reaction to the injection of lipopolysaccharide (LPS) and subconvulsant doses of kainic acid without any acute injury [84]. A recent study has indicated that one of the mechanisms of hyperthermic/febrile seizures is respiratory alkalosis [85]. This indeed opens another line of possible side effects after hyperthermic/febrile seizures as alkalosis is associated with unblocking of NMDA receptors [86] critically involved in learning and memory processes.

Freezing Focus

Freezing focus can be easily created by cooling neocortex by an appropriate source of low temperature (e.g., a metal probe cooled by dry ice or liquid nitrogen). This treatment creates epileptic focus in the neocortex of adult rats, cats and dogs [87]. The focus appears in the matter of hours to days often followed by the development of a secondary focus. The focal findings include ion changes, water disturbances (edema), and decreases in glutamate, glutamine and glutathione, but not GABA. These are the effects of freezing focus in the adult brain, in which the development has been completed. However, if this model is applied early after birth, only dysplastic lesion without any ictal activity is produced [88].

3.2 Seizures in the Abnormal Brain

Several models have produced primary lesions in the brain that predisposes them to develop seizures following subsequent subconvulsant stimuli. These indicate that the impact of the underlying pathologies linked with symptomatic causes of epilepsy on brain development is also important and can confound the effects of seizures [89]. The timing of the lesions can also define outcome. These include:

3.2.1 Freezing Lesion

Freezing lesion uses the finding that a short severe cooling of developing neocortex [90] (transcortical freeze lesion) is associated with a development of microgyric lesions, which may become epileptogenic after a secondary trigger is applied. The freezing lesion has been associated with increases in NR2B and delayed expression of NR2A subunit of the NMDA receptor in the impaired neocortex [91, 92] as well as with downregulation of most of GABA(A) receptor subunits [93]. The microgyrus and its close surroundings (<1 mm) are metabolically hypoactive ([^{14}C]2-deoxyglucose measurements; [94]) suggesting limited inputs in the zone. A decrease in in vitro epileptiform activity in rats at PN40 was seen if lesions were induced at PN0, but not at PN1 [95]. Focal freeze lesion at PN1 predisposes rats to develop hyperthermic seizures at PN10 [96]. Freezing of the neocortex on PN1, but not later on, has been associated with deficits in auditory processing [97], which is more prominent in males that also have more dysplastic neurons at the medial geniculate nucleus than females [98, 99]. Long-term impairments in the performance in the Morris Water Maze and in the Lashley type III maze have also been reported [100].

3.2.2 MAM Lesion

MAM lesion is created by prenatal administration of methylazoxymethanol (MAM), a cycas toxin, which methylates DNA. If injected in the correct period of neuronal division, it leads to dysplasias in the neo- and archi-cortex. MAM lesions induced at gestational days 14–15 (G14–15) were associated with decreased body weight of the offspring and significantly increased seizure susceptibility [101–107]. Spontaneous seizures may eventually occur in about 18% of adult rats after this prenatal impact [108]. Developmentally, rats exposed to MAM at G17 suffer from postpubertal hyperactivity, hypersensitivity to a mild stress, underdevelopment of working memory, and social interaction as well as motor deficits [109, 110]. In rats exposed to MAM on G11–12, there are also deficits in learning in Morris Water Maze [111]. However, since a primary impact with MAM is involved, it may be very difficult to distinguish, what is the contribution of the primary impact and what is caused by seizures. The role of seizures is further discounted by their rare occurrence [108].

3.2.3 Prenatal Corticosteroids

Prenatal corticosteroids may also predispose brain for development of certain seizures. Administration of betamethasone on G15 is associated with postnatal changes, specifically with a decreased weight of the offspring, enhanced transfer latency in the elevated plus maze and increased distance traveled in the central zone of the open field compared to saline-exposed controls. The only morphological finding associated with prenatal betamethasone exposure is increased NPY expression in both dorsal and ventral parts of the hippocampus [112]. This prenatal exposure, if specifically combined with a postnatal administration of *N*-methyl-D-aspartate, has proconvulsant features [113]. Again, the developmental effects are associated with the prenatal impact and not the seizures.

3.3 Genetic Models

There are several animal models with either genetic predisposition to epilepsy or with genetic mutations linked with human epilepsies. These can model focal or generalized types of epilepsies, of the idiopathic or symptomatic type, or may simply be associated with increased susceptibility to seizures. The availability of enriched colonies with specific genetic defects has been advantageous when trying to discern the individual impact of a genetic defect, seizures or environment on brain development. By implementing specific genetic manipulations, it may be possible to establish a cause and effect relationship between the genetic defect and a specific phenotype and study the reversibility or age-dependence of a specific trait. This has best been exemplified in the Rett syndrome model, where mice with MeCP2 mutations develop stereotypies, abnormal motility, deficits in learning and sociability, ataxia, seizures, and early death [114–117]. Two laboratories, however, have independently reported that late restoration of MeCP2 expression in MeCP2 knockout mice results in dramatic reversal of the Rett phenotype. This was achieved by conditional deletion of the inserted Lox-Stop-Lox cassette. Such findings have raised the optimism about the validity of late, restorative therapies [118, 119].

Animal models do not always produce disease phenotypes identical to those seen in humans. Animal models of absence or generalized seizures, like WAG/Rij or GAERS rats, have faster frequencies of spike-wave discharges in their EEGs than human patients have and the age of peak occurrence of seizures lags well into the adult life [120, 121]. In contrast, mice with a MeCP2 mutation are profoundly obese in adulthood, which is unusual for the human patients [115]. Although such findings have raised criticisms on the validity of the existing models, the importance of interactions between genetic background and specific mutations has also been evident in human studies. Siblings with the same mutation linked with the *Autosomal Dominant Epilepsy with Febrile Seizure Plus* syndrome (ADEFS+) may have diverse seizure types, with generalized or focal onset and of variable severity [122, 123]. Genetic interactions have best been demonstrated in mice, in which the fortuitous combination of two genetic mutations of ionic channels, each with opposite effects on neuronal excitability, can attenuate the observed seizures and the otherwise lethal phenotype [124].

Genetic models of epilepsy also offer the advantage of observing the effects of "spontaneous" seizures on brain development, without the exposure to drugs or other means of seizure-induction. However, it is often difficult to accurately log the age of seizure onset in these animal models, as we do in humans, making attempts to determine whether it is the genetic defect or the ensuing seizures that alters brain development a cumbersome task.

4 Effects of Administration of Antiepileptic Drugs

4.1 Studies on Naïve Rats

Normal neuronal activity and plasticity are essential components of brain development [47, 125–129]. Just as seizures may disrupt development by imposing excessive and aberrant patterns of neuronal communication, drugs and interventions that reduce neuronal activity may also have adverse effects. In seizure-naïve animals, both prenatal and early postnatal drug administration may impair brain development, in manners that depend upon the drug, age, and dose administered (Table 2). The timing of exposure is often critical as the vulnerability of each brain region depends upon its maturational stage. For example, valproic acid induces hypomyelination at the corpus callosum if given during the first 10 postnatal days of the rat, but not if given after myelination has been completed [130]. Adverse effects include growth impairment, neurodegeneration, morphological changes involving neuronal structure, myelination and architectural organization of the brain. They pertain to a variety of neuroactive drugs and are not just limited to one mechanism of action. They may eventually lead to a variety of neurodevelopmental deficits, including learning, memory or behavioral problems. Combination therapy increases further the risk of adverse reactions. Co-administration of drugs that have not been associated with neurodegeneration in developing rats, like levetiracetam, lamotrigine or topiramate, with drugs that do, like phenytoin, phenobarbital or carbamazepine, increases the number of apoptotic cells seen in the brain [131, 132].

The animal studies have produced useful models to study the adverse effects of antiepileptics in the developing brain, replicating many aspects of the human clinical experience. However caution is warranted, as in certain cases the outcomes are completely different, in different species. For instance, vigabatrin-induced white matter injury is only observed in rodents but not in humans [133–137]. Furthermore, it is not always possible to conduct the appropriate studies in humans, as would be needed to assess for drug-induced neurodegeneration. Regardless, the experimental data and the clinical observations of functional sequelae in patients taking antiepileptics (reviewed in Chap. 3) emphasize the need for more detailed studies studying the underlying mechanisms and longer follow-ups of patients exposed to these drugs.

Table 2 Effects of antiepileptics on brain development of seizure-naïve rodents

Drug dose/age of exposure	Effect	Reference
Valproic acid		
Prenatal	Malformed embryos (mice>rats)	[149, 188–193]
	Reduced viability	
	Low body weight	
	Impaired cortical migration	
	Reduced forebrain size	
	Abnormalities in monoamine systems (migration, differentiation) (PN50)	
	Decreased leucine incorporation in myelin or nuclear proteins	
	Hippocampal and cortical dysplasias (PN30)	
	Increased apoptosis	
	Impaired LTP, LTD, PPF (PN22–28)	
Early postnatal	Reduced body weight	[130, 131, 158, 176, 193, 194]
	Increased apoptosis if ≤ PN7	
	Hypomyelination of corpus callosum if ≤ PN10	
	Impaired sterol and fatty acid synthesis, if PN2–PN18	
	Impaired LTP, if PN0–PN21	
	Decreased leucine incorporation in myelin or nuclear proteins	
Phenytoin		
Prenatal	Decreased fetal survival rate	[148, 179, 193, 195–203]
	Lower body and brain weight	
	Decrease in benzodiazepine binding sites; Decrease in RO5-4864 displaceable benzodiazepine binding	
	Decreased leucine incorporation in myelin or nuclear proteins	
	Gross motor coordination deficits	
	Sex-specific effects:	
	Male offsprings: hyperactivity; worse eye opening, olfactory orientation, startle response	
	Female offsprings: hypoactivity, increased grooming, worse negative geotaxis, air righting, reactivity, locomotor and maze activity	
	Delayed swimming and acoustic startle ontogeny	
	Impaired water maze performance	
	Impaired passive avoidance retention	
	Abnormal circling behavior	
	Impaired reference memory-based spatial learning deficit, impaired spatial discrimination (PN50–70)	
	Dose-dependent and age-dependent effects (vulnerable period: G11–G14 rodents)	
	No dynamic air righting at PN15–20	
	Enhanced release of catecholamine and corticosterone after stress in adult females	

(continued)

Table 2 (continued)

Drug dose/age of exposure	Effect	Reference
Prenatal and early postnatal	Delayed maturation of Purkinje cells (reversible)	[204]
Early postnatal	Reduced body weight Increased apoptosis (PN7) Decreased neurotrophin signaling (PN7) Decreased leucine incorporation in myelin or nuclear proteins	[138, 158, 176]
Lamotrigine		
Prenatal	Decreased maternal weight and litter size (high dose) Hippocampal and cortical dysplasias (all doses) Probable migration defect	[191]
Early postnatal (PN7)	No effect on apoptosis if given alone in doses thought to correspond to therapeutic levels Increased apoptosis if given with phenobarbital or MK-801 Dose-dependent effects on phenytoin-induced apoptosis	[132]
Levetiracetam		
Prenatal	No change in litter size No cortical/hippocampal dysplasias	[191]
Early postnatal (PN7)	No neurodegeneration Increased neurodegeneration if given with phenytoin or carbamazepine	[131]
Topiramate		
Prenatal	No change in maternal weight or litter size No cortical/hippocampal dysplasias	[191]
Early postnatal (PN7)	No significant cell death with therapeutic doses Increased neurodegeneration if at toxic doses or if given with phenytoin	[131, 176]
Vigabatrin		
Prenatal	Lower maternal weight *At PN30*: Hippocampal and cortical dysplasias Increased apoptosis No change in BrdU+ cells in hippocampus Impaired migration	[147]
Early postnatal	Neurodegeneration (high doses) (PN7) *PN1–7 or PN4–14 administration*: Impaired synaptogenesis, decrease in fEPSPs, delayed sensory and motor reflex development, reduced open field activity, impaired spatial learning and object recognition till adulthood *PN12–16 administration*: decreased myelin staining *PN14–26 administration*: white matter injury (partially reversible), hyperactivity	[133, 134, 158, 205, 206]

(continued)

Table 2 (continued)

Drug dose/age of exposure	Effect	Reference
Benzodiazepines		
Prenatal	Alterations in flunitrazepam binding and adrenergic receptors Deficits in learning, locomotor and explorative activity, increased peak amplitude in acoustic startle reflex, anxiety-driven behavior Impairment in anxiety, learning and memory may depend on genetic substrate Sex-specific effects on the sensitivity to pentylenetetrazole-induced seizures in adulthood: *Male offsprings*: increased sensitivity *Female offsprings*: decreased sensitivity	[150, 151, 154, 177, 207, 208]
Prenatal and early postnatal	Altered exploratory behavior, mesolimbic dopamine turnover, and low affinity $GABA_A$ receptors (PN90)	[155]
Early postnatal	Neurodegeneration (diazepam, clonazepam) Decreased local cerebral metabolic rates for glucose Altered expression of GABA receptor subunits, GABA transporters and GAD (PN90) Increased activity, delayed task acquisition, no learning-memory impairment, reduced level of anxiety to novelty	[153, 154, 158, 209]
Phenobarbital		
Prenatal	Reduced brain weight Delayed swimming ontogeny, if given G7–14: No change in maternal weight or litter size No cortical/hippocampal/dysplasias if given G14–19 Cerebellar degeneration if given G9–18 (at PN50) Deficits in learning, activity, sexual behavior, reproductive function	[147, 193, 198, 210–212]
Early postnatal	Reduced body and brain weight Increases apoptosis (high therapeutic range; if given till PN7) Altered expression of GABA receptor subunits, GABA transporters and GAD (PN90) Cerebellar degeneration, if given PN2–21 (PN14, PN50 assessment) Decreased local cerebral metabolic rates for glucose Delay in auditory function acquisition Decreased neurotrophin signaling Decreased leucine incorporation in myelin or nuclear proteins	[153, 158, 160, 176, 193, 210, 213]
Carbamazepine		
Prenatal	No hippocampal or cortical dysplasias at PN30	[191]
Early postnatal (PN7)	Induces neurodegeneration (thalamus) Increased neurodegeneration if given with phenytoin	[131]

(continued)

Table 2 (continued)

Drug dose/age of exposure	Effect	Reference
Sulthiame		
Early postnatal (PN7)	Increased apoptosis	[214]
Trimethadione		
Prenatal	Delayed swimming ontogeny Increased figure-8 activity and maze times Decreased spontaneous alternation behavior; Sensitive periods: G7–10, G15–18	[198]
Ketogenic diet		
Postnatal (post-weanling)	Decreased brain growth Impaired visual-spatial memory	[145]

4.2 Studies on Rats with Prior Experience of Seizures, Stress or Excitotoxic Insults

Overall, the picture presented by studies on naïve rats appears to be grim, yet it is important to remember that these are obtained from stress, seizure, and disease-naïve developing rats. These observations may relate to the in utero exposure of naïve fetuses to antiepileptics that have been administered to their pregnant mothers. They may also be very relevant to young patients with epilepsies, who are chronically treated with such drugs, albeit their seizures may be focal or not as frequent. But even in such scenarios, we need better answers as to how antiepileptic treatment modifies the adverse effects of seizures and epilepsy, associated stress, and the epigenetic influences due to the already altered interactions between the affected subject and its environment.

Few studies have elaborated on the effects of antiepileptic drugs on the developing brain with seizures or other stressful or excitotoxic insults that may predispose to epilepsy. Optimistic reports of clear benefits of certain therapies for recurrent or prolonged seizures have emerged, but the answers are usually drug and model specific. Unlike their effects in naïve rats, NMDA antagonists do not induce neurodegeneration at PN7 in pups that had experienced electroconvulsive seizures [138], whereas in the neonatal cortical freezing lesion model, they reduce the resulting pathology and impairments in auditory processing [99]. Topiramate administration after recurrent brief neonatal flurothyl-induced seizures decreased mossy fiber sprouting and prevented the seizure-induced impairment in spatial learning and open field activity [139]. Also, topiramate administration after perinatal hypoxia-induced seizures decreased the seizure-induced injury later in life [140]. Chronic treatment with valproic acid, gabapentin or topiramate, starting 1 day after kainic acid-induced status epilepticus at PN35 rats, prevented the appearance of histopathological injury, neurodevelopmental deficits, and epilepsy [141, 142]. In contrast, phenobarbital increased the deficits in spatial learning without altering the histopathological damage [141, 143]. In a model of excitotoxic injury with seizures induced at PN5 with S-bromowillardiine, an agonist on kainate/AMPA receptors, topiramate, but not phenytoin, phenobarbital or carbamazepine, had neuroprotective effect [144].

One of the alternative methods of treating intractable seizures is the ketogenic diet, which has shown efficacy in reducing the number of spontaneous seizures in juvenile rats after an episode of lithium-pilocarpine induced SE at PN20 or kainic acid-induced SE at PN30 [145, 146]. However, not only there was no benefit on subsequent seizure-induced injury but ketogenic diet impaired brain growth and visual-spatial memory in later life [145, 146].

Another important parameter is the stress associated with the disease, both on the subject, but also probably on the embryos prenatally stressed by the pregnant mother's seizures. More studies need to be done to ascertain whether the impact of the fetal stress may depend upon the type and severity of seizures. Repetitive kindled seizures (one per day, between G15–20) did not produce any measurable effects on the birth and migration of cortical and hippocampal progenitors in the offspring of Wistar rats [147]. With more severe seizures, inducing three electroconvulsive seizures daily, at G16, G18, G20, the offspring were observed to have decreased benzodiazepine binding sites in the cerebral cortex till PN21, decreased seizure threshold and delayed onset of acoustic startle and eye opening [148]. In this model, though, the effects may also result from direct electrical stimulation of the fetuses. It is well known, and beyond the scope of this review, that prenatal stressors have an effect in development. Simply the prenatal shipping stress increased the neurodegeneration induced by valproic acid, that lacks anxiolytic effects, in the offsprings of treated pregnant dams [149]. Unlike the experience with naïve rats, a benefit of prenatal administration of anxiolytic drugs, like diazepam, in pregnant dams subjected to forced swim stress has been reported, as it alleviated the adverse effects of maternal stress on the offspring [150–152].

There are many aspects that need further investigation, such as the age and sex specific effects of both older and novel antiepileptic therapies on different models of epilepsy, their mechanisms and methods to abort the deleterious side effects. More animal models are needed to study these interactions within the context of specialized syndromes, like catastrophic epileptic encephalopathies or symptomatic epilepsies, which often require different therapeutic approaches.

4.3 Candidate Mechanisms for the Effects of Antiepileptics on Brain Development

A variety of mechanisms have been implicated and used to develop methods to reverse the deficits. First, by simply altering neuronal activity, antiepileptics may trigger transient or persisting changes in gene expression, affecting the expression of neurotransmitters, their receptors and second messengers, growth factors, influencing cellular proliferation and migration, synaptogenesis and neuronal communication, as well as survival [129, 130, 153–157]. Indeed, early exposure to antiepileptics may decrease the expression and activity of growth factors, such as brain derived neurotrophic factor (BDNF), neurotrophin-3 (NT-3), glial cell line-derived neurotrophic factor (GDNF), events that have been linked to the observed neurodegeneration [131, 158, 159]. Interventions that prevented such drug-induced neurodegeneration, like 17β-estradiol or erythropoietin, restored the activity of these signaling pathways [160, 161]. Alternatively, antiepileptics may affect histone-chromosomal interactions and gene expression, a mechanism proposed for the teratogenic effects and hypomyelination associated with valproic acid, a known histone deacetylase inhibitor [130, 156]. Drug-associated teratogenicity has also been linked to interference with metabolic pathways, like the methionine cycle and folate metabolism [162, 163].

Although such changes may occur in adults, administration of antiepileptic drugs in developing pups may be more detrimental as it may interfere with critical periods of brain development [164–166]. One such period is the early developmental stage when $GABA_A$ signaling is still depolarizing and capable of promoting calcium-sensitive differentiation [47, 167]. Depolarizing GABA may control proliferation, neuronal migration and differentiation, synaptogenesis, and is important for normal patterns of interneuronal communications. In most brain regions, the switch of $GABA_A$ signaling from depolarizing to hyperpolarizing occurs by the 3rd postnatal week and in certain brain regions occurs earlier in females [46, 49, 52, 168–170]. Administration of $GABA_A$ergic agonists alters the expression of chloride cotransporters and consequently the direction of $GABA_A$ergic signaling [50, 52, 171, 172]. As a result, they may potentially prematurely deprive developing neurons from the neurotrophic effects of depolarizing GABA, a mechanism that may partially explain why they may be more vulnerable to drug-induced adverse effects. However, severe early life seizures or stress also alter $GABA_A$ signaling, and as a result, the effects of $GABA_A$-acting drugs may not be the same, or as negative, as in naive brains [46].

Another age-related factor for the vulnerability of the very young naïve brain to neuroactive drugs may relate to its need for a certain level of excitatory NMDA-mediated neurotransmission. Administration of NMDA receptor antagonists, like MK-801, causes diffuse neurodegeneration in the brain of pups younger than PN14, but not in PN21 pups [159]. Indeed, the expression of NMDA receptors is increased during the first 2 postnatal weeks [173] and because of the prevailing NR2B subunit their subunit composition differs from older rats, in which the NR2B subunit is replaced by NR2A [174]. There are currently no available data on felbamate-induced apoptosis, a known inhibitor of NMDA receptors containing the NR2B subunit [175]. Such effects do not extend however to non-NMDA glutamatergic neurotransmission, as topiramate is not known to induce apoptosis [131, 176].

Epigenetic influences may also modify the effects of antiepileptics on brain development. Prenatal shipping stress exacerbated valproic acid induced neurodegeneration [149]. In contrast, environmental modifications implemented during the early postnatal life, like brief intermittent maternal separation or handling, compensated for some of the prenatal effects of antiepileptics on behavior [150, 177].

As we gain more insight into the pathogenetic mechanism of epilepsy and antiepileptic effects, it is becoming increasingly more obvious that there are common pathways of interaction, which need to be explored to design more effective and safe methods of treating seizures.

5 Additional Confounders

Several factors may alter the effects of seizures and their treatment on the developing brain. These include the genetic background, concomitant systemic or metabolic disturbances, stressors and prior experiences, or environmental factors. Among these, sex is an important factor, which can affect outcome of seizures and their therapy, even in developing animals [30, 46, 50, 151, 167, 178–181]. Known explanations are genetic and hormonal factors, sex differences in the timing and patterns of neurogenesis during brain development and in signaling pathways, as well as their different biological responsiveness to environmental stressors and experiences [182]. Indeed, sex may interfere with the conditions priming the brain for increased seizure susceptibility, as occurs with the neonatal freezing lesions [183]. Prenatal betamethasone exposure induces sex-specific behavioral changes [112], that may result in differential seizure susceptibility of both sexes and therefore, differential outcome of brain development. Prenatal stressors or antiepileptics, like phenytoin or benzodiazepines, may enhance the sex differences in seizure susceptibility and other behavioral and functional outcomes [151, 179, 184–186]. Thus, this modifier needs to be taken into account when the seizure consequences are evaluated. The best approach is to investigate both sexes and evaluate the effects of seizures on brain development separately.

6 Conclusions

Early in life, prolonged or recurrent seizures may impair brain development in an age specific fashion with different consequences during the various distinct period that comprise infancy and childhood in humans. Antiepileptic drugs administered during critical periods may also significantly alter brain function, although most of these studies were not carried out in brains with specific epileptic potential. Understanding the impact of seizures and concurrent effects of treatments must take into account the cause of the seizures, whether the brain is already compromised, gender, seizure duration and frequency, mechanism of action of the antiepileptic therapy and the need to treat versus not treating. Further studies of these factors are required to design age and gender appropriate treatments aiming at cures without side effects.

Acknowledgments Supported by grants from NIH/NINDS NS020253, NS041366, NS045243, NS058303, NS059504, NS062947, NS056093, International Rett Syndrome Research Foundation grant, grants from People Against Childhood Epilepsy, Johnson & Johnson, March of Dimes Birth Defect Foundation, and the Heffer Family Foundation. Solomon Moshé is the recipient of a Martin A. and Emily L. Fisher fellowship in Neurology and Pediatrics.

References

1. Hauser W (1994) The prevalence and incidence of convulsive disorders in children. Epilepsia 35:1–6
2. Hauser WA (1997) Incidence and prevalence. In: Engel J Jr, Pedley TA (eds) Epilepsy: a comprehensive textbook. Lippincott-Raven Publishers, Philadelphia, pp 47–57
3. Huttenlocher PR, Hapke RJ (1990) A follow-up study of intractable seizures in childhood. Ann Neurol 28:699–705
4. McNelis AM, Dunn DW, Johnson CS, Austin JK, Perkins SM (2007) Academic performance in children with new-onset seizures and asthma: a prospective study. Epilepsy Behav 10:311–318
5. Wasterlain CG, Duffy TE (1976) Status epilepticus in immature rats. Protective effects of glucose on survival and brain development. Arch Neurol 33:821–827
6. Wasterlain CG (1976) Effects of neonatal status epilepticus on rat brain development. Neurology 26:975–986
7. Holmes GL, Gairsa JL, Chevassus-Au-Louis N, Ben-Ari Y (1998) Consequences of neonatal seizures in the rat: morphological and behavioral effects. Ann Neurol 44:845–857
8. Wasterlain CG (1997) Recurrent seizures in the developing brain are harmful. Epilepsia 38:728–734
9. Wasterlain CG, Shirassaka Y (1994) Seizures, brain damage and brain development. Brain Dev 16:279–295
10. Kaindl AM, Asimiadou S, Manthey D, Hagen MV, Turski L, Ikonomidou C (2006) Antiepileptic drugs and the developing brain. Cell Mol Life Sci 63:399–413
11. Jensen FE (2006) Pediatric epilepsy models. Epilepsy Res 68:28–31
12. Holmes GL (1997) Epilepsy in the developing brain: lessons from the laboratory and clinic. Epilepsia 38:12–30
13. Holmes GL, Moshé SL (1990) Consequences of seizures in the developing brain. J Epilepsy 3(suppl):13
14. Lado FA, Laureta EC, Moshe SL (2002) Seizure-induced hippocampal damage in the mature and immature brain. Epileptic Disord 4:83–97
15. Haut SR, Velíšková J, Moshé SL (2004) Susceptibility of immature and adult brains to seizure effects. Lancet Neurol 3:608–617
16. Swann JW, Baram TZ, Jensen FE, Moshé SL (2008) Seizure mechanisms and vulnerability in the developing brain. In: Engel J Jr, Pedley TA (eds) Epilepsy: a comprehensive textbook, vol 1. Wolters Kluwer/Lippincot Williams & Wilkins, Philadelphia, pp 469–479
17. Scantlebury MH, Heida JG, Hasson HJ, Veliskova J, Velisek L, Galanopoulou AS, Moshe SL (2007) Age-dependent consequences of status epilepticus: animal models. Epilepsia 48(Suppl 2): 75–82
18. Holmes GL, Khazipov R, Ben-Ari Y (2002) Seizure-induced damage in the developing human: relevance of experimental models. Prog Brain Res 135:321–334
19. Haut SR, Shinnar S, Moshe SL (2005) Seizure clustering: risks and outcomes. Epilepsia 46:146–149
20. Velíšek L, Moshé SL (2002) Effects of brief seizures during development. In: Sutula T, Pitkanen A (eds) Do seizures damage the brain, vol 135. Elsevier, Amsterdam, pp 355–364
21. Sperber EF (1996) The relationship between seizures and damage in the maturing brain. In: Heinemann U, Engel J Jr, Avanzini G, Meldrum BS, Mouritzen-Dam A, Wasterlain C (eds) Progressive nature of epileptogenesis. Elsevier, Amsterdam, pp 365–376
22. Haas KZ, Sperber EF, Opanashuk LA, Stanton PK, Moshé SL (2001) Resistance of the immature hippocampus to morphologic and physiologic alterations following status epilepticus or kindling. Hippocampus 11:615–625
23. Dobbing J, Sands J (1979) Comparative aspects of the brain growth spurt. Early Hum Dev 3:79–83
24. Avishai-Eliner S, Brunson KL, Sandman CA, Baram TZ (2002) Stressed-out, or in (utero)? Trends Neurosci 25:518–524

25. Gottlieb A, Keydar I, Epstein HT (1977) Rodent brain growth stages. An analytical review. Biol Neonate 32:166–176
26. Nehlig A (1993) Imaging and the ontogeny of brain metabolism. Baillieres Clin Endocrinol Metab 7:627–642
27. Majak K, Pitkanen A (2004) Do seizures cause irreversible cognitive damage? Evidence from animal studies. Epilepsy Behav 5(Suppl 1):S35–S44
28. Stafstrom CE (2007) Neurobiological mechanisms of developmental epilepsy: translating experimental findings into clinical application. Semin Pediatr Neurol 14:164–172
29. Faverjon S, Silveira DC, Fu DD, Cha BH, Akman C, Hu Y, Holmes GL (2002) Beneficial effects of enriched environment following status epilepticus in immature rats. Neurology 59:1356–1364
30. Velíšková J, Moshé SL (2001) Sexual dimorphism and developmental regulation of substantia nigra function. Ann Neurol 50:596–601
31. Liu Z, Yang Y, Silveira DC, Sarkisian MR, Tandon P, Huang LT, Stafstrom CE, Holmes GL (1999) Consequences of recurrent seizures during early brain development. Neuroscience 92:1443–1454
32. Huang L, Cilio MR, Silveira DC, McCabe BK, Sogawa Y, Stafstrom CE, Holmes GL (1999) Long-term effects of neonatal seizures: a behavioral, electrophysiological, and histological study. Brain Res Dev Brain Res 118:99–107
33. Hoffmann AF, Zhao Q, Holmes GL (2004) Cognitive impairment following status epilepticus and recurrent seizures during early development: support for the "two-hit hypothesis". Epilepsy Behav 5:873–877
34. Morgan JI, Cohen DR, Hempstead JL, Curran T (1987) Mapping patterns of c-fos expression in the central nervous system after seizure. Science 237:192–197
35. Szot P, White SS, McCarthy EB, Turella A, Rejniak SX, Schwartzkroin PA (2001) Behavioral and metabolic features of repetitive seizures in immature and mature rats. Epilepsy Res 46:191–203
36. Kim DK, Jang TJ (2006) Cyclooxygenase-2 expression and effect of celecoxib in flurothyl-induced neonatal seizure. Int J Exp Pathol 87:73–78
37. Swann JW, Le JT, Lee CL (2007) Recurrent seizures and the molecular maturation of hippocampal and neocortical glutamatergic synapses. Dev Neurosci 29:168–178
38. Ni H, Jiang YW, Bo T, Wang JM, Wu XR (2005) c-Fos, N-methyl-d-aspartate receptor 2C, GABA-A-alpha1 immonoreactivity, seizure latency and neuronal injury following single or recurrent neonatal seizures in hippocampus of Wistar rat. Neurosci Lett 380:149–154
39. Schmid R, Tandon P, Stafstrom CE, Holmes GL (1999) Effects of neonatal seizures on subsequent seizure-induced brain injury. Neurology 53:1754–1761
40. Albala BJ, Moshé SL, Okada R (1984) Kainic-acid-induced seizures: a developmental study. Dev Brain Res 13:139–148
41. Tremblay E, Nitecka L, Berger M, Ben-Ari Y (1984) Maturation of kainic acid seizure-brain damage syndrome in the rat. I. Clinical, electrographic and metabolic observations. Neuroscience 13:1051–1072
42. Nitecka L, Tremblay E, Charton G, Bouillot JP, Berger ML, Ben-Ari Y (1984) Maturation of kainic acid seizure-brain damage syndrome in the rat. II Histopathological sequelae. Neuroscience 13:1073–1094
43. Sarkisian MR, Tandon P, Liu Z, Yang Y, Hori A, Holmes GL, Stafstrom CE (1997) Multiple kainic acid seizures in the immature and adult brain: ictal manifestations and long-term effects on learning and memory. Epilepsia 38:1157–1166
44. de Feo MR, Mecarelli O, Palladini G, Ricci GF (1986) Long-term effects of early status epilepticus on the acquisition avoidance behavior in rats. Epilepsia 27:476–482
45. Lynch M, Sayin U, Bownds J, Janumpalli S, Sutula T (2000) Long-term consequences of early postnatal seizures on hippocampal learning and plasticity. Eur J Neurosci 12:2252–2264
46. Galanopoulou AS (2008) Dissociated gender-specific effects of recurrent seizures on GABA signaling in CA1 pyramidal neurons: role of GABAA receptors. J Neurosci 28:1557–1567
47. Ben-Ari Y (2002) Excitatory actions of GABA during development: the nature of the nurture. Nat Rev Neurosci 3:728–739

48. Rivera C, Voipio J, Payne JA, Ruusuvuori E, Lahtinen H, Lamsa K, Pirvola U, Saarma M, Kaila K (1999) The K+/Cl- co-transporter KCC2 renders GABA hyperpolarizing during neuronal maturation. Nature 397:251–255
49. Kyrozis A, Chudomel O, Moshe SL, Galanopoulou AS (2006) Sex-dependent maturation of GABAA receptor-mediated synaptic events in rat substantia nigra reticulata. Neurosci Lett 398:1–5
50. Galanopoulou AS, Kyrozis A, Claudio OI, Stanton PK, Moshe SL (2003) Sex-specific KCC2 expression and GABA(A) receptor function in rat substantia nigra. Exp Neurol 183:628–637
51. Lauren HB, Lopez-Picon FR, Korpi ER, Holopainen IE (2005) Kainic acid-induced status epilepticus alters GABA receptor subunit mRNA and protein expression in the developing rat hippocampus. J Neurochem 94:1384–1394
52. Galanopoulou AS (2006) Sex- and cell-type-specific patterns of GABAA receptor and estradiol-mediated signaling in the immature rat substantia nigra. Eur J Neurosci 23:2423–2430
53. Avallone J, Gashi E, Magrys B, Friedman LK (2006) Distinct regulation of metabotropic glutamate receptor (mGluR1 alpha) in the developing limbic system following multiple early-life seizures. Exp Neurol 202:100–111
54. Nagaraja RY, Becker A, Reymann KG, Balschun D (2005) Repeated administration of group I mGluR antagonists prevents seizure-induced long-term aberrations in hippocampal synaptic plasticity. Neuropharmacology 49(Suppl 1):179–187
55. Shannon HE, Peters SC, Kingston AE (2005) Anticonvulsant effects of LY456236, a selective mGlu1 receptor antagonist. Neuropharmacology 49(Suppl 1):188–195
56. Shannon HE, Eberle EL, Peters SC (2005) Comparison of the effects of anticonvulsant drugs with diverse mechanisms of action in the formalin test in rats. Neuropharmacology 48:1012–1020
57. More L, Gravius A, Pietraszek M, Belozertseva I, Malyshkin A, Shekunova E, Barberi C, Schaefer D, Schmidt WJ, Danysz W (2007) Comparison of the mGluR1 antagonist A-841720 in rat models of pain and cognition. Behav Pharmacol 18:273–281
58. Kornblum HI, Sankar R, Shin DH, Wasterlain CG, Gall CM (1997) Induction of brain derived neurotrophic factor mRNA by seizures in neonatal and juvenile rat brain. Mol Brain Res 44:219–228
59. Tandon P, Yang Y, Das K, Holmes GL, Stafstrom CE (1999) Neuroprotective effects of brain-derived neurotrophic factor in seizures during development. Neuroscience 91:293–303
60. Scharfman HE, Goodman JH, Sollas AL, Croll SD (2002) Spontaneous limbic seizures after intrahippocampal infusion of brain-derived neurotrophic factor. Exp Neurol 174:201–214
61. Walz C, Jungling K, Lessmann V, Gottmann K (2006) Presynaptic plasticity in an immature neocortical network requires NMDA receptor activation and BDNF release. J Neurophysiol 96:3512–3516
62. Xiu-Yu S, Ruo-Peng S, Ji-Wen W (2007) Consequences of pilocarpine-induced recurrent seizures in neonatal rats. Brain Dev 29:157–163
63. Crusio WE, Schwegler H, Lipp HP (1987) Radial-maze performance and structural variation of the hippocampus in mice: a correlation with mossy fibre distribution. Brain Res 425:182–185
64. Porter BE, Maronski M, Brooks-Kayal AR (2004) Fate of newborn dentate granule cells after early life status epilepticus. Epilepsia 45:13–19
65. Cha BH, Akman C, Silveira DC, Liu X, Holmes GL (2004) Spontaneous recurrent seizure following status epilepticus enhances dentate gyrus neurogenesis. Brain Dev 26:394–397
66. Liu Z, Gatt A, Werner SJ, Mikati MA, Holmes GL (1994) Long-term behavioral deficits following pilocarpine seizures in immature rats. Epilepsy Res 19:191–204
67. Cilio MR, Sogawa Y, Cha BH, Liu X, Huang LT, Holmes GL (2003) Long-term effects of status epilepticus in the immature brain are specific for age and model. Epilepsia 44:518–528
68. Santos NF, Marques RH, Correia L, Sinigaglia-Coimbra R, Calderazzo L, Sanabria ER, Cavalheiro EA (2000) Multiple pilocarpine-induced status epilepticus in developing rats: a long-term behavioral and electrophysiological study. Epilepsia 41:S57–S63
69. da Silva AV, Regondi MC, Cavalheiro EA, Spreafico R (2005) Disruption of cortical development as a consequence of repetitive pilocarpine-induced status epilepticus in rats. Epilepsia 46(Suppl 5):22–30

70. Goddard GV, McIntyre DC, Leech CK (1969) A permanent change in brain function resulting from daily electrical stimulation. Exp Neurol 25:295–330
71. Kaada B (1951) Somato-motor, autonomic and electrocorticographic responses to electrical stimulation of "rhinencephalic" and other structures in primates, cat and dog. Acta Physiol Scand 24:1–285
72. Haas K, Sperber EF, Moshé SL (1990) Kindling in developing animals: expression of severe seizures and enhanced development of bilateral foci. Dev Brain Res 56:275–280
73. Haas KZ, Sperber EF, Moshe SL (1992) Kindling in developing animals: interactions between ipsilateral foci. Dev Brain Res 68:140–143
74. Moshé SL (1981) The effects of age on the kindling phenomenon. Dev Psychobiol 14:75–81
75. Moshé SL, Albala BJ (1982) Kindling in developing rats: persistence of seizures into adulthood. Dev Brain Res 4:67–71
76. Moshé SL, Sharpless NS, Kaplan J (1981) Kindling in developing rats: afterdischarge thresholds. Brain Res 211:190–195
77. Holmes GL, Chronopoulos A, Stafstrom CE, Mikati MA, Thurber SJ, Hyde PA, Thompson JL (1993) Effects of kindling on subsequent learning, memory, behavior, and seizure susceptibility. Brain Res Dev Brain Res 73:71–77
78. Sarkisian MR, Holmes GL, Carmant L, Liu Z, Yang Y, Stafstrom CE (1999) Effects of hyperthermia and continuous hippocampal stimulation on the immature and adult brain. Brain Dev 21:318–325
79. Thurber S, Chronopoulos A, Stafstrom CE, Holmes GL (1992) Behavioral effects of continuous hippocampal stimulation in the developing rat. Dev Brain Res 68:35–40
80. Baram TZ, Gerth A, Schultz L (1997) Febrile seizures: an appropriate-aged model suitable for long-term studies. Dev Brain Res 98:265–270
81. Toth Z, Yan XX, Haftoglou S, Ribak CE, Baram TZ (1998) Seizure-induced neuronal injury: vulnerability to febrile seizures in an immature rat model. J Neurosci 18:4285–4294
82. Dube C, Richichi C, Bender RA, Chung G, Litt B, Baram TZ (2006) Temporal lobe epilepsy after experimental prolonged febrile seizures: prospective analysis. Brain 129:911–922
83. Chang YC, Huang AM, Kuo YM, Wang ST, Chang YY, Huang CC (2003) Febrile seizures impair memory and cAMP response-element binding protein activation. Ann Neurol 54:706–718
84. Heida JG, Boisse L, Pittman QJ (2004) Lipopolysaccharide-induced febrile convulsions in the rat: short-term sequelae. Epilepsia 45:1317–1329
85. Schuchmann S, Schmitz D, Rivera C, Vanhatalo S, Salmen B, Mackie K, Sipila ST, Voipio J, Kaila K (2006) Experimental febrile seizures are precipitated by a hyperthermia-induced respiratory alkalosis. Nat Med 12:817–823
86. Ou-Yang Y, Kristian T, Mellergard P, Siesjo BK (1994) The influence of pH on glutamate- and depolarization-induced increases of intracellular calcium concentration in cortical neurons in primary culture. Brain Res 646:65–72
87. Lewin E (1972) The production of epileptogenic cortical foci in experimental animals by freezing. In: Purpura DP, Penry JK, Tower D, Woodbury DM, Walter R (eds) Experimental models of epilepsy: a manual for the laboratory worker. Raven, New York, pp 37–49
88. Luhmann HJ, Raabe K (1996) Characterization of neuronal migration disorders in neocortical structures: I. Expression of epileptiform activity in an animal model. Epilepsy Res 26:67–74
89. Luhmann HJ (2006) The cortical freeze lesion model. In: Pitkanen A, Schwartzkroin PA, Moshé SL (eds) Models of seizures and epilepsy. Elsevier, Amsterdam
90. Dvorak K, Feit J (1977) Migration of neuroblasts through partial necrosis of the cerebral cortex in newborn rats. Contribution of the problems of morphological developmental period of cerebral microgyria. Acta Neuropathologica 38:203–212
91. DeFazio RA, Hablitz JJ (2000) Alterations in NMDA receptors in a rat model of cortical dysplasia. J Neurophysiol 83:315–321
92. Hablitz JJ, DeFazio RA (2000) Altered receptor subunit expression in rat neocortical malformations. Epilepsia 41(Suppl 6):S82–S85

93. Redecker C, Luhmann HJ, Hagemann G, Fritschy JM, Witte OW (2000) Differential down-regulation of GABAA receptor subunits in widespread brain regions in the freeze-lesion model of focal cortical malformations. J Neurosci 20:5045–5053
94. Kraemer M, Roth-Haerer A, Bruehl C, Luhmann HJ, Witte OW (2001) Metabolic and electrophysiological alterations in an animal model of neocortical neuronal migration disorder. JNeuroreport 12:2001–2006
95. Jacobs KM, Hwang BJ, Prince DA (1999) Focal epileptogenesis in a rat model of polymicrogyria. J Neurophysiol 81:159–173
96. Scantlebury MH, Ouellet PL, Psarropoulou C, Carmant L (2004) Freeze lesion-induced focal cortical dysplasia predisposes to atypical hyperthermic seizures in the immature rat. Epilepsia 45:592–600
97. Threlkeld SW, McClure MM, Rosen GD, Fitch RH (2006) Developmental timeframes for induction of microgyria and rapid auditory processing deficits in the rat. Brain Res 1109:22–31
98. Herman AE, Galaburda AM, Fitch RH, Carter AR, Rosen GD (1997) Cerebral microgyria, thalamic cell size and auditory temporal processing in male and female rats. Cereb Cortex 7:453–464
99. Rosen GD, Herman AE, Galaburda AM (1999) Sex differences in the effects of early neocortical injury on neuronal size distribution of the medial geniculate nucleus in the rat are mediated by perinatal gonadal steroids. Cereb Cortex 9:27–34
100. Rosen GD, Waters NS, Galaburda AM, Denenberg VH (1995) Behavioral consequences of neonatal injury of the neocortex. Brain Res 681:177–189
101. de Feo MR, Mecarelli O, Ricci GF (1995) Seizure susceptibility in immature rats with microencephaly induced by prenatal exposure to methylazoxymethanol acetate. Pharmacol Res 31:109–114
102. Germano IM, Zhang YF, Sperber EF, Moshe SL (1996) Neuronal migration disorders increase susceptibility to hyperthermia-induced seizures in developing rats. Epilepsia 37:902–910
103. Germano IM, Sperber EF (1998) Transplacentally induced neuronal migration disorders: an animal model for the study of the epilepsies. J Neurosci Res 51:473–488
104. Baraban SC, Schwartzkroin PA (1996) Flurothyl seizure susceptibility in rats following prenatal methylazoxymethanol treatment. Epilepsy Res 23:189–194
105. Chevassus-au-Louis N, Baraban SC, Gaiarsa J-L, Ben-Ari Y (1999) Cortical malformations and epilepsy: new insights from animal models. Epilepsia 40:811–821
106. Choi IS, Cho JH, Kim JH, Jung SC, Bae YC, Lee MG, Choi BJ (2004) Excitability of CA1 neurons in the model of malformation-associated epilepsy. Neuroreport 15:1639–1642
107. Choi IS, Cho JH, Lee MG, Choi BJ (2005) Pilocarpine-induced seizure susceptibility in rats following prenatal methylazoxymethanol treatment. Biol Pharm Bull 28:1408–1413
108. Harrington EP, Moddel G, Najm IM, Baraban SC (2007) Altered glutamate receptor - transporter expression and spontaneous seizures in rats exposed to methylazoxymethanol in utero. Epilepsia 48:158–168
109. Le Pen G, Gourevitch R, Hazane F, Hoareau C, Jay TM, Krebs MO (2006) Peri-pubertal maturation after developmental disturbance: a model for psychosis onset in the rat. Neuroscience 143:395–405
110. Balduini W, Lombardelli G, Peruzzi G, Cattabeni F (1991) Treatment with methylazoxymethanol at different gestational days: physical, reflex development and spontaneous activity in the offspring. Neurotoxicology 12:179–188
111. Fiore M, Korf J, Antonelli A, Talamini L, Aloe L (2002) Long-lasting effects of prenatal MAM treatment on water maze performance in rats: associations with altered brain development and neurotrophin levels. Neurotoxicol Teratol 24:179–191
112. Velíšek L (2006) Prenatal exposure to betamethasone decreases anxiety in developing rats: Hippocampal neuropeptide Y as a target molecule. Neuropsychopharmacology 31:2140–2149
113. Velíšek L, Jehle K, Asche S, Velíšková J (2007) Model of infantile spasms induced by N-methyl-D-aspartic acid in prenatally impaired brain. Ann Neurol 61:109–119
114. Chahrour M, Zoghbi HY (2007) The story of Rett syndrome: from clinic to neurobiology. Neuron 56:422–437

115. Guy J, Hendrich B, Holmes M, Martin JE, Bird A (2001) A mouse Mecp2-null mutation causes neurological symptoms that mimic Rett syndrome. Nat Genet 27:322–326
116. Chen RZ, Akbarian S, Tudor M, Jaenisch R (2001) Deficiency of methyl-CpG binding protein-2 in CNS neurons results in a Rett-like phenotype in mice. Nat Genet 27:327–331
117. Pelka GJ, Watson CM, Radziewic T, Hayward M, Lahooti H, Christodoulou J, Tam PP (2006) Mecp2 deficiency is associated with learning and cognitive deficits and altered gene activity in the hippocampal region of mice. Brain 129:887–898
118. Guy J, Gan J, Selfridge J, Cobb S, Bird A (2007) Reversal of neurological defects in a mouse model of Rett syndrome. Science 315:1143–1147
119. Giacometti E, Luikenhuis S, Beard C, Jaenisch R (2007) Partial rescue of MeCP2 deficiency by postnatal activation of MeCP2. Proc Natl Acad Sci USA 104:1931–1936
120. Marescaux C, Vergnes M, Depaulis A (1992) Genetic absence epilepsy in rats from Strasbourg – a review. J Neural Transm Suppl 35:37–69
121. Coenen AM, Van Luijtelaar EL (1987) The WAG/Rij rat model for absence epilepsy: age and sex factors. Epilepsy Res 1:297–301
122. Wallace RH, Scheffer IE, Parasivam G, Barnett S, Wallace GB, Sutherland GR, Berkovic SF, Mulley JC (2002) Generalized epilepsy with febrile seizures plus: mutation of the sodium channel subunit SCN1B. Neurology 58:1426–1429
123. Scheffer IE, Harkin LA, Grinton BE, Dibbens LM, Turner SJ, Zielinski MA, Xu R, Jackson G, Adams J, Connellan M, Petrou S, Wellard RM, Briellmann RS, Wallace RH, Mulley JC, Berkovic SF (2007) Temporal lobe epilepsy and GEFS+ phenotypes associated with SCN1B mutations. Brain 130:100–109
124. Glasscock E, Qian J, Yoo JW, Noebels JL (2007) Masking epilepsy by combining two epilepsy genes. Nat Neurosci 10:1554–1558
125. Karmarkar UR, Dan Y (2006) Experience-dependent plasticity in adult visual cortex. Neuron 52:577–585
126. Leinekugel X (2003) Developmental patterns and plasticities: the hippocampal model. J Physiol Paris 97:27–37
127. Zito K, Svoboda K (2002) Activity-dependent synaptogenesis in the adult Mammalian cortex. Neuron 35:1015–1017
128. Kempermann G, van Praag H, Gage FH (2000) Activity-dependent regulation of neuronal plasticity and self repair. Prog Brain Res 127:35–48
129. Zhou Z, Hong EJ, Cohen S, Zhao WN, Ho HY, Schmidt L, Chen WG, Lin Y, Savner E, Griffith EC, Hu L, Steen JA, Weitz CJ, Greenberg ME (2006) Brain-specific phosphorylation of MeCP2 regulates activity-dependent Bdnf transcription, dendritic growth, and spine maturation. Neuron 52:255–269
130. Shen S, Li J, Casaccia-Bonnefil P (2005) Histone modifications affect timing of oligodendrocyte progenitor differentiation in the developing rat brain. J Cell Biol 169:577–589
131. Kim J, Kondratyev A, Gale K (2007) Antiepileptic drug-induced neuronal cell death in the immature brain: effects of carbamazepine, topiramate, and levetiracetam as monotherapy versus polytherapy. J Pharmacol Exp Ther 323:165–173
132. Katz I, Kim J, Gale K, Kondratyev A (2007) Effects of lamotrigine alone and in combination with MK-801, phenobarbital, or phenytoin on cell death in the neonatal rat brain. J Pharmacol Exp Ther 322:494–500
133. Qiao M, Malisza KL, Del Bigio MR, Kozlowski P, Seshia SS, Tuor UI (2000) Effect of long-term vigabatrin administration on the immature rat brain. Epilepsia 41:655–665
134. Sidhu RS, Del Bigio MR, Tuor UI, Seshia SS (1997) Low-dose vigabatrin (gamma-vinyl GABA)-induced damage in the immature rat brain. Exp Neurol 144:400–405
135. Jackson GD, Grunewald RA, Connelly A, Duncan JS (1994) Quantitative MR relaxometry study of effects of vigabatrin on the brains of patients with epilepsy. Epilepsy Res 18:127–137
136. Sivenius J, Paljarvi L, Vapalahti M, Nousiainen U, Riekkinen PJ (1993) Vigabatrin (gamma-vinyl-GABA): neuropathologic evaluation in five patients. Epilepsia 34:193–196
137. Van Paesschen W, Duncan JS, Connelly A (1998) A comparison of the neuropathological effects of vigabatrin and carbamazepine in patients with newly diagnosed localization-

related epilepsy using MR-based cerebral T2 relaxation time measurements. Epilepsy Res 29:155–160
138. Kim JS, Kondratyev A, Tomita Y, Gale K (2007) Neurodevelopmental impact of antiepileptic drugs and seizures in the immature brain. Epilepsia 48(Suppl 5):19–26
139. Zhao Q, Hu Y, Holmes GL (2005) Effect of topiramate on cognitive function and activity level following neonatal seizures. Epilepsy Behav 6:529–536
140. Koh S, Tibayan FD, Simpson JN, Jensen FE (2004) NBQX or topiramate treatment after perinatal hypoxia-induced seizures prevents later increases in seizure-induced neuronal injury. Epilepsia 45:569–575
141. Bolanos AR, Sarkisian M, Yang Y, Hori A, Helmers SL, Mikati M, Tandon P, Stafstrom CE, Holmes GL (1998) Comparison of valproate and phenobarbital treatment after status epilepticus in rats. Neurology 51:41–48
142. Cilio MR, Bolanos AR, Liu Z, Schmid R, Yang Y, Stafstrom CE, Mikati MA, Holmes GL (2001) Anticonvulsant action and long-term effects of gabapentin in the immature brain. Neuropharmacology 40:139–147
143. Mikati MA, Holmes GL, Chronopoulos A, Hyde P, Thurber S, Gatt A, Liu Z, Werner S, Stafstrom CE (1994) Phenobarbital modifies seizure-related brain injury in the developing brain. Ann Neurol 36:425–433
144. Sfaello I, Baud O, Arzimanoglou A, Gressens P (2005) Topiramate prevents excitotoxic damage in the newborn rodent brain. Neurobiol Dis 20:837–848
145. Zhao Q, Stafstrom CE, Fu DD, Hu Y, Holmes GL (2004) Detrimental effects of the ketogenic diet on cognitive function in rats. Pediatr Res 55:498–506
146. Su SW, Cilio MR, Sogawa Y, Silveira DC, Holmes GL, Stafstrom CE (2000) Timing of ketogenic diet initiation in an experimental epilepsy model. Brain Res Dev Brain Res 125:131–138
147. Manent JB, Jorquera I, Mazzucchelli I, Depaulis A, Perucca E, Ben-Ari Y, Represa A (2007) Fetal exposure to GABA-acting antiepileptic drugs generates hippocampal and cortical dysplasias. Epilepsia 48:684–693
148. Gallager DW, Wakeman EA (1982) Prenatal exposure to electroconvulsive seizures and phenytoin: development of benzodiazepine binding sites, reflex behaviors and seizure thresholds in exposed offspring. Eur J Pharmacol 85:143–153
149. Ogawa T, Kuwagata M, Hori Y, Shioda S (2007) Valproate-induced developmental neurotoxicity is affected by maternal conditions including shipping stress and environmental change during early pregnancy. Toxicol Lett 174:18–24
150. Cannizzaro E, Martire M, Gagliano M, Plescia F, La Barbera M, Mantia G, Mineo A, Cannizzaro G, Cannizzaro C (2005) Reversal of prenatal diazepam-induced deficit in a spatial-object learning task by brief, periodic maternal separation in adult rats. Behav Brain Res 161:320–330
151. Nicosia A, Giardina L, Di Leo F, Medico M, Mazzola C, Genazzani AA, Drago F (2003) Long-lasting behavioral changes induced by pre- or neonatal exposure to diazepam in rats. Eur J Pharmacol 469:103–109
152. Drago F, Di Leo F, Giardina L (1999) Prenatal stress induces body weight deficit and behavioural alterations in rats: the effect of diazepam. Eur Neuropsychopharmacol 9:239–245
153. Raol YH, Zhang G, Budreck EC, Brooks-Kayal AR (2005) Long-term effects of diazepam and phenobarbital treatment during development on GABA receptors, transporters and glutamic acid decarboxylase. Neuroscience 132:399–407
154. Rothe T, Bigl V (1989) The ontogeny of benzodiazepine receptors in selected regions of the rat brain: effect of perinatal exposure to diazepam. Neuropharmacology 28:503–508
155. Gruen RJ, Deutch AY, Roth RH (1990) Perinatal diazepam exposure: alterations in exploratory behavior and mesolimbic dopamine turnover. Pharmacol Biochem Behav 36:169–175
156. Menegola E, Di Renzo F, Broccia ML, Giavini E (2006) Inhibition of histone deacetylase as a new mechanism of teratogenesis. Birth Defects Res C Embryo Today 78:345–353
157. Faiella A, Wernig M, Consalez GG, Hostick U, Hofmann C, Hustert E, Boncinelli E, Balling R, Nadeau JH (2000) A mouse model for valproate teratogenicity: parental effects, homeotic transformations, and altered HOX expression. Hum Mol Genet 9:227–236

158. Bittigau P, Sifringer M, Ikonomidou C (2003) Antiepileptic drugs and apoptosis in the developing brain. Ann N Y Acad Sci 993:103–114 discussion 123–104
159. Ikonomidou C, Bosch F, Miksa M, Bittigau P, Vockler J, Dikranian K, Tenkova TI, Stefovska V, Turski L, Olney JW (1999) Blockade of NMDA receptors and apoptotic neurodegeneration in the developing brain. Science 283:70–74
160. Asimiadou S, Bittigau P, Felderhoff-Mueser U, Manthey D, Sifringer M, Pesditschek S, Dzietko M, Kaindl AM, Pytel M, Studniarczyk D, Mozrzymas JW, Ikonomidou C (2005) Protection with estradiol in developmental models of apoptotic neurodegeneration. Ann Neurol 58:266–276
161. Dzietko M, Felderhoff-Mueser U, Sifringer M, Krutz B, Bittigau P, Thor F, Heumann R, Buhrer C, Ikonomidou C, Hansen HH (2004) Erythropoietin protects the developing brain against N-methyl-D-aspartate receptor antagonist neurotoxicity. Neurobiol Dis 15:177–187
162. Wegner C, Nau H (1992) Alteration of embryonic folate metabolism by valproic acid during organogenesis: implications for mechanism of teratogenesis. Neurology 42:17–24
163. Ubeda N, Alonso-Aperte E, Varela-Moreiras G (2002) Acute valproate administration impairs methionine metabolism in rats. J Nutr 132:2737–2742
164. Hooks BM, Chen C (2007) Critical periods in the visual system: changing views for a model of experience-dependent plasticity. Neuron 56:312–326
165. Knudsen EI (2004) Sensitive periods in the development of the brain and behavior. J Cogn Neurosci 16:1412–1425
166. Berardi N, Pizzorusso T, Maffei L (2000) Critical periods during sensory development. Curr Opin Neurobiol 10:138–145
167. Galanopoulou AS (2005) GABA receptors as broadcasters of sexually differentiating signals in the brain. Epilepsia 46(Suppl 5):107–112
168. Banke TG, McBain CJ (2006) GABAergic input onto CA3 hippocampal interneurons remains shunting throughout development. J Neurosci 26:11720–11725
169. Khazipov R, Khalilov I, Tyzio R, Morozova E, Ben-Ari Y, Holmes GL (2004) Developmental changes in GABAergic actions and seizure susceptibility in the rat hippocampus. Eur J Neurosci 19:590–600
170. Tyzio R, Holmes GL, Ben-Ari Y, Khazipov R (2007) Timing of the developmental switch in GABA(A) mediated signaling from excitation to inhibition in CA3 rat hippocampus using gramicidin perforated patch and extracellular recordings. Epilepsia 48(Suppl 5):96–105
171. Ganguly K, Schinder AF, Wong ST, Poo M (2001) GABA itself promotes the developmental switch of neuronal GABAergic responses from excitation to inhibition. Cell 105:521–532
172. Leitch E, Coaker J, Young C, Mehta V, Sernagor E (2005) GABA type-A activity controls its own developmental polarity switch in the maturing retina. J Neurosci 25:4801–4805
173. Sanchez RM, Jensen FE (2001) Maturational aspects of epilepsy mechanisms and consequences for the immature brain. Epilepsia 42:577–585
174. Williams K, Russell SL, Shen YM, Molinoff PB (1993) Developmental switch in the expression of NMDA receptors occurs in vivo and in vitro. Neuron 10:267–278
175. Harty TP, Rogawski MA (2000) Felbamate block of recombinant N-methyl-D-aspartate receptors: selectivity for the NR2B subunit. Epilepsy Res 39:47–55
176. Glier C, Dzietko M, Bittigau P, Jarosz B, Korobowicz E, Ikonomidou C (2004) Therapeutic doses of topiramate are not toxic to the developing rat brain. Exp Neurol 187:403–409
177. Cannizzaro C, Martire M, Steardo L, Cannizzaro E, Gagliano M, Mineo A, Provenzano G (2002) Prenatal exposure to diazepam and alprazolam, but not to zolpidem, affects behavioural stress reactivity in handling-naive and handling-habituated adult male rat progeny. Brain Res 953:170–180
178. Moshe SL (1997) Sex and the substantia nigra: administration, teaching, patient care, and research. J Clin Neurophysiol 14:484–494
179. McCartney MA, Scinto PL, Wang SS, Altan S (1999) Developmental effects of phenytoin may differ depending on sex of offspring. Neurotoxicol Teratol 21:119–128
180. Lemmens EM, Lubbers T, Schijns OE, Beuls EA, Hoogland G (2005) Gender differences in febrile seizure-induced proliferation and survival in the rat dentate gyrus. Epilepsia 46:1603–1612

181. Persad V, Cortez MA, Snead OC III (2002) A chronic model of atypical absence seizures: studies of developmental and gender sensitivity. Epilepsy Res 48:111–119
182. Pfaff DW, Arnold AA, Etgen AM, Fahrbach SE, Rubin RT (eds) (2002) Hormones, brain and behavior. Academic, San Diego
183. Fitch RH, Brown CP, Tallal P, Rosen GD (1997) Effects of sex and MK-801 on auditory-processing deficits associated with developmental microgyric lesions in rats. Behav Neurosci 111:404–412
184. Bowman RE, MacLusky NJ, Sarmiento Y, Frankfurt M, Gordon M, Luine VN (2004) Sexually dimorphic effects of prenatal stress on cognition, hormonal responses, and central neurotransmitters. Endocrinology 145:3778–3787
185. Schmitz C, Rhodes ME, Bludau M, Kaplan S, Ong P, Ueffing I, Vehoff J, Korr H, Frye CA (2002) Depression: reduced number of granule cells in the hippocampus of female, but not male, rats due to prenatal restraint stress. Mol Psychiatry 7:810–813
186. Frye CA, Wawrzycki J (2003) Effect of prenatal stress and gonadal hormone condition on depressive behaviors of female and male rats. Horm Behav 44:319–326
187. Ojeda SR, Urbanski HF (1994) Puberty in the rat. In: Knobil E (ed) The physiology of reproduction, vol I. Raven, New York, pp 363–411
188. Menegola E, Broccia ML, Nau H, Prati M, Ricolfi R, Giavini E (1996) Teratogenic effects of sodium valproate in mice and rats at midgestation and at term. Teratog Carcinog Mutagen 16:97–108
189. Miyazaki K, Narita N, Narita M (2005) Maternal administration of thalidomide or valproic acid causes abnormal serotonergic neurons in the offspring: implication for pathogenesis of autism. Int J Dev Neurosci 23:287–297
190. Narita N, Kato M, Tazoe M, Miyazaki K, Narita M, Okado N (2002) Increased monoamine concentration in the brain and blood of fetal thalidomide- and valproic acid-exposed rat: putative animal models for autism. Pediatr Res 52:576–579
191. Manent JB, Jorquera I, Franco V, Ben-Ari Y, Perucca E, Represa A (2008) Antiepileptic drugs and brain maturation: fetal exposure to lamotrigine generates cortical malformations in rats. Epilepsy Res 78(2–3):131–139
192. Zhang MM, Yu K, Xiao C, Ruan DY (2003) The influence of developmental periods of sodium valproate exposure on synaptic plasticity in the CA1 region of rat hippocampus. Neurosci Lett 351:165–168
193. Patsalos PN, Wiggins RC (1982) Brain maturation following administration of phenobarbital, phenytoin, and sodium valproate to developing rats or to their dams: effects on synthesis of brain myelin and other subcellular membrane proteins. J Neurochem 39:915–923
194. Bolanos JP, Medina JM, Williamson DH (1990) Inhibition of sterol but not fatty acid synthesis by valproate in developing rat brain in vivo. Biochem J 272:251–253
195. Elmazar MM, Sullivan FM (1981) Effect of prenatal phenytoin administration on postnatal development of the rat: a behavioral teratology study. Teratology 24:115–124
196. Mullenix P, Tassinari MS, Keith DA (1983) Behavioral outcome after prenatal exposure to phenytoin in rats. Teratology 27:149–157
197. Vorhees CV (1987) Fetal hydantoin syndrome in rats: dose-effect relationships of prenatal phenytoin on postnatal development and behavior. Teratology 35:287–303
198. Vorhees CV (1985) Fetal anticonvulsant syndrome in rats: effects on postnatal behavior and brain amino acid content. Neurobehav Toxicol Teratol 7:471–482
199. Mach M, Dubovicky M, Navarova J, Kovacovsky P, Ujhazy E (2006) Vitamin E supplementation in phenytoin induced developmental toxicity in rats: postnatal study. Neuro Endocrinol Lett 27(Suppl 2):69–73
200. Navarova J, Ujhazy E, Dubovicky M, Mach M (2005) Phenytoin induced oxidative stress in pre- and postnatal rat development – effect of vitamin E on selective biochemical variables. Biomed Pap Med Fac Univ Palacky Olomouc Czech Repub 149:325–328
201. Makatsori A, Michal D, Eduard U, Bakos J, Jezova D (2005) Neuroendocrine changes in adult female rats prenatally exposed to phenytoin. Neurotoxicol Teratol 27:509–514

202. Schilling MA, Inman SL, Morford LL, Moran MS, Vorhees CV (1999) Prenatal phenytoin exposure and spatial navigation in offspring: effects on reference and working memory and on discrimination learning. Neurotoxicol Teratol 21:567–578
203. Weisenburger WP, Minck DR, Acuff KD, Vorhees CV (1990) Dose-response effects of prenatal phenytoin exposure in rats: effects on early locomotion, maze learning, and memory as a function of phenytoin-induced circling behavior. Neurotoxicol Teratol 12:145–152
204. Biggiogera M, Scherini E, Barni S, Bernocchi G (1983) Effects of diphenylhydantoin on postnatal ontogenesis in the rat. Ultrastructural analysis of Purkinje neurons and hepatocytes. J Submicrosc Cytol 15:1027–1040
205. Levav T, Wirthaim O, Weiss R, Grossman Y, Golan H (2008) Impaired synaptogenesis and long-term modulation of behavior following postnatal elevation of GABA levels in mice. Neuropharmacology 54:387–398
206. Levav T, Saar T, Berkovich L, Golan H (2004) Perinatal exposure to GABA-transaminase inhibitor impaired psychomotor function in the developing and adult mouse. Int J Dev Neurosci 22:137–147
207. Rayburn WF, Christensen HD, Gold KM, Gonzalez CL (2002) Neurobehavior effects in four strains of mice offspring exposed prenatally to alprazolam. Am J Obstet Gynecol 187:968–972
208. Rothe T, Langer M (1988) Prenatal diazepam exposure affects beta-adrenergic receptors in brain regions of adult rat offspring. J Neurochem 51:1361–1366
209. Schroeder H, Humbert AC, Desor D, Nehlig A (1997) Long-term consequences of neonatal exposure to diazepam on cerebral glucose utilization, learning, memory and anxiety. Brain Res 766:142–152
210. Fishman RH, Ornoy A, Yanai J (1983) Ultrastructural evidence of long-lasting cerebellar degeneration after early exposure to phenobarbital in mice. Exp Neurol 79:212–222
211. Gupta C, Yaffe SJ, Shapiro BH (1982) Prenatal exposure to phenobarbital permanently decreases testosterone and causes reproductive dysfunction. Science 216:640–642
212. Zemp JW, Middaugh LD (1975) Some effects of prenatal exposure to d-amphetamine sulfate and phenobarbital on developmental neurochemistry and on behavior. Addict Dis 2:307–331
213. Pereira de Vasconcelos A, Boyet S, Nehlig A (1990) Consequences of chronic phenobarbital treatment on local cerebral glucose utilization in the developing rat. Brain Res Dev Brain Res 53:168–178
214. Manthey D, Asimiadou S, Stefovska V, Kaindl AM, Fassbender J, Ikonomidou C, Bittigau P (2005) Sulthiame but not levetiracetam exerts neurotoxic effect in the developing rat brain. Exp Neurol 193:497–503

Overview of Neural Mechanisms in Developmental Disorders

Ayush Batra, Erin Carlton, and Kathleen Franco

Abstract The development of the mammalian brain is both a highly organized and chaotic process that is nothing short of a miracle. The complexity with which the human brain develops from but a mere plate of ectodermal cells is still not fully understood. Numerous disorders have their etiology seeded in abnormal development during the prenatal and perinatal stages of central nervous system (CNS) development. Various signaling molecules and the precisely orchestrated expression and silencing of neural-specific genes may be disrupted in the CNS development of patients suffering from developmental disorders. In this chapter we present an overview of the normal neural mechanisms of mammalian brain development and how errors in this process contribute to developmental disorders. Developmental disorders are defined as those disorders of the CNS that impair normal functioning and development in early childhood. Developmental pathologies encompass various learning disabilities, dyslexias, dyspraxias, metabolic disorders, and anatomical abnormalities, including those that may result from teratogen exposure. Developmental disorders may affect one specific area of development, or may involve numerous areas as in the case of pervasive developmental disorders. More global developmental disorders result from inherent metabolic abnormalities and more recently described immune mediated mechanisms.

Keywords Developmental disorders • Neurulation • Neural migration • Neural pruning • Neural tube defects • Pervasive developmental disorders • Attention deficit hyperactive disorder • Anxiety disorders • Prenatal infections

---

A. Batra
Cleveland Clinic Lerner College of Medicine, 9500 Euclid Avenue, Cleveland, OH, 44195, USA
e-mail: batraa@ccf.org

D. Janigro (ed.), *Mammalian Brain Development*, Contemporary Neuroscience,
DOI 10.1007/978-1-60761-287-2_7,

1 Mechanisms of Development

1.1 Neurulation

As a part of organogenesis, neurulation occurs during the first 4 weeks of gestation [1]. The notochord, a collection of mesodermal cells that define the early axis of development of the embryo, will induce the ectoderm germ layer to form the neural plate. This plate will then fold lengthwise along itself, creating the neural tube, eventually becoming the spinal cord and brain. It was previously thought that between days 18 and 29 of gestation the lateral edges of the neural plate fused rostrally and caudally, creating the neural tube [1]. However, studies have shown that neurulation occurs at four sites of the neural tube: midcervical, cranial junction, stomadeum, and the rhombencephalon region [2, 3]. Most Neural Tube Disorders (NTDs) can be explained by failure to close at one or more of these sites.

Two processes, primary and secondary neurulation, occur in distinct morphologic and molecular routes. Primary neurulation creates the cranial and upper spinal regions, while secondary neurulation arises only in the lower sacral and caudal regions of the body axis [4–6].

The anterior, or primary, tube forms from the neural plate. Eventually, this tube will extend into the cervicothoracic area. The epithelial tissue from which the neural plate is formed is a continuous layer of cells held together by tight junctions [7].

Although the key characteristic of primary neurulation is the bending of the epithelial sheet into a tube, additional steps are involved in this process. Initially, the ectoderm becomes columnar, creating the neural plate. The edges of the newly formed plate begin to thicken, giving rise to neural folds. Simultaneously the plate extends allowing for elongation of the tube and establishment of a groove. Finally, this groove closes, creating the neural tube [4]. Cytoskeletal genes promoting cell adhesion are key to the development of the neural tube [4, 8].

The secondary or posterior neurulation occurs in undifferentiated tissue after the development of anterior regions [5]. Unlike primary neurulation, the secondary process occurs in mesenchymal cells. These cells unite to give a rod-like medullary cord, eventually becoming neuroepithelium and lumen which forms a neural tube. Distinct molecular mechanisms regulate secondary neurulation [7]. Figure 1 describes the steps in primary and secondary neurulation. While low levels of sialic acid N-CAM has been suggested in primary neurulation, a highly sialiated form is believed necessary for the secondary process of mesenchymal to epithelial transformation [5, 9].

Because the genetic basis for many NTDs is not yet known, attention has been directed to non-genetic influences such as metabolic factors. Evidence that neural tube formation depends on particular biochemical pathways comes from NTDs that follow the use of teratogenic substances or maternal metabolic disorders.

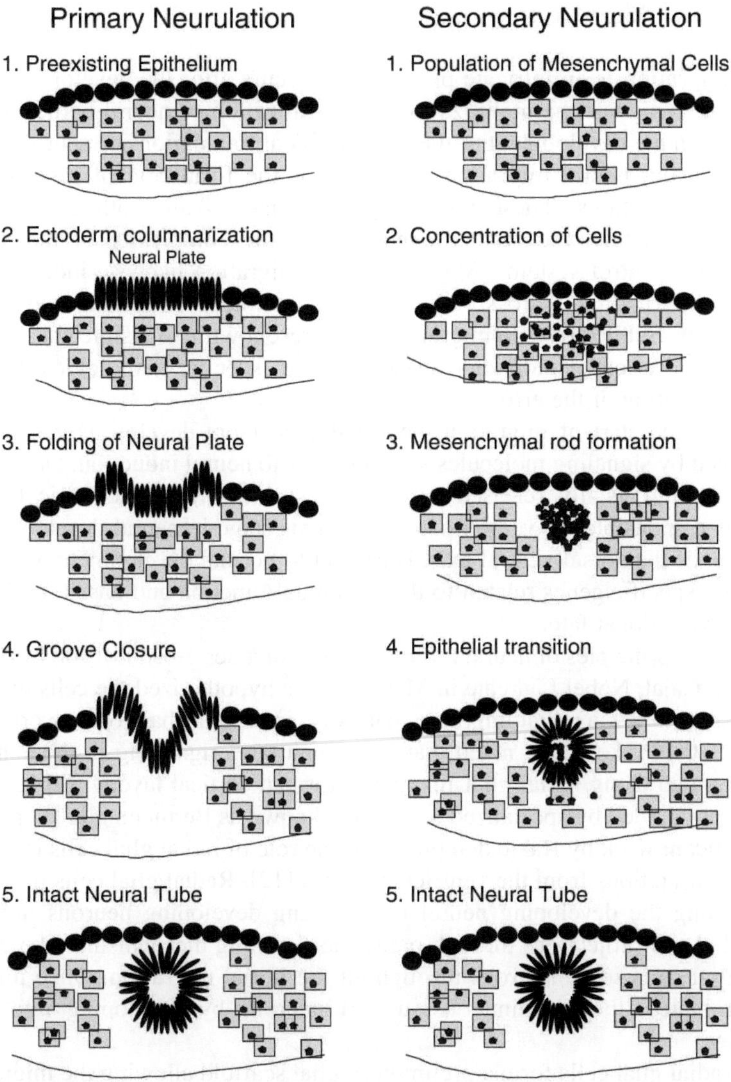

Fig. 1 Primary and secondary neurulation. Primary neurulation originates from an epithelial layer. The preexisting cells undergo columnarization to form the neural plate. The plate thickens, forming the neural folds, which bends configuring the neural groove. This structure, which allows for lengthening of the final formation, closes completing the neural tube. Secondary neurulation is developed from mesenchymal cells. After congregating, cells merge to form the medullary cord. This cord becomes epithelium and later lumen which finally develops into the neural tube. Adapted from [7]

1.2 Neural Migration

Neural migration is an intricate process that occurs after the development of the notochord. It signals the mobilization and localization of neural bodies and glial cells early in the development of the CNS. Critical foundations of neural networks are laid during neural migration allowing for the formation of more complex synaptic connections. These multi-origined synapses are both excitatory and inhibitory in nature and further contribute to the complexity of the dynamic higher-order control systems. Major errors in migration likewise induce massive cortical abnormalities, grossly disfigured anatomical structures and multiple impairments. Subtle errors in migration may present with no visible abnormalities, yet leave the affected individual predisposed to specific pathologies dependent upon the location of the error.

Prior to the start of migration, neuronal precursors develop axonal identities determined by signaling molecules simultaneous to neural induction. Dorsalization occurs through locally released peptide growth factors, while sonic hedgehog (SHH) is responsible for ventral patterning. This stage in neural development, the formation of the dorsal-ventral axis, is critical to normal development. Cells begin to express specific genes related to their eventual function and based on their predetermined regional fate.

The basic principles of neural migration were first described in 1906 by Santiago Ramon Y Cajal, Nobel Laureate in Medicine. He hypothesized the cells of the nervous system follow a radial migratory pathway, creating a chain of networked cells. In the 1970s, Pasko Rakic noted that radial neurons originating in the ventricular zone migrated along radial glial fibers and formed cortical layers [10]. Cerebellar neurons are guided by specialized astrocytes, known as Bermann glia [11].

Significant work by Rakic demonstrated the role of radial glial cells in directing neuronal migrations from the ventricular zones [12]. Radial glial cells project processes along the developing neural tube. Young developing neurons induce the radial glial cell projections through paracrine signaling mechanisms. The dynamic radial glial cells serve neurons throughout the initial migration. Once migration has terminated, they continue to support neurons by developing into stellate astrocytes.

The radial glial cells form a preliminary glial scaffold allowing the migration of neurons of the ventricular zone to form layers of the cortex. This vast migratory process proceeds through waves in a highly organized pattern. Recent literature suggests the role of local calcium fluxes modulating the waves of neuronal migration [13]. The terminal position of migrating neurons is determined in a unique inside-out pattern, with cells migrating at the onset, forming the inner most laminar layer, while the final cells to migrate form the outer layer of the cortex.

As the initial framework of neurons is laid radially in various regions of the cortex and cerebellum, a unique tangential migratory process occurs with neurons throughout the CNS. Interneurons, defined as "tangential migrators", establish necessary inhibitory connections initially. These interneurons form additional synapses throughout post natal development and are the basis for many learning theories.

Tangential migration by neurons other than interneurons occurs to allow cells to populate discrete regions within the ventricular zone [14].

1.3 Neural Network Pruning

No sooner is the foundation laid for the developing CNS, than does the process of neural pruning begins. A majority of the neurons are formed by the second trimester of prenatal life. However, over 90% of these neurons will undergo apoptosis during the process of neural pruning. This programmed neuronal death is a fundamental component of normal development. An excess of region specific neurons result from germinal precursors. These neurons compete amongst each other for various trophic factors and synapse formation [15]. In essence, this survival of the fittest model of neuronal selection allows for the best qualified neurons to survive, while the lesser ones fade into the background. In addition to selecting the fit neurons, pruning plays an essential role in creating complex connection patterns. In the developing cortex, neurons from layer 5 undergo a great deal of selection and modification through programmed cell death [16]. This process is transcriptionally regulated with specific transcription factors serving unique roles among the different layers, and cortical systems. The homeodomain transcription factor Otx1 regulates an essential part of the pruning activity witnessed in layer 5 cortical neurons, and Otx1 null mice demonstrate major errors in synapse generation [17].

As the framework for the developing CNS is formed by the second trimester of prenatal life, glial cells play an important role in neural pruning as well. Initially, they help regulate the number of neural precursors that are able to populate a given region. They are also capable of pruning neuronal projections through direct axonal engulfment or receptor mediated dendritic spine retraction [18].

The process of neurulation, neural migration, and neural pruning are the fundamental components of normal mammalian brain development. Table 1 highlights the developmental disorder that may arise when brain development may go awry. The following section discusses specific pathologies which involve specific stages of development, while certain pathologies are combinations of multiple stages. Figure 2 broadly simplifies the types of errors that result within the developmental process.

Table 1 Developmental Disorders within the International Classification of Disease, 10th edition

Title	Description	Blocks
Mental and behavioral disorders	Pervasive developmental disorders, affective disorders, disorders of psychological development, hyperkinetic disorders, conduct disorders, mental retardation	F00–F99
Diseases of the nervous system	Extrapyramidal and movement disorders, cerebral palsy and other paralytic syndromes	G00–G99
Congenital anomalies	Neural tube defects, chromosomal anomalies	Q00–Q99
Neoplasms	Brain and other CNS tumors	C00–D48

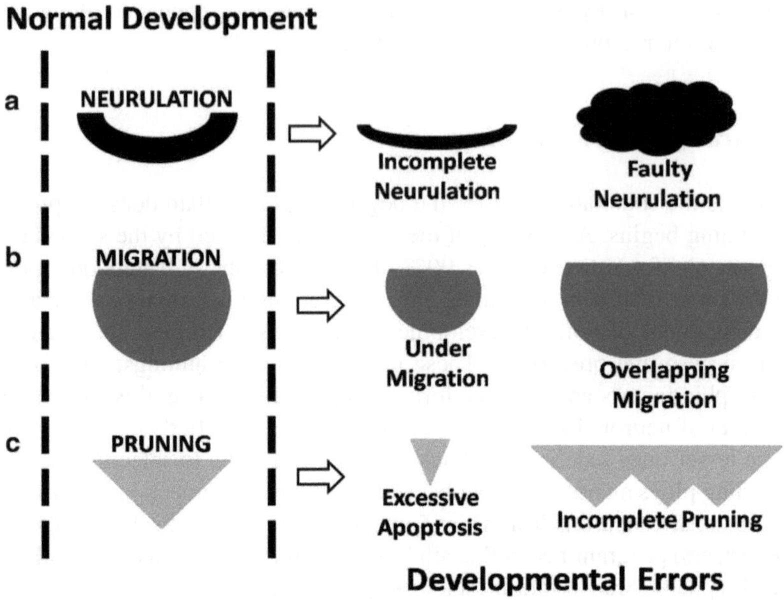

Fig. 2 Normal development and developmental errors. (**a**) Neurulation is the first step of mammalian brain development. Incomplete neurulation is responsible for many diverse developmental errors, which may present as minor neural tube defects if incomplete, or in utero death if neurulation never occurs properly. (**b**) Neural migration is the second major step of mammalian brain development and is the process by which functional neuronal regions are organized. Inappropriate migration results in areas of under population and underdevelopment or areas of overlapping migration resulting in neural death and regional dysfunction. (**c**) Neural pruning is the final major step of mammalian brain development and is chiefly responsible for creating neural networks and circuits. Excessive apoptosis results in inadequate neural communication while incomplete pruning may result in inefficient neural circuits again leading to neuronal dysfunction

2 Developmental Disorder Pathology

2.1 Neural Tube Defects

Neural Tube Defects (NTD), an incomplete closure of the cranium and/or the spinal cord, are some of the most common congenital abnormalities occurring in 1.4–2% of pregnancies. These malformations, caused by incorrect neurulation or abnormal development of the neural tube, are classified as open or closed, depending on the presence of an overlying skin cover. Multiple etiologies may induce different abnormalities consistent with the various sites of neural tube malclosure. Table 2 summarizes the various neural tube defects.

Table 2 Classification of neural tube defects

Open NTDs	Closed NTDs	Secondary defects
Spinal rachisis	Lipoma	Chiari malformations
Spina bifida aperta	*Lipomeningocele*	*Type I*
Myeloschisis	Split Cord	*Type II (Arnold Chiari)*
Myelomeningocele	*Diastematomyelia*	*Type III*
Iniencephaly	*Diplomyelia*	*Type IV*
Encephalocele	Neurenteric cysts	Tethered Cord
Anencephaly	Spina bifida occulta	
	Dermal Sinus	
	Sacral agenesis	
	Sirenomyelia	

Note: For a description of each disease see [1]

2.1.1 Anencephaly

Occurring in approximately 0.1–0.7 per 1,000 births, anencephaly is the failure of the cephalic region of the spinal cord to close or the absence of a brain, specifically the cerebral cortex and white matter. Nerve, glial and connective tissues, as well as brainstem, cerebellum and spinal cord, are present, however, hemorrhagic and severely malformed. Approximately 65% of anencephalic fetuses will die in utero and nearly all by postnatal week one. Anencephaly, specifically frontal and parietal defects, originates from the cranial junction of the prosencephalon and mesencephalon [19].

2.1.2 Spina Bifida

Spina Bifida (SB), an abnormal closure of the vertebral arches, is classified in three forms; spina bifida occulta, meningocele, and meningomyelocele. Spina bifida occulta, a closed NTD affecting 10% of the population, is the failed closure of one or more posterior vertebral arches. Both the meninges and spinal cord remain intact and the patient characteristically presents with a small tuft of hair on the lower back. As the mildest form of SB, many individuals are asymptomatic and the disorder is only discovered upon evaluation for back pain. Neurologic dysfunction may result, however, from tethering of the spinal cord causing weakness, scoliosis, urinary or bowel issues, gait instability and pain.

Meningocele, the least common form of SB, can occur in the posterior or anterior. In posterior menigocele, the vertebrae remain unfused and the damaged meninges appear as a cyst containing CSF. Spinal cord and nerve function remains intact. A cyst protrudes the unfused inner vertebrae into the retroperitoneum in anterior meningocele. Neurologic function typically is normal.

Myelomeningocele is the most severe form of SB. The spinal column fails to close, allowing the protrusion of the spinal cord. A sac or cyst may form from the meningeal membrane to cover the cord. When associated with myeloschisis, the neural fold does not fuse, causing a flat mass of exposed neural tissue and spinal cord. Symptoms of lower extremity weakness and atrophy, urinary incontinence, and foot malformations are often seen. When occurring in the high lumbar region, myelomeningocele causes total paralysis and incontinence. Patients often present with hydrocephalus and Arnold Chiari II malformations, an abnormal herniation of the cerebellum and brain stem.

The mechanism behind myelomeningocele, like other NTDs, is described by von Recklinhausen as a 'non-closure' defect. In this widely accepted theory, the NTD results from a failure of neural tube closure [19]. In addition to the initial failure of neural tube formation, erroneous midline axial integration during gastrulation leads to myelomeningocele [19]. It is believed that the Henson node does not properly place the neuroepithelium surrounded notochord, causing the formation of two heminotocords and inevitably giving rise to two hemineural plates. Eventually, two hemicords are located on each side of the node and interrupt neurulation causing myelomeningocele [19].

2.1.3 Folic Acid Deficiency

Folate, a coenzyme, is necessary for the development of red and white blood cells, function of the CNS, and normal mitosis and meiosis. Folic acid is essential for the development of methionine, required for methylation. These reactions are essential in the production of lipids, proteins, and myelin.

The association between folic acid deficiency and NTDs is most likely multifactorial. Anti-Epileptic Drugs (AEDs), such as valproic acid (VPA) and carbamezapine have been implicated as antagonists to folate absorption. VPA and carbamezapine have been independently associated with spina bifida, specifically. A 1–2% prevalence is seen with VPA exposure compared to 0.5% in carbamezapine [20–22]. When metabolized by microsomes to arene-oxides or epoxides, these drugs often accumulate in fetal tissue causing toxic effects. The enzyme 5,10-methylene tetrahydrofolate reductase (MTHFR) transfers a methyl group from folic acid and converts homocysteine to methionine. When mutated, MTHFR has reduced enzymatic activity and has been implicated in neural tube defects. The mechanism and biochemical pathway of folic acid deficiency and neural tube defects is reviewed in Fig. 3.

To overcome possible MTHFR deficiency, folic acid supplementation is prescribed and reportedly prevents 50–70% of NTDs when taken daily before conception and throughout the first trimester of pregnancy. In 1992, the United States Public Health Service recommended that all women capable of becoming pregnant take 400 μg of folic acid daily through improved diet, fortification of foods or dietary supplements.

2.2 Neural Migration Errors

Neural migration is perhaps one of the most important aspects of CNS development, second only to proper cell proliferation during neurulation. Genetic defects in essential migratory proteins, such as Reelin, GFAP, and BLP2, result in abnormal

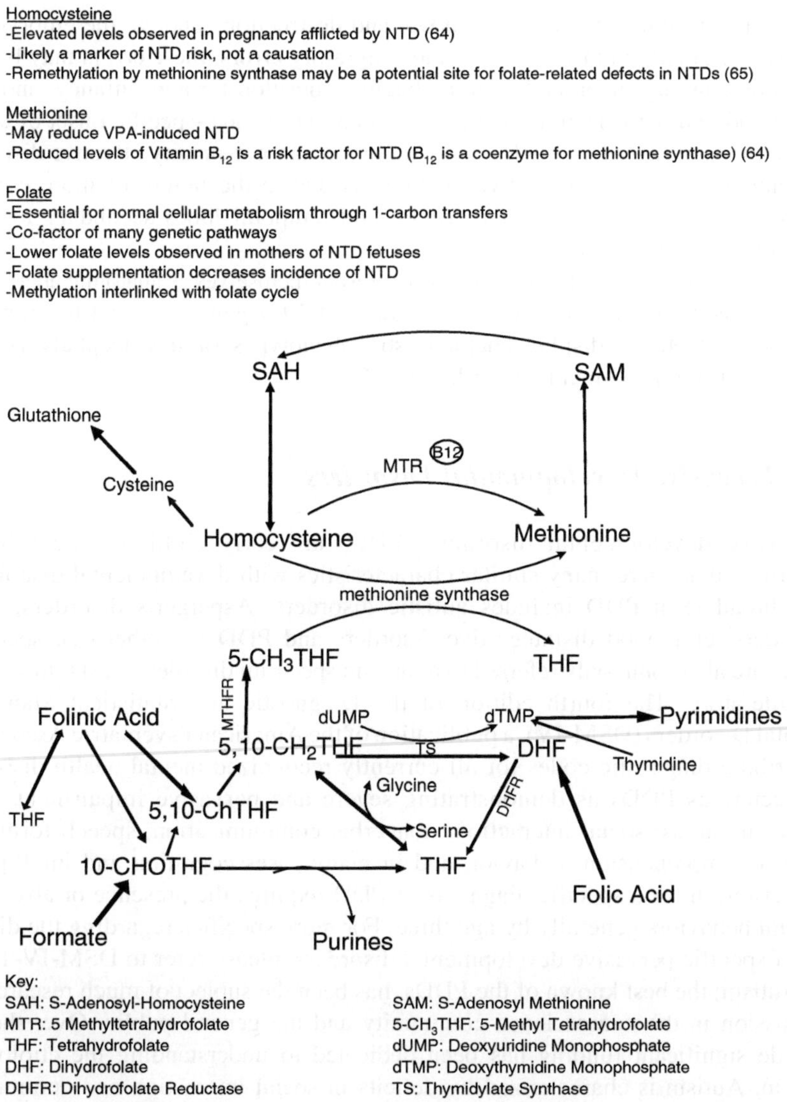

Fig. 3 Metabolic pathways of homocysteine, methionine and folate. Reduction in the efficacy of methionine synthase or Methylene tetrahydrofolate reductase (MTR) and the co-enzyme B12 may inhibit methylation of homocysteine, thereby causing an increase in homocysteine. This methylating site has been proposed as a possible location for folate-related NTDs

positioning of the initial neural networks [23, 24]. Nonetheless, the remarkable resilience of the immature brain allows neurons to bypass abnormally placed cells early on, creating the necessary functional circuits. Unfortunately, these abnormal cell populations remain in non-native locations permanently, and the immune privileged environment inhibits their easy removal and destruction. Though the embryo may progress normally in instances of minor migratory errors, these errors may become pathologic as higher order functions begin to develop through infancy, and early childhood. Numerous studies have speculated that the age-specific onset of certain developmental disorders, such as autism, mood and personality effective disorders, and attention deficit hyperactive disorder, is due to the failure of more complex neuronal circuits to form in the frontal and temporal lobes secondary to initial errors in the groundwork [25].

More serious neuronal migration errors present themselves as anatomic aberrations. Individuals with mutations or deletions in the DXC1 gene, essential for migration dependent proteins, display characteristic phenotypes of lissencephaly (smooth brain syndrome) and mental retardation [26].

2.3 Pervasive Developmental Disorders

Pervasive developmental disorders (PDD) are early childhood neurological disorders that share many similar characteristics with developmental disabilities. The broad term PDD includes autistic disorders, Asperger's disorders, Rett's disorders, childhood disintegrative disorders and PDD not otherwise specified. PDDs are also commonly referred to as autism spectrum disorders (ASD) throughout the literature. The fourth edition of the Diagnostic and Statistical Manual of Mental Disorders (DSM-IV), a publication of the American Psychiatric Association describing diagnostic codes for all currently recognized mental health disorders, characterizes PDDs as demonstrating severe and pervasive impairment in the following areas: social interaction, nonverbal communication, speech formation, speech comprehension, behavior, and in many cases cognition and intelligence. The criterion for a specific diagnosis of PDD requires the presence or absence of certain behaviors generally by age three. For more specifics regarding the diagnosis of specific pervasive developmental disorders, please refer to DSM-IV-TR.

Autism, the best known of the PDDs, has been the subject of much research and discussion in the neuroscience community and the general public. Over the past decade significant funding has been dedicated to understanding the etiology of autism. Autism is characterized by deficits in social interaction and communication. Usually by the age of three, patients display stereotypic repetitive behaviors and restricted interests. There is increasing evidence reporting autism as a genetic disorder. Mutations for several voltage gated and ligand gated ion channels have been identified with autism and include MET receptor tyrosine kinase genes [27] and CACNA1C (an L-type calcium channel gene) [28]. Recent studies from the Dolmetsch group point to the role of disrupted calcium signaling in the development of autism [29]. Behavioral characteristics similar to PDDs in animal models incorporate the use of valproic acid [30] and thalidomide treatment *in utero*, suggesting a prenatal origin. However, the environmental influences on autism cannot be totally ignored. Though this issue remains a consistent area of debate among geneticists, developmental biologists, and the general public, our final understanding may indeed include degrees of both. Other PDD's such as Asperger's and Rett Syndrome also have strong associations with genetic mutations, including mutations in the X-linked gene methyl CpG-binding protein 2 (MECP2), that impacts neuronal development and synapse formation [31]. These proteins are generally expressed throughout neural development, but their greatest impact lies in the postnatal period, explaining the apparent cognitive decline in children suffering from Rett Syndrome at an early age.

2.4 *Attention Deficit Hyperactivity Disorder*

Attention Deficit Hyperactivity Disorder (ADHD) affects approximately 4% of all children with estimates between 3 and 11% [32]. This disorder, described in DSM-IV-TR, is characterized by a child's inability to pay attention, with hyperactivity and/or impulsivity. The onset is often before the age of 7, and impairments are present in at least two different settings such as home and school. Imaging studies performed over the past 15 years have shed new light on its neurophysiological basis. Magnetic resonance imaging demonstrated that cerebellum and cerebrum volumes are significantly smaller in children with ADHD [33]. Studies demonstrate that children with ADHD show lower developmental trajectories for specific behaviors influenced by the cerebrum and cerebellum [33]. These anatomical abnormalities found in ADHD patients suggest no errors in many developmental processes later in childhood and adolescence, since growth curves lie parallel to healthy children. Rather, the atypical volumes observed in ADHD patients points to an early developmental event delaying the appropriate growth of the cerebrum and cerebellum at an early crucial time point. Detailed genetic analysis of ADHD over the past decade focused on genes involved in dopamine transmission [34]. Specific genes involved include the dopamine transporter gene DAT1, and the dopamine receptors DRD4 and DRD5. Detailed gene mapping studies have been conducted in an effort to correlate classical "psychiatric" diseases with defects in specific genes relating to the observed pathology [35]. The role of environmental influences on the development of ADHD is less clear, though there appears to be at least an association with the pathogenesis of the disease. The effect of a stable parenting environment can drastically alter the expression and course of the disease. Environmental studies demonstrating the improvement of ADHD symptoms by teaching more effective parenting strategies demonstrate an environmental influence on the course of the disease [34].

2.5 *Anxiety Disorders*

Anxiety disorders have an incidence of 18.1%, and a disease prevalence of just over 28%. These disorders are grouped with obsessive compulsive disorders (OCD), post traumatic stress disorders (PTSD), separation and social anxiety disorders, specific phobia disorders, and panic disorders. The DSM-IV criteria classify 12 different types of anxiety that include the persistence of illogical fear as one of the characteristics. Numerous studies have demonstrated precise anatomical origins for anxiety disorders, specifically the amygdala within the medial temporal lobe [36]. The amygdala is involved in the conditioning of fear through Pavlov's conditioning models, and inborn errors within this region may predispose individuals to anxiety disorders. The molecular mechanisms behind these physiological changes remain a promising area for future discoveries.

2.6 Prenatal Infectious Disease Exposure or Inflammation Mediated Response

Throughout the intrauterine period, the embryo or fetus is vulnerable to many infectious diseases or inflammatory responses. These infections may be caused by viral or microbial microorganisms. The fetus is susceptible to agents which cross the placenta and those located in the genital tract. Because some agents must cross the placenta to reach the fetus, subsequent infections depend on permeability of the virus or bacteria. Due to immaturity of the fetal blood brain barrier, adequate defense mechanisms may not be available once infected.

The most common sources of neonatal infection, toxoplasma gondii, rubella, cytomegalovirus, and herpes (TORCH), can cause low birth weight, microcephaly and cerebral calcification among others. Such prenatal exposures are associated with various neurological sequelae, including schizophrenia, neuronal migration disorders, cerebral palsy and epilepsy. Table 3 summarizes the most common fetal infections that result in developmental disorders.

2.6.1 Cytomegalovirus Infection

Cytomegalovirus (CMV), a DNA herpes virus infection can be transmitted through body fluids, intimate sexual contact, fetal intrauterine infection, intrapartum infection, or postpartum infection from breast feeding [37]. After acute infection, CMV, like other herpesviruses, becomes latent. Despite the presence of an IgG antibody, viral shedding occurs upon reentering the lytic cycle. Although most infections are asymptomatic, some individuals may experience fever and other mononucleosis symptoms.

CMV infects approximately 0.2–2% of live births and is the most frequent congenital infection worldwide [38]. Only 10% of infected newborns are symptomatic at birth. Of these, 20% will not survive while 90% of the survivors will have some severe neurological sequelae [38, 39].

Mothers with a primary infection will transmit the virus to their child in approximately 40% of cases, often causing severe morbidity [38, 39]. Acute maternal infection creates a more significant risk than secondary infection. Maternal immunity does not prevent congenital infection, however if measurable antibody titers are detected at conception, risk of transmission decreases significantly. CMV may be transmitted though cervicitis, however, placental infection is much more common. Frequency of fetal infection increases later in pregnancy. Severity, however, is higher when maternal

Table 3 Commonly acquired fetal infections and the role they may play in developmental disorders

Virus	Transmission (fetal/child)	Mechanism	Subsequent defects
CMV	Fetal Intrauterine Infections Intrapartum Infection Breast Feeding	Placental Infection Viral Replication Fetal Transmission Viral Replication (in renal tubular epithelium) Viral excretion to amniotic fluid	Cerebral Malformation Neurodegeneration Deafness Cerebral Palsy Epilepsy Optic Atrophy Microcephaly Hydrocephalus Mental Retardation
Rubella	Respiratory Intrauterine	Placental Infection (during maternal viremia) Viral transport to fetal circulation (suggested by necrotic cells in endothelial lumen) Eventual damage to organ systems (Exact overall mechanism unknown)	Congenital Cataracts Hearing Defects Congenital Heart Disease Mental Retardation Encephalitis Ventriculomegaly Psychiatric Disorders
HIV	Maternal-Infant (Vertical) Vaginal Delivery Labor	Viral attachment (To T-Helper lymphocyte) Immune cell activation by T-Helper Viral RNA conversion to DNA DNA incorporation into cell DNA replication within cell Cell destruction Subsequent infection of other cells	Encephalopathy Microcephaly Brain Atrophy Spasticity Cognitive Delay Hypertonicity Hyperreflexia

infection occurs during the first 18 weeks of pregnancy [40]. This trend is also reported in maternal rubella and toxoplasma gondii infection [40].

Congenital CMV may cause prenatal damage, including cerebral malformation or postnatal neurodegeneration. Cytomegalic inclusion disease, caused by CMV, can induce pre-term birth, pneumonitis and neurological symptoms. The latter may involve deafness, cerebral palsy, epilepsy, optic atrophy, microcephaly, hydrocephalus, delayed psychomotor skills and mental retardation [41].

Because primary and secondary maternal CMV causes congenital infection, prevention is difficult and is complicated by CMV's asymptomatic nature [41]. At present, no definitive treatment is available.

2.6.2 Rubella Infection

Also known as German Measles, Rubella is a single-stranded RNA (ssRNA) virus contracted through respiratory transmission. Viraemia occurs after a 13–20 day incubation period. Rubella was first noted to affect fetal health in 1941 when McAllister Gregg described its association with congenital cataracts. As one of the first viruses understood to lead to congenital abnormalities after in utero exposure, Rubella continues to be a model of viral teratogenic effect.

In addition to congenital cataracts, this ssRNA virus has been implicated in hearing defects, congenital heart disease, mental retardation, encephalitis, ventriculomegaly and various psychiatric disorders [42]. Congenital rubella infections occur within the first 8 weeks of pregnancy, posing approximately 10–58% chance of major malformation. One recent study suggests, however, this risk may, in fact, be dramatically higher. Though a risk remains into gestational week 24, an earlier infection dramatically increases risk to the fetus due to organogenesis events occurring throughout the first 8 weeks [43]. Cardiac and optical malformations are most prevalent at this time. If infection occurs at weeks 16–24, retinopathy and hearing defects prevail.

Prenatal rubella infection occurs across the placenta [40]. Rapid cell death and constant viral infection causes damage to almost all germ layers. Additionally, reduced cell division and chromosomal aberrations are observed. Fetal immune mechanisms can defend against rubella infection if exposure occurs after the first trimester. However, if maternal infection occurs during the earliest gestational period, fetal infection is imminent.

Prenatal rubella infection can be diagnosed in two ways. Rubella specific IgM antibody detection in fetal blood can be done through a fetoscopy. Because the fetus does not produce enough IgM antibody until week 22, abnormalities may already be present [40]. Improved sensitivity of diagnostic tests for fetal rubella infections will help identify early infections, which opens the window for early treatment options. Rubella can also be isolated through percutaneous umbilical cord blood sampling. This technique, however, remains unreliable due to poor sensitivity and risks placed on the mother. Currently, a third option, detection of rubella RNA or viral proteins through chorionic villus biopsies, is being explored but efficacy is yet to be established [40].

2.6.3 Human Immunodeficiency Virus Infection

Vertical or maternal-infant transmission is the primary way that children are infected with Human Immunodeficiency Virus type 1 (HIV) [44]. This occurs in approximately 15–40% of infants born to seropositive mothers [45, 46]. Although not conclusive, studies suggest vertical transmission occurs late in pregnancy, during labor or vaginal delivery. The risk of transmission increases with preterm delivery, fetal skin trauma, and maternal bleeding. Infection with high maternal viral load, low CD4+ T cell count, and the presence of other infections such as syphilis or CMV may also increase this risk [46–49].

Shortly upon infection, HIV enters the nervous system. With early exposure (i.e. in utero), HIV often manifests itself as encephalopathy. The incidence increases with advanced maternal disease due to elevated viral load. As the predominant feature of child HIV, encephalopathy may occur prior to immunological compromise and contributes to risk of mortality and morbidity in children. One study comparing HIV-positive infants to HIV-positive adults found that within the first year of infection 9.9% of infants presented with encephalopathy compared to 0.3% of adults [50]. In addition to encephalopathy, microcephaly, brain atrophy, motor dysfunction in lower extremities, spasticity and cognitive delay may be displayed [51].

Because the virus is not found in neurons but rather in macrophages and microglia, its neuronal effect is most likely indirect. For example, infected microglia cause immune-activated macrophages to cross the blood brain barrier, while neurotoxic macrophages may create pro-inflammatory cytokines, chemokines and regulatory proteins [52]. Such cytokines, specifically found in patients with HIV-encephalopathy, have exhibited pro-apoptotic behaviors. Additionally, Caspase-3, a pro-apoptotic enzyme has been found up-regulated in the neurons, microglia and macrophages of these patients [53]. Glycoproteins gp120 and gp41, contained in the viral coat, may have direct neurotoxic effects. Gp 120 has been shown in the presence of macrophages, to act as a toxin, mediating excitotoxicity [54]. This, however, is repressed by N-methyl-D-aspartate and calcium channel blockers [54].

HIV-positive infants resulting from vertical transmission typically present with early onset and progressive encephalopathy. Children account for severe HIV-related neurodevelopmental issues [51]. A comorbidity of severe neurological illness, delayed brain growth, and hypotonia followed by severe hypertonicity and hyperreflexia is seen. Although this type of encephalopathy may be indistinguishable from those caused by prenatal damage, these infants can be separated by a high viral systematic load [51]. Because encephalopathy is often the only presenting HIV-defining factor in children, viral load should be determined.

2.7 Cerebral Palsy

Cerebral palsy (CP) is a static, non-progressive disorder of movement and posture, resulting from CNS or brain injury or insult during the prenatal, perinatal or postnatal periods. As the major developmental disability of function in children, CP is distinguished by an inability to normally control motor function, which in turn retards the affected child's capacity for exploration, speaking and overall learning [55]. In addition to the motor function retardation, cognitive ability may also be limited. United States and worldwide prevalence stands at 1.5–2.0 cases per 1,000 births. Despite technological and medical advances, this rate has remained as such for approximately 40 years [56, 57].

CP syndromes vary in severity and type, with this directly related to the portion of the brain which was damaged. Its classification is considered multiaxial. It is characterized according to "extremities involved (monoplegia, hemiplegia, diplegia and quadriplegia) and the characteristics of neurologic dysfunction (spastic, hypotonic, dystonic, athetotic or a combination)" [58]. Specific pathological and imaging patterns have been observed throughout the different types of CP. In hemiplegic CP, periventricular atrophy is often present, while gross malformation of cerebral development is evident in approximately one-sixth of affected children [59]. Because approximately one-third of these children present with a normal CT or MRI, such pathology can be attributed to microscopic malformations of development rather than injury to the already-developed brain. Cystic lesions, cortical atrophy and hydrocephalus have been associated with quadriplegia type CP [60]. Marble-like basal ganglia, impaired cerebellar pathways and an enlarged ventricle system have also been seen in various forms of CP.

Although a multifactorial etiology of CP has been established, only a limited number of cases are related to recognized causes [60]. Postnatal causes have been recognized, however, in most instances birth injury or hypoxic-ischemic insults at delivery are not to blame. The increased occurrence of congenital abnormalities in children with CP versus those without supports the theory of an underlying prenatal cause. A Californian study found 19.2% of children with CP also had other congenital defects while this occurred in only 4.3% of controls [61]. In fact, the Collaborative Perinatal Project lists congenital malformations and birth weight below 2,100 g to be the main predictors of Cerebral Palsy. In addition, minor irregularities outside of the nervous system and neuronal migration are considered related to CP [62, 63].

3 Conclusion

As our understanding of developmental disorders grows, we must continue to focus on the fundamental neural processes that have gone wrong. The key to successfully treating developmental disorders lies in understanding the mechanisms by which normal CNS developmental processes are affected in these pathological states. Though presented as three discrete processes, neurulation, neural migration and neural pruning proceed in a non-discreet manner. The processes of neural migration and neural pruning show the greatest overlap, and thus further complicate our understanding of the mechanisms behind developmental disorders. Nutritional deficiencies, viruses, and teratogens all affect the developmental process in non-specific ways, affecting each step of development to some degree. Yet fundamental genetic alterations in developmentally expressed genes may have specific targets. By identifying these triggers, whether they are specific or non-specific in nature, and by successfully correlating their impact on the fundamental processes of development we move forward in our quest for novel treatment strategies.

References

1. Yerby MS (2003) Clinical care of pregnant women with epilepsy: neural tube defects and folic acid supplementation. Epilepsia 44(Suppl 3):33–40
2. Golden JA, Chernoff GF (1995) Multiple sites of anterior neural tube closure in humans: evidence from anterior neural tube defects (anencephaly). Pediatrics 95(4):506–510
3. Van Allen MI, Kalousek DK, Chernoff GF, Juriloff D, Harris M, McGillivray BC et al (1993) Evidence for multi-site closure of the neural tube in humans. Am J Med Genet 47(5):723–743
4. Colas JF, Schoenwolf GC (2001) Towards a cellular and molecular understanding of neurulation. Dev Dyn 221(2):117–145
5. Griffith CM, Wiley MJ, Sanders EJ (1992) The vertebrate tail bud: three germ layers from one tissue. Anat Embryol (Berl) 185(2):101–113
6. Criley BB (1969) Analysis of embryonic sources and mechanims of development of posterior levels of chick neural tubes. J Morphol 128(4):465–501
7. Lowery LA, Sive H (2004) Strategies of vertebrate neurulation and a re-evaluation of teleost neural tube formation. Mech Dev 121(10):1189–1197
8. Copp AJ, Greene ND, Murdoch JN (2003) The genetic basis of mammalian neurulation. Nat Rev Genet 4(10):784–793
9. Sunshine J, Balak K, Rutishauser U, Jacobson M (1987) Changes in neural cell adhesion molecule (NCAM) structure during vertebrate neural development. Proc Natl Acad Sci USA 84(16):5986–5990
10. Rakic P (2006) A century of progress in corticoneurogenesis: from silver impregnation to genetic engineering. Cereb Cortex 16(Suppl 1):i3–i17
11. Hatten ME, Liem RK (1981) Astroglial cells provide a template for the positioning of developing cerebellar neurons in vitro. J Cell Biol 90(3):622–630
12. Rakic P (1971) Neuron-glia relationship during granule cell migration in developing cerebellar cortex. A Golgi and electronmicroscopic study in Macacus Rhesus. J Comp Neurol 141(3): 283–312
13. Weissman TA, Riquelme PA, Ivic L, Flint AC, Kriegstein AR (2004) Calcium waves propagate through radial glial cells and modulate proliferation in the developing neocortex. Neuron 43(5):647–661
14. O'Rourke NA, Dailey ME, Smith SJ, McConnell SK (1992) Diverse migratory pathways in the developing cerebral cortex. Science 258(5080):299–302
15. Stahl S (2002) Essential psychopharmacology neuroscientific basis and practical applications, 2nd edn. Cambridge University Press, Cambridge
16. Kantor DB, Kolodkin AL (2003) Curbing the excesses of youth: molecular insights into axonal pruning. Neuron 38(6):849–852
17. Weimann JM, Zhang YA, Levin ME, Devine WP, Brulet P, McConnell SK (1999) Cortical neurons require Otx1 for the refinement of exuberant axonal projections to subcortical targets. Neuron 24(4):819–831
18. Freeman MR (2006) Sculpting the nervous system: glial control of neuronal development. Curr Opin Neurobiol 16(1):119–125
19. Dias MS, Partington M (2004) Embryology of myelomeningocele and anencephaly. Neurosurg Focus 16(2):E1
20. Hiilesmaa VK (1992) Pregnancy and birth in women with epilepsy. Neurology 42(4 Suppl 5): 8–11

21. Lindhout D, Schmidt D (1986) In-utero exposure to valproate and neural tube defects. Lancet 1(8494):1392–1393
22. Rosa FW (1991) Spina bifida in infants of women treated with carbamazepine during pregnancy. N Engl J Med 324(10):674–677
23. Cicchetti D, Cannon TD (1999) Neurodevelopmental processes in the ontogenesis and epigenesis of psychopathology. Dev Psychopathol 11(3):375–393
24. Palmini A, Andermann E, Andermann F (1994) Prenatal events and genetic factors in epileptic patients with neuronal migration disorders. Epilepsia 35(5):965–973
25. Kriegstein AR (2005) Constructing circuits: neurogenesis and migration in the developing neocortex. Epilepsia 46(Suppl 7):15–21
26. Gleeson JG, Allen KM, Fox JW, Lamperti ED, Berkovic S, Scheffer I et al (1998) Doublecortin, a brain-specific gene mutated in human X-linked lissencephaly and double cortex syndrome, encodes a putative signaling protein. Cell 92(1):63–72
27. Campbell DB, Sutcliffe JS, Ebert PJ, Militerni R, Bravaccio C, Trillo S et al (2006) A genetic variant that disrupts MET transcription is associated with autism. Proc Natl Acad Sci USA 103(45):16834–16839
28. Splawski I, Yoo DS, Stotz SC, Cherry A, Clapham DE, Keating MT (2006) CACNA1H mutations in autism spectrum disorders. J Biol Chem 281(31):22085–22091
29. Krey JF, Dolmetsch RE (2007) Molecular mechanisms of autism: a possible role for Ca2+ signaling. Curr Opin Neurobiol 17(1):112–119
30. Schneider T, Przewlocki R (2005) Behavioral alterations in rats prenatally exposed to valproic acid: animal model of autism. Neuropsychopharmacology 30(1):80–89
31. Williamson SL, Christodoulou J (2006) Rett syndrome: new clinical and molecular insights. Eur J Hum Genet 14(8):896–903
32. Zametkin AJ, Ernst M (1999) Problems in the management of attention-deficit-hyperactivity disorder. N Engl J Med 340(1):40–46
33. Castellanos FX, Lee PP, Sharp W, Jeffries NO, Greenstein DK, Clasen LS et al (2002) Developmental trajectories of brain volume abnormalities in children and adolescents with attention-deficit/hyperactivity disorder. JAMA 288(14):1740–1748
34. Daley D (2006) Attention deficit hyperactivity disorder: a review of the essential facts. Child Care Health Dev 32(2):193–204
35. Asherson PJ, Curran S (2001) Approaches to gene mapping in complex disorders and their application in child psychiatry and psychology. Br J Psychiatry 179:122–128
36. Garakani A, Mathew SJ, Charney DS (2006) Neurobiology of anxiety disorders and implications for treatment. Mt Sinai J Med 73(7):941–949
37. Cheeseman SH, McGraw BR (1983) Studies on cytomegalovirus in human milk. J Infect Dis 148(3):615–616
38. Stagno S, Pass RF, Cloud G, Britt WJ, Henderson RE, Walton PD et al (1986) Primary cytomegalovirus infection in pregnancy. Incidence, transmission to fetus, and clinical outcome. JAMA 256(14):1904–1908
39. Demmler GJ (1991) Infectious Diseases Society of America and Centers for Disease Control. Summary of a workshop on surveillance for congenital cytomegalovirus disease. Rev Infect Dis 13(2):315–329
40. Enders G, Bader U, Lindemann L, Schalasta G, Daiminger A (2001) Prenatal diagnosis of congenital cytomegalovirus infection in 189 pregnancies with known outcome. Prenat Diagn 21(5):362–377
41. Ahlfors K, Ivarsson SA, Harris S, Svanberg L, Holmqvist R, Lernmark B et al (1984) Congenital cytomegalovirus infection and disease in Sweden and the relative importance of primary and secondary maternal infections. Preliminary findings from a prospective study. Scand J Infect Dis 16(2):129–137
42. Scheld WM, Whitely RJ, Durack DT (1991) Infections of the central nervous system. Raven, New York
43. Hardy JB (1973) Fetal consequences of maternal viral infections in pregnancy. Arch Otolaryngol 98(4):218–227

44. Newell ML, Peckham C (1993) Risk factors for vertical transmission of HIV-1 and early markers of HIV-1 infection in children. AIDS 7(Suppl 1):S91–S97
45. Oxtoby MJ (1994) Pediatric AIDS: the challenge of HIV in infants, children, adolescents, 2nd edn. Williams & Wilkins, Baltimore
46. Mofenson LM, Burns DN (1991) Passive immunization to prevent mother-infant transmission of human immunodeficiency virus: current issues and future directions. Pediatr Infect Dis J 10(6):456–462
47. de Rossi A, Ometto L, Mammano F, Zanotto C, del Mistro A, Giaquinto C et al (1993) Time course of antigenaemia and seroconversion in infants with vertically acquired HIV-1 infection. AIDS 7(11):1528–1529
48. Goedert JJ, Duliege AM, Amos CI, Felton S, Biggar RJ (1991) High risk of HIV-1 infection for first-born twins. The International Registry of HIV-exposed Twins. Lancet 338(8781): 1471–1475
49. Bryson YJ, Luzuriaga K, Sullivan JL, Wara DW (1992) Proposed definitions for in utero versus intrapartum transmission of HIV-1. N Engl J Med 327(17):1246–1247
50. Tardieu M, Le Chenadec J, Persoz A, Meyer L, Blanche S, Mayaux MJ (2000) HIV-1-related encephalopathy in infants compared with children and adults. French Pediatric HIV Infection Study and the SEROCO Group. Neurology 54(5):1089–1095
51. Mitchell W (2001) Neurological and developmental effects of HIV and AIDS in children and adolescents. Ment Retard Dev Disabil Res Rev 7(3):211–216
52. Persidsky Y, Ghorpade A, Rasmussen J, Limoges J, Liu XJ, Stins M et al (1999) Microglial and astrocyte chemokines regulate monocyte migration through the blood-brain barrier in human immunodeficiency virus-1 encephalitis. Am J Pathol 155(5):1599–1611
53. Gelbard HA, Boustany RM, Schor NF (1997) Apoptosis in development and disease of the nervous system: II. Apoptosis in childhood neurologic disease. Pediatr Neurol 16(2):93–97
54. Rafalowska J (1998) HIV-1-infection in the CNS. A pathogenesis of some neurological syndromes in the light of recent investigations. Folia Neuropathol 36(4):211–216
55. Boyle CA, Decoufle P, Yeargin-Allsopp M (1994) Prevalence and health impact of developmental disabilities in US children. Pediatrics 93(3):399–403
56. Clark SL, Hankins GD (2003) Temporal and demographic trends in cerebral palsy – fact and fiction. Am J Obstet Gynecol 188(3):628–633
57. Gibson CS, MacLennan AH, Goldwater PN, Haan EA, Priest K, Dekker GA (2006) Neurotropic viruses and cerebral palsy: population based case-control study. BMJ 332(7533):76–80
58. Kuban KC, Leviton A (1994) Cerebral palsy. N Engl J Med 330(3):188–195
59. Wiklund LM, Uvebrant P (1991) Hemiplegic cerebral palsy: correlation between CT morphology and clinical findings. Dev Med Child Neurol 33(6):512–523
60. Nelson KB (2003) Can we prevent cerebral palsy? N Engl J Med 349(18):1765–1769
61. Croen LA, Grether JK, Curry CJ, Nelson KB (2001) Congenital abnormalities among children with cerebral palsy: More evidence for prenatal antecedents. J Pediatr 138(6):804–810
62. Stanley F, Blair E, Alberman E (2000) Cerebral palsies: epidemiology and causal pathways. MacKeith Press, London
63. Coorssen EA, Msall ME, Duffy LC (1991) Multiple minor malformations as a marker for prenatal etiology of cerebral palsy. Dev Med Child Neurol 33(8):730–736
64. Mills JL, McPartlin JM, Kirke PN, Lee YJ, Conley MR, Weir DG et al (1995) Homocysteine metabolism in pregnancies complicated by neural-tube defects. Lancet 345(8943):149–151
65. Lucock MD, Daskalakis I, Lumb CH, Schorah CJ, Levene MI (1998) Impaired regeneration of monoglutamyl tetrahydrofolate leads to cellular folate depletion in mothers affected by a spina bifida pregnancy. Mol Genet Metab 65(1):18–30

Drug Permeation Across the Fetal Maternal Barrier

Chaitali Ghosh and Nicola Marchi

Abstract Mammalians have barriers that regulate and limit the distribution of a broad variety of compounds and xenobiotics. Among these, the placental barrier, which forms during embryogenesis, protects the developing fetus. The placenta is composed of several layers of cells acting as a barrier for the diffusion of substances between the maternal and fetal circulatory systems. Lipid-soluble molecules can readily cross while the transfer of large-molecular-weight molecules are limited. Anatomical and functional properties of the placental barrier and similarities to the blood-brain barrier (BBB) will be discussed.

Keywords Pre-natal • Fetus protection • Epigenetic mechanism • Drug transporters

1 Introduction

2 Anatomy and Development of the Placenta

Embryonic development starts with the formation of the zygote, containing single diploid nucleus holding paternal and maternal chromosomes. As it moves into the oviduct, a newly formed embryo undergoes a series of cell division known as cleavage. The cleavage process early differentiates the embryo into two groups of cells. The outer cell mass called trophoblast, will form the placenta and the associated membranes while the inner cell mass, called embryoblast, will form the embryo and the amnion. At this stage the embryo (or *morula*) undergoes further cleavage developing into the blastocyst. The blastocyst enters the uterine cavity and implant on the endometrial inside layer of the uterine wall. The contact with the endometrium

N. Marchi
Cleveland Clinic, Cerebrovascular Research, Lerner Research Institute, 9500 Euclid Avenue, NB2-137, Cleveland, OH, 44195, USA
e-mail: marchin@ccf.org

D. Janigro (ed.), *Mammalian Brain Development*, Contemporary Neuroscience,
DOI 10.1007/978-1-60761-287-2_8,

triggers the proliferation of the trophoblasts and embryoblasts, leading to the formation of the amniotic cavity and yolk sac, thus setting the basic structure for the development of the embryo.

In humans the gestation period is approximately 38 weeks. Most of the organs systems are formed during the embryonic period, spanning from the third to the eighth week. The fetal period is characterized by the maturation of the organs and growth. From the end of the third week until birth, the fetus receives nutrients and eliminates its metabolic wastes *via* the placenta, an organ that has both maternal (*decidual* endometrium) and fetal (*chorion*) components. The maternal face of the placenta, called *basal plate*, consists of syncitiotrophoblasts and a supporting layer of *decidua basalis*. On the fetal side, the layers of the chorion form the *chorionic plate* of the placenta.

As the blastocyst implants, it stimulates a response in the uterine wall called decidual reaction. This reaction involves the cells of the endometrial stroma, which underlie the endometrial epithelium. These cells start to accumulate lipid and glycogen becoming *decidual* cells. The stroma becomes thicker and highly vascularized and the endometrium as a whole is then called deciduas and represents the maternal portion of the placenta. In particular, the *decidua basalis* will be involved in the future formation of this organ. The formation of embryo/fetal part of the placenta starts with the differentiation of the trophoblasts into cytotrophoblasts and syncitiotrophobalsts. Between days 6 and 9, the embryo is completely implanted in the endometrium, mostly as result of enzymes such as metalloproteinase secreted by the cytotrophoblasts to break down the extracellular matrix between endometrial cells. The subsequential development of a functional placenta starts from this stage and evolves with the formation of the amniotic cavity, coalescence of the extraembryonic mesoderm to form the chorionic cavity and migration at day 12 of the primary yolk sac at the end of the chorionic cavity. It is not until day 16 that the extraembryonic mesoderm associated with the cytotrophoblasts penetrates the core of the primary stem *villi*, thus transforming them into the secondary stem *villi*. By the end of the third week, this villous mesoderm gives rise to blood vessels that connect with the vessels forming the embryo, thus establishing a working utero-placental circulation. Such extraembryonic mesoderm proliferates to shape the tertiary stem *villi* that project into the trophoblastic lacunae that become blood-filled after 10 weeks. Starting near the end of the second trimester branching of the *villi* will complete the placental villous tree. During the fourth and fifth months, walls of decidual tissues grow into the *intervillous space* from the maternal side of the placenta, separating the *villi* into 15–25 groups called *cotyledons*. Maternal blood freely flows across cotyledons.

The placental starts functioning early during pregnancy. For instance, during early gestation the primary function is to mediate the implantation of the embryo at the uterus. After implantation, the main placental function is to regulate nutrients and oxygen uptake from the mother. Generally, the placenta offers protection to the fetus, strictly regulating the entry of xenobiotics of maternal origin while assisting the passage of essential nutrients (Fig. 1). The placenta also

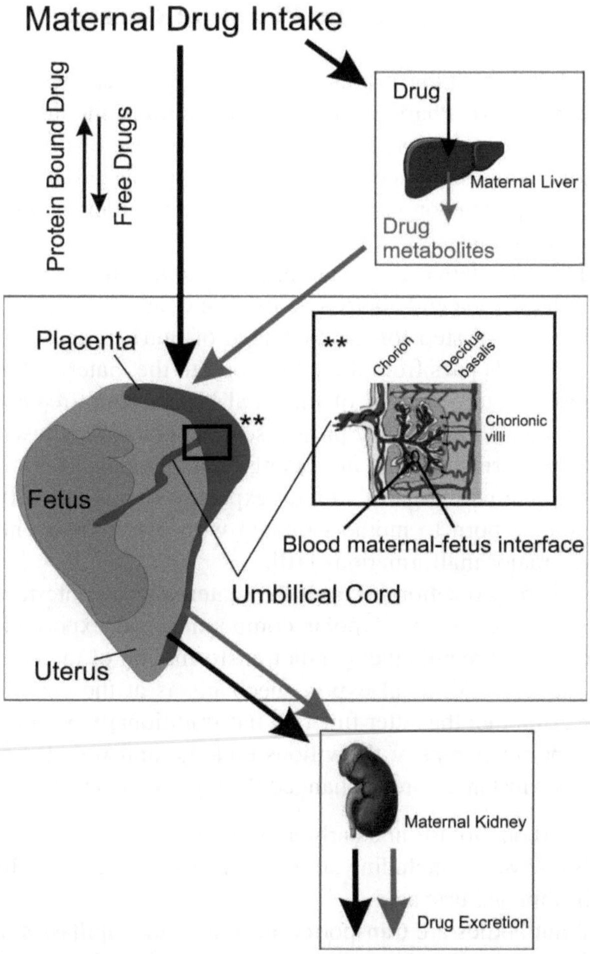

Fig. 1 Drug disposition in mother and fetus through the placental barrier. Drug is administrated to the mother. After absorption, the drug will circulate in blood as free or bound to plasma proteins. The drug can (1) directly reach the fetus through the placenta (*black arrow*) or (2) reach the fetus after metabolization at the maternal liver (*green arrow*). The overall pathway of the drug is also shown. The placenta itself is capable to biotransform drugs due to the presence of metabolic enzymatic machinery. Through the umbilical vein oxygenated blood is transferred from mother to the fetus and through umbilical arteries deoxygenated blood containing both the drug and its metabolites is released by the fetus back to the maternal compartment where final excretion takes place through maternal kidney

acts much the same as the kidney in the elimination of xenobiotics [1–3]. The placenta also plays an active role in regulating maternal physiology to the nutritional benefit of the fetus, being involved in the production of angiogenic factors, vasodilators and hormones [4].

3 Levels of Protection at the Placental Barrier

Clinically, when managing a pregnant patient, the treatment of two individuals should be considered. Pharmacological interventions should be based on a risk/benefit assessment encompassing both. While the use of drugs has increased the risk of teratogenicity, some medical conditions such as gestational diabetes, hyperthyroidism or hypertension require drug therapy in order to ensure optimal health of the mother and fetus.

Maternal blood enters the placenta passing thought the spiral arteries, bathes the *villi* and then leaves via endometrial veins. The contact of maternal blood with *villi* represents the critical step for the exchange of nutrient and oxygen. At the same time, waste products pass from the fetal blood to the maternal blood. The placenta contains approximately 150 ml of maternal blood, being replaced approximately 3–4 times/minute. As pregnancy progresses, the placenta adjusts its structural and functional characteristics to the increasing fetal demands [5–9]. The first trimester is the most critical with regards to fetal exposure to noxious substances. Thus, up to 20% of infants born to mothers treated with cytotoxic agents during the first trimester bear major malformations [10].

Three levels of protection occur at the maternal-fetus interface (Fig. 2). In particular: (1) restricted entry of polar compounds and export mechanisms at the syncitiotrophobalsts membrane; (2) biotransformation of molecules at the cytotrophoblasts; and, (3) additional export mechanisms at the endothelial/mesodermal membranes. Although the latter function has traditionally been neglected, the presence of transporter pumps at the villous endothelium was demonstrated [11–13]. The following substances are exchanged through the placenta:

1. Oxygen, carbon dioxide and carbon monoxide.
2. Nutrients and waste, including amino acids, fatty acids, carbohydrates, vitamins, urea, bilirubin and uric acid.
3. Maternal antibodies are transported into the fetal capillaries, thereby providing passive immunity to various diseases. Maternal lymphocytes also pass into fetal circulation. However, certain antibodies (e.g., against chicken pox and whooping cough) do not cross.
4. Most peptide hormones do not cross the placenta. However, steroid or synthetic hormones can pass. Most peptide hormones do not cross the placenta. Only some steroid or synthetic hormones can pass. The fetus is almost self-sufficient in its hormonal requirements. Exception is made for some steroid hormones, which are produced by the feto-placental unit. In this case, the hormones can cross the placental barrier despite of the presence of drug transporter proteins and carry out integrated functions in both the fetus and the mother. Hormones produced by the feto-placental unit include progesterone, estradiol, chronic gonadotropin (hCG), and placental lactogens.
5. Infectious agents penetrating the placenta include cytomegalovirus, rubella, measles and syphilis bacteria. The human immunodeficiency virus (HIV) can also be transmitted.

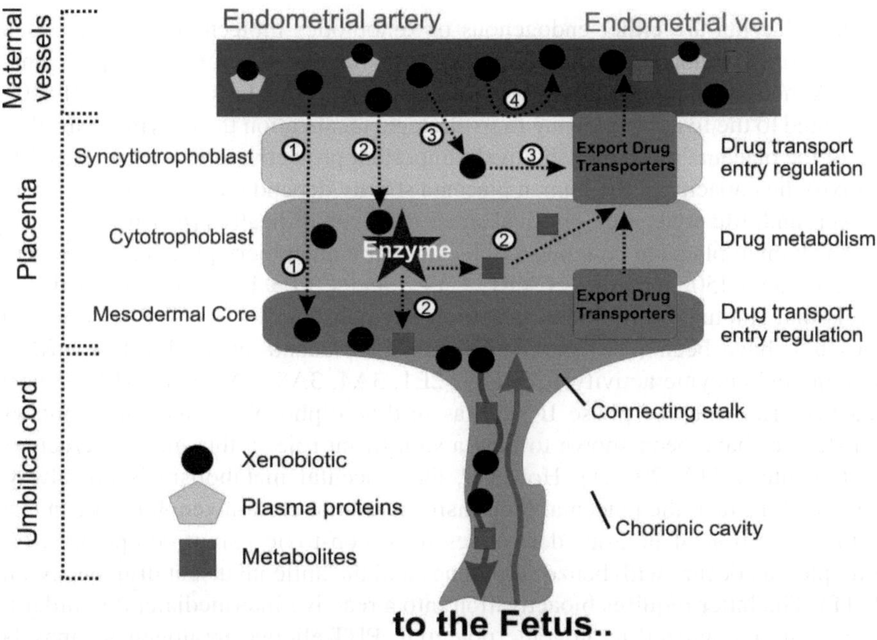

Fig. 2 Placental barrier: three levels of protection at the maternal fetus interface. The drug reaches the placenta through the endometrial artery. Three biological layers constituting the placenta will determine the route of the free drug. In particular a given drug can cross directly and reach the umbilical cord and the fetus (1), be metabolized at the cytotrophoblasts level (2) and then either reach the umbilical cord or be pumped back in the endometrial vein by specific drug transporters present at the syncitiotrophobalsts or at the mesodermal core (3 and 4). Wastes and extruded drugs will leave through the endometrial vein. The diagrams depict the distribution of drug transporters and metabolic enzymes at the placenta

6. Many drugs or abuse substances can cross the placental barrier leading to cardiovascular and cerebral defects and increasing the incidence of spontaneous abortion. Teratogenic drugs include ethanol, cocaine, LSD and heroin. Other teratogens crossing the placenta include excessive vitamin A, anticoagulants, anti-depressant and chemotherapy drugs.

Drugs can cross the placenta via specific transport mechanisms present in the maternal-facing apical (brush border) membrane and fetal-facing basal membrane of the syncitiotrophobalsts [1]. It is also well established that the placenta metabolizes a variety of pharmacologically active molecules.

4 Metabolic Properties of the Placenta

Generally, the metabolization of xenobiotics and endogenous molecules occur following a two steps process: phase I (oxido-reduction reactions) and phase II (conjugation with polyatomic groups). Compounds that can be metabolized by the human placenta are either endogenous or xenobiotics molecules (e.g., medicinal drugs, drugs of abuse, food-contaminating toxins and environmental pollutants) [14]. Although the metabolic capacity of the human placenta is approximately 10% compared to the liver [15], owing to its strategic localization the enzymatic machinery of the placenta plays a qualitatively important protective role for the fetus. The detoxifying capacity of the human placenta strictly depend on the maternal genetic background, life style, diet, physical stress and state of health [16–19]).

The human placenta contains a rich enzymatic machinery [17, 20, 21]. Several cytochrome P450, including CYP1, CYP2 and CYP3 have been identified at levels that fluctuate throughout gestation. Cytochrome P450 (CYP) enzymes in particular have been well characterized in the placenta at the level of mRNA, protein, and enzyme activity. CYP1A1, 2E1, 3A4, 3A5, 3A7 and 4B1 have been detected. Enzymes of phase II such as uridine-diphosphate and glucuronosyltransferases have been shown to play a significant role in this metabolic/detoxification phase [17, 20, 21]. However, the placental metabolism is not always beneficial. In fact, the placental biotransformation of certain xenobiotics can lead to the formation of noxious derivatives from non-toxic parent compounds. For example this occurs with benzo (a)pyrenes and the anticonvulsant drug phenytoin (PHT). The latter requires bioactivation into a reactive intermediate(s) in order to achieve its recognized teratogenic potential. PHT-elicited teratogenesis may be due to the in situ generation of reactive intermediates, resulting from the enzymatic bioactivation of PHT [22] or as a consequence of PHT induced episodes of hypoxia/reoxygenation [23].

5 Mechanisms of Drug Passage Across the Placenta

5.1 Drug Distribution: Similarities Between BBB and Blood-Placental Barrier

Blood brain barrier: The BBB contributes to brain homeostasis by protecting the brain from potentially harmful endogenous and exogenous substances. Multi-drug resistance proteins (e.g., P-gp) have been demonstrated to play a key role in influencing the permeability properties of the BBB. Thus, such proteins can actively transport a variety of lipophilic drugs out of the brain capillary endothelial cells that form the BBB. Multi-drug transporter proteins limit the distribution of drugs to the CNS. Such mechanism is ontologically related to protection of the CNS by hampering the passage of potential toxic molecules present in the blood stream. Modulation of efflux transporters at the BBB forms a novel strategy to enhance the penetration of drugs into the brain and may yield new therapeutic options for drug-resistant CNS diseases. Transporters inhibitors include cyclosporin A, verapamil, tetrandrine, and doxorubicin. While, one hand, the use of such inhibitors could improve the passage of drugs across the BBB in pathological states (e.g., brain tumor or epilepsy), on the other it could facilitate the harmful action of toxins present in the peripheral blood [24].

Placental barrier: It has been shown that nearly all drugs that are administered during pregnancy will enter to some degree in the circulation of the fetus via passive diffusion. In addition, some drugs are pumped across the placenta by various active drug transporters located on both the fetal and maternal side of the trophoblast layer. It is only in recent years that the impact of active transporters (e.g., P-gp) on the disposition of drugs has been demonstrated. These transporters are similar to the ones present at the BBB. Presence of P-gp in the rat chorioallantoic placenta starting soon after its development signifies the involvement of P-gp in transplacental pharmacokinetics during the whole period of placental maturation. P-gp limits the entry of various potentially toxic drugs and xenobiotics into the fetus and is thus considered a placental protective mechanism. However, such transporters limit the entrance of drug, reducing the overall pharmacological potency. For example, indinavir may be used in the management of HIV infection during pregnancy. Poor maternal-to-fetal transfer of indinavir has been reported, but the mechanisms remain unknown. Recently, involvement of placental P-gp has been suggested and the use of transport modulators can increase the maternal-to-fetal transfer [25].

5.2 Drug Physical Properties Affecting Placental Permeability

The passage of molecules across physiological barriers occurs according to mechanisms dictated by the physical properties of a specific compound. Moreover, both the placental and the BBB regulate the entry of endogenous compound and xenobiotics by drug transporter proteins or facilitate/passive drug diffusion. The placenta resembles a lipid bi-layer membrane, therefore only non-protein-bound drugs can diffuse across it. Understanding drug binding is important to determine the profile of fetal/maternal total drug level at steady state [26, 27]. Usually, lipophilic drugs with a molecular weight <500 Da can cross the placenta. Increase in the molecular weight leads to a progressive decrease of permeability [28]. Lipid solubility is an important factor determining the permeability of a drug across a biological membrane. Drugs that are lipid-soluble will pass easily through the placenta, whereas hydrophilic drugs are generally impermeable. Fetal exposure to drugs also depends on maternal pharmacokinetics parameters, including distribution volume and hemodynamic changes. Ideally, one may wish to identify drugs that will not cross the placental barrier. However, with the exception of drugs with large molecular weights, such as heparin or insulin, most drugs appear to cross the placenta and are associated with varying degrees of fetal exposure [29].

Normally, transplacental exchanges are known to involve passive transfer, active transport, facilitated diffusion, phagocytosis and pinocytosis [30]. Only the passive transfer of drugs across the placenta has received much attention by investigators, although active transport systems are being identified [1]. In particular:

5.3 Passive Diffusion

Passive diffusion of drugs through the maternal-fetal unit is the predominant form of permeation of molecules driven by concentration gradient. Input of energy and competitive inhibition are not involved in passive diffusion. In this case, the direction and the rate by which a drug crosses a biological membrane depend on the concentration and drug lipophilicity.

5.4 Facilitated Diffusion

It requires the presence of a carrier. This system can be saturated at high concentrations of drugs as predicted by the Michaelis-Menten law. Such mechanism of transport does not require energy. Transport stops when the drug concentrations in the maternal and fetal circulations are equal. Facilitated diffusion provides a means to increase transport rate when the functional and metabolic needs of the fetus would not be met by passive diffusion alone (e.g., increasing demand of carbohydrates) [31].

5.5 Active Transport

Active transport across the placental membrane occurs *via* protein pumps adenosine triphosphate (ATP) dependent. Active transporters work against concentration gradient and are saturable systems. Active transport systems include those for nutrients such as amino acids, vitamins and glucose. Less is known about the active transport of xenobiotics across the placental barrier [32–34]. Drugs that are actively transported are structurally similar to endogenous substrates [5]. Thus, the original function of such carriers is to transfer endogenous compounds that are essential for fetal development. However, the exogenous drugs that are substrates, can use these transporters as a vehicle, gaining access into the fetal circulation [5]. Transporters are located in the following regions:

- *Brush border membrane* (maternal-side, Fig. 2)

  Serotonin transporter (SERT)
  Norepinephrine transporter (NET)
  Carnitine transporters (OCTN2)
  Equilibrative nucleotide transporter-1 (ENT1)
  Monocarboxylate transportera (MCTs)
  P-glycoprotein (MDR1)
  Placenta-specific ABC transporter, Breast cancer resistant protein (BCRP)
  Multidrug resistance-related proteins (MRP1, MRP2, and MRP3)

- *Basal membrane* (fetal-side, Fig. 2)

  Reduced folate transporter (RFT-1)
  Monocarboxylate transporters (MCTs)
  Organic anion transporter-4 (OAT4)
  Organic anion transporting polypeptide-B (OATP-B)

- *Unknown localization*

  Organic cation transporter-3 (OCT3) (basal membrane?)
  OCTN1 (brush border membrane?)

Changes in the levels of expression of drug transporters can affect the entry of drugs into the placenta [35]. In addition, when xenobiotics are recognized as substrates by placental transporters, it is possible that the transfer of endogenous substrates could be compromised. This however, depends on the relative concentrations as well as relative affinities of drugs and endogenous substrates to a given transporter (Table 1).

5.6 Efflux Transporters

Depending on their localization and orientation, placental drug transporters will direct a substrate inward or outward toward the fetus. Drug transporters include the ABC-cassette family: P-glycoprotein (P-gp), Multidrug Resistance-associated Proteins (MRPs) and Breast Cancer Resistance Protein (BCRP). See also Table 1 and Figs. 2 and 3.

P-Glycoprotein (P-gp). P-gp has broad substrate selectivity. It transports lipophilic drugs (neutral or cationic) including antimicrobials (e.g., rifampin), antivirals (e.g., anti-HIV protease inhibitors), antiarrythmics (e.g., verapamil) and antineoplastics (e.g., vincristine). P-gp is a transmembrane glycoprotein encoded by the human multidrug resistance gene *MDR1*. It is expressed on the maternal side of the placental trophoblast layer in the brush border membrane [12], where it mediates the active efflux of lipophilic drugs from the fetal compartment, using energy derived by ATP hydrolysis [36] (Figs. 2 and 3). It was demonstrated that in CF-1 mouse, a naturally occurring knock-out of the *mdr1a* gene, fetal exposure to avermectin caused cleft palate at much lower doses than those needed to cause the same birth defect in control animals [37]. Lack of P-gp expression in *mdr1a/b*-deficient mice was also associated with increased exposure of the fetus to digoxin, saquinavir or paclitaxel [38]. These studies also showed that placental P-gp function can be blocked by the P-gp inhibitors PSC833 (acyclosporine A analog) and GG120918, resulting in increased placental passage and fetal exposure to drugs that are P-gp substrates [38].

Multidrug Resistance-associated Proteins (MRPs). MRPs constitute another member of ATP-dependent efflux transporters (Fig. 3) [39]. Unlike P-gp, MRPs appear to transport polar compounds. The presence of MRPs in the placenta is somewhat controversial. Recent findings show that the human placenta expresses at least three members of the MRP family: MRP1, MRP2, and MRP3 [12]. MRP2 is expressed on the apical membrane of the syncitiotrophobalsts, while MRP1 and MRP3 are expressed on the basolateral membrane and the blood vessel endothelia (Fig. 2) [40, 41]. The role of MRPs in the placenta has not been fully clarified, but it is likely to be related to efflux of polar conjugates of xenobiotics or metabolites of endogenous compounds. When compared with the wild-type mice, the combined deficiency of P-gp and MRP1 in mice results in a dramatically increased bone

Table 1 Drug transporters at the Blood-Placental Barrier

Type of transporter	Location in the placenta	Type and direction of transport	Function	Prototypical drug substrates
P-glycoprotein (ABCB1, MDR1)	Brush-border membrane	Efflux	Remove xenobiotics from fetal compartment	Anticancer drugs, protease inhibitors, drugs of abuse, steroids
MRPs (ABCC1-ABCC6)	Brush-border membrane or basolateral membrane	Efflux	Remove xenobiotics from the placenta	Organic anions, glutathione conjugates, nucleotide analogs
BCRP (ABCG2 or MXR/ABCP)	Brush-border membrane	Efflux	Remove xenobiotics and estrone sulfates from fetal compartment	Quercetin
hENT1 (SLC29A1)	Brush-border membrane	Influx/Efflux	Mediate nucleotide transport	Anticancer and antiviral drugs, nucleotide analogs
hENT2 (SLC29A2)	Unknown	Influx/Efflux	Mediate nucleotide transport	Anticancer and antiviral drugs, nucleotide analogs
SERT (SLC6A4)	Brush-border membrane	Influx	Transport serotonin	Amphetamines
NET (SLC6A2)	Brush-border membrane	Influx	Transport norepinephrine and dopamine	Amphetamines
OCTN2 (SLC22A5)	Brush-border membrane	Influx/Efflux	Transport carnitine	Verapamil quinidine, TEA
MCT (SLC16A5)	Brush-border membrane	Influx/Efflux	Transport monocarboxylates	Aspirin, cefdinir, valproic acid
Sodium/multivitamin transporter (SMVT)	Brush-border membrane	Efflux	Transports biotin and pantothenate	Carbamazepine, primidone
OCT3 (EMT, SLC22A3)	Unknown	Influx/Efflux	Transport catecholamines	Amphetamines, antidepressants
OCTN1 (SLC22A4)	Unknown	Influx/Efflux	Transport organic cations	Verapamil, quinidine

marrow and gastrointestinal toxicity to intraperitoneally administered vincristine (up to 128-fold) and etoposide (3.5-fold). In P-gp-deficient mice, the corresponding increase in toxicity is lower, only 16- and 1.75-fold respectively.

Breast Cancer Resistance Protein (BCRP). BCRP (encoded by ABCG2) is referred as an ABC half-transporter because it has a secondary structure that has 6 transmembrane domains as opposed to the 12 of other ABC transporters (e.g., P-gp, Fig. 3) [42]. Immunohistochemical studies of BCRP localize the transporter to the syncytiotrophoblastic plasma membrane of the chorionic *villi* [43]. Although the

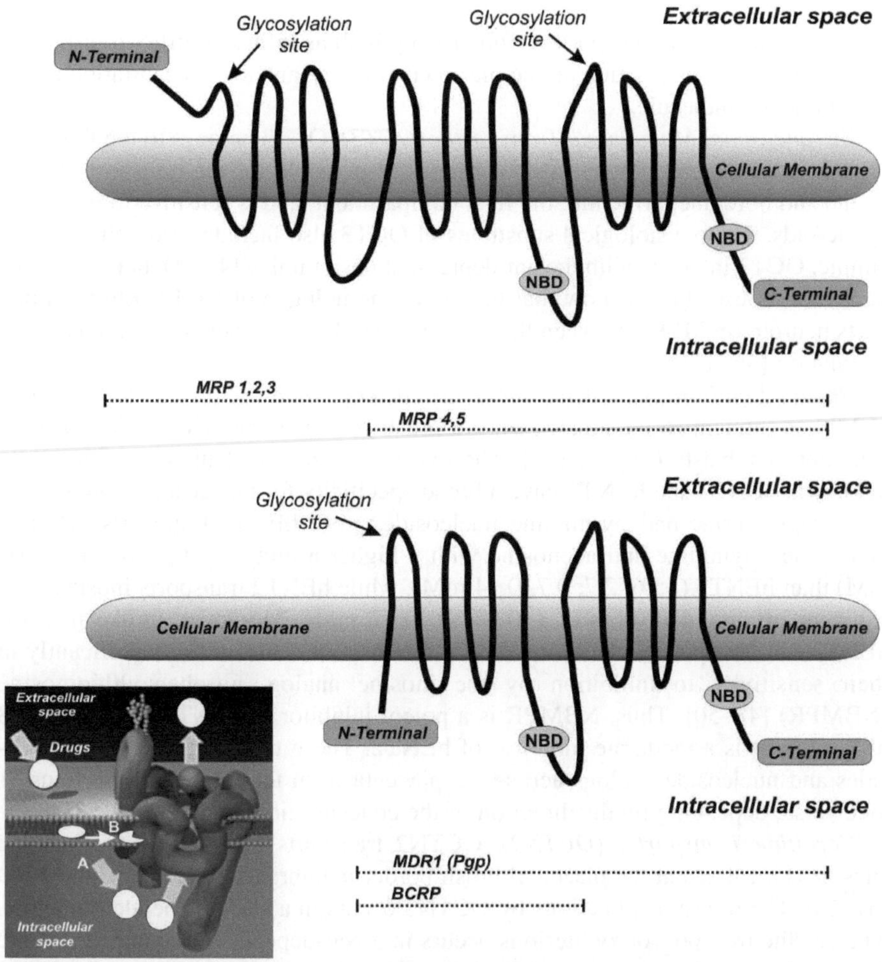

Fig. 3 Drug transporters are membrane proteins regulating the penetration across biological systems of a large variety of clinically relevant drugs (see also Table 2). These proteins have been initially discovered and characterized for their ability to confer resistance of cancer cells to chemotherapeutics [28]. Among these drug transporters, members of the ATP binding cassette (ABC) family includes P-glycoprotein (P-gp), multidrug resistance-associated proteins (MRP) and breast cancer-resistance protein (BCRP)

substrate selectivity of BCRP has not been completely elucidated, there is considerable overlap in substrates between BCRP and P-gp. Real-time RT-PCR expression profiling demonstrated a high level of BCRP expression in female reproductive tissues including the placenta [44]. BCRP can be induced in cell lines after the continuous administration of chemotherapeutic drugs, thus developing drug resistance in response to chemotherapy.

5.7 Influx/Efflux Transporters

These are facilitative transporters, functioning both as influx or efflux transporters depending on the orientation on the plasma membrane and the concentration gradient of a given molecule.

Extraneuronal Monoamine Transporter (OCT3). OCT3 belongs to the family of organic cationic transporters. This carrier is responsible for the transport of dopamine and norepinephrine into the fetal compartment, and is sensitive to inhibition by steroids. The physiological substrates of OCT3 also include serotonin and histamine. OCT3 interacts with the antidepressant desipramine [45]. Data from OCT3 knockout mouse studies show that the mouse homologue of OCT3 (*Orct3*) transports neurotoxin MPP+ between the placenta and the fetus, but not to the maternal circulation [46, 47].

Human Equilibrative Nucleoside Transporters 1 and 2 (hENT1 and hENT2). hENT1 is thought to be situated on the brush border membrane of the placental syncitiotrophobalsts (Fig. 2) [48]. The exact localization of hENT2 is currently unknown. hENT1 and hENT2 have a broad specificity for nucleosides and nucleoside drugs, purine and pyrimidine nucleosides [49]. hENT1 transports cytidine, guanosine, thymidine and adenosine with a higher affinity (~0.6, 0.14, 0.3, 0.04 mM) than hENT2 (~5.6, 2.7, 0.7, 0.14 mM), while hENT2 transports inosine with a higher affinity than hENT1 (~0.05 vs. 0.17 mM). Moreover, unlike hENT1, hENT2 can transport nucleobases. The two transporters also differ significantly in their sensitivity to inhibition by the inosine analog, nitrobenzylthioinosine (NBMPR) [48–50]. Thus, NBMPR is a potent inhibitor of hENT1 (IC50 ~ 0.4–8 nM) while it is a moderate inhibitor of hENT2. These transporters move nucleosides and nucleoside analogs across the placenta from the mother to the fetus or vice versa, depending on the direction of the concentration gradient [51, 52].

Carnitine Transporter (OCTN2). OCTN2 transports carnitine into privileged sites. It is localized at the placental brush border membrane as well as in the BBB [51, 53]. The transport of cations by OCTN2 occurs in a Na^+-independent manner, whereas the transport of zwitterions occurs in a Na^+-dependent manner. A genetic disorder known as primary carnitine deficiency is caused by mutations in OCTN2 zwitterions transport domains, which results in a loss of carnitine transport function [53, 54]. Because OCTN2 transports organic cations, its activity can be inhibited by cationic drug substrates. Several zwitterionic and cationic drugs inhibit OCTN2 in human embryonic kidney cells, including pyrilamine, quinidine, verapamil and valproate [55].

Acetylcholine is translocated via OCTN2 transporters, and that this activity can be inhibited by OCTN2 substrate drugs such as verapamil [56].

6 Fate of Drugs in Pregnancy

6.1 Does Serum Protein Play a Role in Drug Delivery During Pregnancy?

Protein binding of antiepileptic drugs like PHT and phenobarbitone was found to be reduced in pregnant women affected by epilepsy. Increase in unbound fraction of diazepam and valproic acid in pregnancy was reported [57]. The possibility of a transient increase of free drug level in serum could favor a toxic effects in the mother and in the fetus. This possibility is particularly relevant in view of the evidence that relatively large doses of diazepam are given to the mother during pregnancy. This may adversely affect the vital parameters of the newborn. Moreover, during pregnancy, protein binding affinity changes along with changes in the concentration of specific proteins. Decrease in maternal serum albumin may lead to corresponding increase in free fraction of drug [29].

Patho-physiological changes occurring during pregnancy may affect maternal drug protein binding, increase drug distribution volume or alter liver metabolism and renal clearance. Changes in the plasma peak drug level and in half-life time can ultimately affect the passage of drug into fetus. Drugs whose dosage needs to be monitored include anticonvulsants, lithium, digoxin, beta-blockers, amphicillin and cefuroxime. Very few drugs given in pregnancy fail to cross the placenta and, as a rule, drug molecules not bound to plasma protein diffuse along a concentration gradient to establish equilibrium. The pharmacological response of the fetus to drugs depends on the concentration of free drug in fetal blood. Adverse effects seldom occur if the maternal dose is appropriate. Exceptions include tetracycline, antithyroid drugs, coumarin anticoagulants, aspirin, indomethacin and cerebral depressant drugs (opiates, barbiturates and phenothiazines). The total (bound+free) drug concentration on the fetal side of the placenta is usually lower than on the maternal side except when differences in blood pH and/or protein binding favors accumulation of drugs in the fetus [58].

7 Xenobiotics Percolation Through the Placenta

Several types of chemicals have been documented to cross into the fetal environment. These chemicals include substance derived from tobacco smoke such as nicotine and cotinine (nicotine's major metabolite) [59, 60], illicit drugs [61], prescription pharmaceuticals [62], alcohol [63], and environmental chemicals such as pesticides [64] (Table 2). In general these chemicals can be divided into two

Table 2 Xenobiotic percolation through the tissue/matrix

Chemical class	Type of chemical	Tissue/matrix
Pharmaceuticals	Anesthetics, Analgesics, Antihistamines, Adrenergics, Expectorants, Antidepressants, Anticonvulsants	Meconium
Illicit drugs	Cocaine, Opiates, Cannabinoids, Morphine, Methadone, Stimulants	Meconium
Alcohol	Fatty acid ethyl esters	Meconium
Tobacco	Cotinine	Meconium
Metals	Manganese, magnesium, iron, copper, zinc, selenium, rubidium, strontium, cadmium, cesium, methylmercury, other metals	Cord Blood
Pesticides	Arochlor , Chlordane, Chlorpyrifos, Organophosphorus metabolites, DDT, Lindane, Malathion, Parathion	Meconium
	Chlorpyrifos, Diazinon, Bendiocarb, DDE, Hexachlorobenzene	Cord Blood
	Pesticide metabolites, chlorinated phenols	Amniotic fluids
PCBs	Sum PCB, Sum TEQ, OH-PCBs, Sum PCBs	Cord Blood
PBDEs	Sum PBDE, BDE 47	Cord Blood
Phytoestrogens	Daidzein, Genistein	

PCB polychlorinated biphenyls; *TEQ* toxic equivalents (of dioxins); *PBDEs* polybrominated diphenyl ethers; *OH-PCBs* hydroxylated PCBs

classes: (1) chemicals that have short biological half lives (non-persistant) and (2) chemicals that bioaccumulate in the body (persistant). Chemicals with short biological half lives are rapidly metabolized by the maternal liver and then excreted as polar metabolites. In most instances, the metabolites do not possess the same degree of toxicity as the parent chemical or may not be toxic at all [65]. Persistent chemicals are generally lipophilic and bind strongly to proteins [65]. These substances cross into the fetal system and can accumulate in the fetal environment either by depositing in fetal matrices or by circulating in the fetal bloodstream. Moreover, the timing of drug exposure is critical because of the developing nature of the placenta. Weeks 1–14 of gestation represents the critical period for exposures to chemicals [66].

8 Fetal-Maternal Unit Along with Developing Blood-Brain Barrier

The xenobiotics which enter through the placental barrier and reach the developing fetus also can easily access the nervous system of the fetus by permeating through the developing blood-brain barrier (BBB) (see also chapter by L. Cucullo) [67]. The BBB is a specialized system of brain microvascular endothelial cells that shields the brain from toxic substances in the blood, supplies brain tissues with nutrients and filter harmful compounds from the brain back to the bloodstream. The developing brain is susceptible to various toxicants as the BBB is not fully developed. These neurotoxicants interfere with the functional integrity of BBB of the developing nervous system and to delay its maturation [68, 69]. Lack of active efflux from brain to blood at immature BBB enhances the effect and therefore gestational period is a vital time to protect the fetus from xenobiotics [70].

Given its physiological and developmental purpose, it is likely that the placental barrier undertakes the functional and protective role that the BBB will acquire at a later stage of embryogenesis and postnatally. During the early enbryogenic stage of brain development, the placenta protects the developing fetal central nervous system from those molecules that in adulthood will be excluded by the BBB. Both the placental and the BBBs possess drug transporters and metabolic systems regulating the passage of molecules from a "peripheral" into a "central" (fetal or brain) compartment. Even though ontogenically not directly related, the two barriers perform analogous tasks and share evolutionary-driven protective functions.

References

1. Syme MR, Paxton JW, Keelan JA (2004) Drug transfer and metabolism by the human placenta. Clin Pharmacokinet 43:487–514
2. Myllynen P, Pasanen M, Pelkonen O (2005) Human placenta: a human organ for developmental toxicology research and biomonitoring. Placenta 26:361–371
3. Myren M, Mose T, Mathiesen L, Knudsen LE (2007) The human placenta – an alternative for studying foetal exposure. Toxicol In Vitro 21:1332–1340
4. Cross JC (2006) Placental function in development and disease. Reprod Fertil Dev 18:71–76
5. Ganapathy V, Prasad PD, Ganapathy ME, Leibach FH (2000) Placental transporters relevant to drug distribution across the maternal-fetal interface. J Pharmacol Exp Ther 294:413–420
6. Marin JJ, Macias RI, Serrano MA (2003) The hepatobiliary-like excretory function of the placenta. A review. Placenta 24:431–438
7. van der Aa EM, Peereboom-Stegeman JH, Noordhoek J, Gribnau FW, Russel FG (1998) Mechanisms of drug transfer across the human placenta. Pharm World Sci 20:139–148
8. Mihaly GW, Morgan DJ, Marshall AW, Smallwood RA, Cockbain S, MacLellan D, Hardy KJ (1982) Placental transfer of ranitidine during steady-state infusions of maternal and fetal sheep. J Pharm Sci 71:1008–1010
9. Mihaly GW, Morgan DJ (1983) Placental drug transfer: effects of gestational age and species. Pharmacol Ther 23:253–266
10. Reynoso EE, Shepherd FA, Messner HA, Farquharson HA, Garvey MB, Baker MA (1987) Acute leukemia during pregnancy: the Toronto Leukemia Study Group experience with long-term follow-up of children exposed in utero to chemotherapeutic agents. J Clin Oncol 5:1098–1106
11. Dye JF, Jablenska R, Donnelly JL, Lawrence L, Leach L, Clark P, Firth JA (2001) Phenotype of the endothelium in the human term placenta. Placenta 22:32–43
12. St Pierre MV, Serrano MA, Macias RI, Dubs U, Hoechli M, Lauper U, Meier PJ, Marin JJ (2000) Expression of members of the multidrug resistance protein family in human term placenta. Am J Physiol Regul Integr Comp Physiol 279:R1495–R1503
13. Patel P, Weerasekera N, Hitchins M, Boyd CA, Johnston DG, Williamson C (2003) Semi quantitative expression analysis of MDR3, FIC1, BSEP, OATP-A, OATP-C, OATP-D, OATP-E and NTCP gene transcripts in 1st and 3rd trimester human placenta. Placenta 24:39–44

14. Pastrakuljic A, Derewlany LO, Koren G (1999) Maternal cocaine use and cigarette smoking in pregnancy in relation to amino acid transport and fetal growth. Placenta 20:499–512
15. Hakkola J, Pelkonen O, Pasanen M, Raunio H (1998) Xenobiotic-metabolizing cytochrome P450 enzymes in the human feto-placental unit: role in intrauterine toxicity. Crit Rev Toxicol 28:35–72
16. Paakki P, Stockmann H, Kantola M, Wagner P, Lauper U, Huch R, Elovaara E, Kirkinen P, Pasanen M (2000) Maternal drug abuse and human term placental xenobiotic and steroid metabolizing enzymes in vitro. Environ Health Perspect 108:141–145
17. Pasanen M (1999) The expression and regulation of drug metabolism in human placenta. Adv Drug Deliv Rev 38:81–97
18. Uszynski M (1992) Heparin neutralization by an extract of the human placenta: measurements and the concept of placental barrier to heparin. Gynecol Obstet Invest 33:205–208
19. Atalla A, Maser E (2001) Characterization of enzymes participating in carbonyl reduction of 4-methylnitrosamino-1-(3-pyridyl)-1-butanone (NNK) in human placenta. Chem Biol Interact 130–132:737–748
20. Pasanen M, Pelkonen O (1994) The expression and environmental regulation of P450 enzymes in human placenta. Crit Rev Toxicol 24:211–229
21. Marin JJ, Briz O, Serrano MA (2004) A review on the molecular mechanisms involved in the placental barrier for drugs. Curr Drug Deliv 1:275–289
22. Parman T, Chen G, Wells PG (1998) Free radical intermediates of phenytoin and related teratogens. Prostaglandin H synthase-catalyzed bioactivation, electron paramagnetic resonance spectrometry, and photochemical product analysis. J Biol Chem 273:25079–25088
23. Azarbayjani F, Danielsson BR (1998) Pharmacologically induced embryonic dysrhythmia and episodes of hypoxia followed by reoxygenation: a common teratogenic mechanism for antiepileptic drugs? Teratology 57:117–126
24. Bauer B, Hartz AM, Fricker G, Miller DS (2005) Modulation of p-glycoprotein transport function at the blood-brain barrier. Exp Biol Med (Maywood) 230:118–127
25. Ushigome F, Takanaga H, Matsuo H, Yanai S, Tsukimori K, Nakano H, Uchiumi T, Nakamura T, Kuwano M, Ohtani H, Sawada Y (2000) Human placental transport of vinblastine, vincristine, digoxin and progesterone: contribution of P-glycoprotein. Eur J Pharmacol 408:1–10
26. Dickinson RG, Fowler DW, Kluck RM (1989) Maternofetal transfer of phenytoin, p-hydroxy-phenytoin and p-hydroxy-phenytoin-glucuronide in the perfused human placenta. Clin Exp Pharmacol Physiol 16:789–797
27. Hill MD, Abramson FP (1988) The significance of plasma protein binding on the fetal/maternal distribution of drugs at steady-state. Clin Pharmacokinet 14:156–170
28. Ward RM (1993) Drug therapy of the fetus. J Clin Pharmacol 33:780–789
29. Gedeon C, Koren G (2006) Designing pregnancy centered medications: drugs which do not cross the human placenta. Placenta 27:861–868
30. Pacifici GM, Nottoli R (1995) Placental transfer of drugs administered to the mother. Clin Pharmacokinet 28:235–269
31. Folkart GR, Dancis J, Money WL (1960) Transfer of carbohydrates across guinea pig placenta. Am J Obstet Gynecol 80:221–223
32. Hahn T, Desoye G (1996) Ontogeny of glucose transport systems in the placenta and its progenitor tissues. Early Pregnancy 2:168–182
33. Moe AJ (1995) Placental amino acid transport. Am J Physiol 268:C1321–C1331
34. Bissonnette JM (1981) Placental transport of carbohydrates. Mead Johnson Symp Perinat Dev Med 18:21–23
35. Unadkat JD, Dahlin A, Vijay S (2004) Placental drug transporters. Curr Drug Metab 5:125–131
36. Fricker G, Miller DS (2002) Relevance of multidrug resistance proteins for intestinal drug absorption in vitro and in vivo. Pharmacol Toxicol 90:5–13
37. Lankas GR, Wise LD, Cartwright ME, Pippert T, Umbenhauer DR (1998) Placental P-glycoprotein deficiency enhances susceptibility to chemically induced birth defects in mice. Reprod Toxicol 12:457–463

38. Smit JW, Huisman MT, van Tellingen O, Wiltshire HR, Schinkel AH (1999) Absence or pharmacological blocking of placental P-glycoprotein profoundly increases fetal drug exposure. J Clin Invest 104:1441–1447
39. Borst P, Evers R, Kool M, Wijnholds J (2000) A family of drug transporters: the multidrug resistance-associated proteins. J Natl Cancer Inst 92:1295–1302
40. Pascolo L, Fernetti C, Pirulli D, Crovella S, Amoroso A, Tiribelli C (2003) Effects of maturation on RNA transcription and protein expression of four MRP genes in human placenta and in BeWo cells. Biochem Biophys Res Commun 303:259–265
41. Leazer TM, Klaassen CD (2003) The presence of xenobiotic transporters in rat placenta. Drug Metab Dispos 31:153–167
42. Scheffer GL, Kool M, Heijn M, de Haas M, Pijnenborg AC, Wijnholds J, van Helvoort A, de Jong MC, Hooijberg JH, Mol CA, van der Linden M, de Vree JM, van der Valk P, Elferink RP, Borst P, Scheper RJ (2000) Specific detection of multidrug resistance proteins MRP1, MRP2, MRP3, MRP5, and MDR3 P-glycoprotein with a panel of monoclonal antibodies. Cancer Res 60:5269–5277
43. Litman T, Jensen U, Hansen A, Covitz KM, Zhan Z, Fetsch P, Abati A, Hansen PR, Horn T, Skovsgaard T, Bates SE (2002) Use of peptide antibodies to probe for the mitoxantrone resistance-associated protein MXR/BCRP/ABCP/ABCG2. Biochim Biophys Acta 1565:6–16
44. Langmann T, Mauerer R, Zahn A, Moehle C, Probst M, Stremmel W, Schmitz G (2003) Real-time reverse transcription-PCR expression profiling of the complete human ATP-binding cassette transporter superfamily in various tissues. Clin Chem 49:230–238
45. Shang T, Uihlein AV, Van Asten J, Kalyanaraman B, Hillard CJ (2003) 1-Methyl-4-phenylpyridinium accumulates in cerebellar granule neurons via organic cation transporter 3. J Neurochem 85:358–367
46. Zwart R, Verhaagh S, Buitelaar M, Popp-Snijders C, Barlow DP (2001) Impaired activity of the extraneuronal monoamine transporter system known as uptake-2 in Orct3/Slc22a3-deficient mice. Mol Cell Biol 21:4188–4196
47. Verhaagh S, Barlow DP, Zwart R (2001) The extraneuronal monoamine transporter Slc22a3/Orct3 co-localizes with the Maoa metabolizing enzyme in mouse placenta. Mech Dev 100:127–130
48. Barros LF, Yudilevich DL, Jarvis SM, Beaumont N, Young JD, Baldwin SA (1995) Immunolocalisation of nucleoside transporters in human placental trophoblast and endothelial cells: evidence for multiple transporter isoforms. Pflugers Arch 429:394–399
49. Ward JL, Sherali A, Mo ZP, Tse CM (2000) Kinetic and pharmacological properties of cloned human equilibrative nucleoside transporters, ENT1 and ENT2, stably expressed in nucleoside transporter-deficient PK15 cells. Ent2 exhibits a low affinity for guanosine and cytidine but a high affinity for inosine. J Biol Chem 275:8375–8381
50. Griffith DA, Jarvis SM (1996) Nucleoside and nucleobase transport systems of mammalian cells. Biochim Biophys Acta 1286:153–181
51. Friedrich A, Prasad PD, Freyer D, Ganapathy V, Brust P (2003) Molecular cloning and functional characterization of the OCTN2 transporter at the RBE4 cells, an in vitro model of the blood-brain barrier. Brain Res 968:69–79
52. Margolis G, Kilham L (1978) Cerebellar, epidermal, and dental defects induced by ribavirin in perinatal hamsters and rats. Exp Mol Pathol 29:44–54
53. Wu X, George RL, Huang W, Wang H, Conway SJ, Leibach FH, Ganapathy V (2000) Structural and functional characteristics and tissue distribution pattern of rat OCTN1, an organic cation transporter, cloned from placenta. Biochim Biophys Acta 1466:315–327
54. Horiuchi M, Kobayashi K, Yamaguchi S, Shimizu N, Koizumi T, Nikaido H, Hayakawa J, Kuwajima M, Saheki T (1994) Primary defect of juvenile visceral steatosis (jvs) mouse with systemic carnitine deficiency is probably in renal carnitine transport system. Biochim Biophys Acta 1226:25–30
55. Ohashi R, Tamai I, Yabuuchi H, Nezu JI, Oku A, Sai Y, Shimane M, Tsuji A (1999) Na(+)-dependent carnitine transport by organic cation transporter (OCTN2): its pharmacological and toxicological relevance. J Pharmacol Exp Ther 291:778–784

56. Wessler I, Roth E, Deutsch C, Brockerhoff P, Bittinger F, Kirkpatrick CJ, Kilbinger H (2001) Release of non-neuronal acetylcholine from the isolated human placenta is mediated by organic cation transporters. Br J Pharmacol 134:951–956
57. Chen SS, Perucca E, Lee JN, Richens A (1982) Serum protein binding and free concentration of phenytoin and phenobarbitone in pregnancy. Br J Clin Pharmacol 13:547–552
58. Mucklow JC (1986) The fate of drugs in pregnancy. Clin Obstet Gynaecol 13:161–175
59. Economides D, Braithwaite J (1994) Smoking, pregnancy and the fetus. J R Soc Health 114:198–201
60. Sastry BV (1991) Placental toxicology: tobacco smoke, abused drugs, multiple chemical interactions, and placental function. Reprod Fertil Dev 3:355–372
61. Ostrea EM Jr (1999) Testing for exposure to illicit drugs and other agents in the neonate: a review of laboratory methods and the role of meconium analysis. Curr Probl Pediatr 29:37–56
62. VanBlerk GA, Majerus TC, Myers RA (1980) Teratogenic potential of some psychopharmacologic drugs: a brief review. Int J Gynaecol Obstet 17:399–402
63. Ostrea EM Jr, Hernandez JD, Bielawski DM, Kan JM, Leonardo GM, Abela MB, Church MW, Hannigan JH, Janisse JJ, Ager JW, Sokol RJ (2006) Fatty acid ethyl esters in meconium: are they biomarkers of fetal alcohol exposure and effect? Alcohol Clin Exp Res 30:1152–1159
64. Whyatt RM, Barr DB (2001) Measurement of organophosphate metabolites in postpartum meconium as a potential biomarker of prenatal exposure: a validation study. Environ Health Perspect 109:417–420
65. Little BB, VanBeveren TT (1996) Placental transfer of selected substances of abuse. Semin Perinatol 20:147–153
66. Barr DB, Wang RY, Needham LL (2005) Biologic monitoring of exposure to environmental chemicals throughout the life stages: requirements and issues for consideration for the National Children's Study. Environ Health Perspect 113:1083–1091
67. Rodier PM (1994) Vulnerable periods and processes during central nervous system development. Environ Health Perspect 102(Suppl 2):121–124
68. Abbott N (2002) Astrocyte-endothelial interactions and blood-brain barrier permeability. J Anat 200:527
69. Abbott NJ (2005) Dynamics of CNS barriers: evolution, differentiation, and modulation. Cell Mol Neurobiol 25:5–23
70. Sinha C, Agrawal AK, Islam F, Seth K, Chaturvedi RK, Shukla S, Seth PK (2004) Mosquito repellent (pyrethroid-based) induced dysfunction of blood-brain barrier permeability in developing brain. Int J Dev Neurosci 22:31–37

In Vivo Imaging of Brain Development: Technologies, Models, Applications, and Impact on Understanding the Etiology of Mental Retardation

Vicko Gluncic

Abstract Development of the mammalian brain proceeds in a precisely controlled sequence of cell divisions, migration, differentiation, and synaptogenesis. It is a process of precise dynamic assembly, and time lapse in vivo imaging of these processes is fundamental for the multidisciplinary endeavor to merge and understand the morphological, physiological, and regulatory processes of neurogenesis.

Modern optical and non-optical imaging technologies enable us to achieve 5-dimensional (5D) imaging of neurogenesis: 3-dimensional (3D) images of neuronal structures, collected over time (4D) in the living tissue or specimen, with simultaneous spectral information and data indicating specific structures, cellular phenotypes, functions, or genes expression patterns (5D). These imaging technologies make it possible to simultaneously monitor structure and function in vivo and merge morphological and physiological perspectives on brain development.

Modern in vivo imaging technologies combined with advances in molecular biology definitely have the potential to unravel basic mechanisms of neurogenesis. Further advances will enable these technologies to elucidate the pathogenesis of developmental neurological disorders and eventually make critical steps of neurogenesis clinically detectable and therapeutically approachable.

Keywords Time laps in vivo imaging • Neurogenesis • Microscopy • Optics • Non optical imaging technologies • Functional imaging • Developmental neurological disorders

1 Introduction

Brain development proceeds in precisely controlled sequence of cell divisions, migration, differentiation, and synaptogenesis, which is well preserved across mammalian species. Precise deposition of neurons into horizontal and vertical arrays

V. Gluncic, MD, PhD
Section of Neurosurgery, The University of Chicago Medical Center, University of Chicago, 5841 South Maryland Avenue, Chicago, IL, 60637, USA
e-mail: vicko.gluncic@gmail.com

D. Janigro (ed.), *Mammalian Brain Development*, Contemporary Neuroscience, DOI 10.1007/978-1-60761-287-2_9,

during neurogenesis, define their connectivity and ultimately the normal function of the mature brain [1–6]. Due to the complexity of cellular and molecular interactions, neurogenesis is highly sensitive to a variety of biological, physical, and chemical agents, as well as to specific genetic mutations [1, 6–15]. When the rate of neuronal proliferation, migration, or subsequent development of axonal projections is distorted, various consequences, including movement abnormalities, epilepsy, mental retardation, and abnormal behavior, have been observed [4, 6, 16–22].

New imaging technologies quickly became an essential tool in developmental neurobiology allowing in vivo observation of the neurogenetic processes, at length-scales from molecules to cortical regions, and over time-scales from milliseconds to days [15, 23]. Given that development of the brain is essentially a process of precise dynamic assembly, this vantage was fundamentally important. It has enabled us to develop a realistic model of how the sequence of gene expression and cascade of multiple molecular pathways establish the number of neurons, guide their allocation into correct areas and layers, determine their differentiation into specific phenotypes, and establish synaptic connections and columnar organization in the developing brain [1, 6]. In vivo imaging combined with advances in molecular biology provided the possibility to unravel basic mechanisms of neurogenesis and elucidate the pathogenesis of developmental neurological disorders.

2 Rational for In Vivo Imaging

Initially, the process of brain development was reconstructed from microscopic images of fixed brain tissue, taken at various points in development [6]. However, neurogenetic processes couldn't be entirely understood from static images. The microscope evolved from a tool used to visualize static histological specimens to one that allows prolonged observation of living cells [15]. Live imaging is essential to determine dynamic parameters or stability of a structure. It offers a unique insight into how neurons in the intact nervous system change over time in relation to different manipulations or behavioral adaptations [1, 15, 23–38] The major advantages of in vivo imaging are:

(a) Only direct observation reveals the dynamism of neurons and developing brain.
(b) Observation of the same neurons over time is more sensitive than assuming how neurons change from static images of fixed tissue, taken from different specimens at various time points.
(c) Observation of the sequence of neurogenetic events can establish their causative relationship.
(d) In vivo imaging of the same area of the developing brain can be compared before, through, and after certain molecular manipulation, media changes or therapeutic interventions.

Furthermore, recent advances of in vivo imaging technologies enabled us to simultaneously monitor structure and function ultimately merging morphological and physiological perspectives on brain development [15, 23, 29, 33].

3 In Vivo Imaging Technologies

In the studies of brain development, we preferably aim to achieve 5-dimensional (5D) imaging: 3-dimensional (3D) images of structure, collected over time (4D) in the living tissue, with simultaneous spectral information and data indicating specific structures, functions, or genes expression patterns (5D) [23, 33]. In addition, we would like to simultaneously image and refocus from the level of the whole brain area to the cellular level. This requires a combination of high spatial and temporal resolutions. In general, high spatial resolution, generally 1 μm or better, is required for imaging neurons. Some applications also require simultaneously high temporal resolution of 1 ms or better. Techniques such as magnetic resonance imaging (MRI), computerized tomography (CT), high resolution ultrasound, and confocal and multiphoton microscopy combined with advanced neuronal labeling techniques are able to provide high-spatial and/or temporal resolution images of the developing brain in vivo. However, technologies with higher 3D spatial resolution take longer to collect images, thereby creating a trade off between spatial and temporal resolution. The fact that these technologies vary widely in their degree of spatial and temporal resolution is the major obstacle in achieving 5D imaging (Fig. 1). Other limitations of in vivo imaging are that specimens generally need to

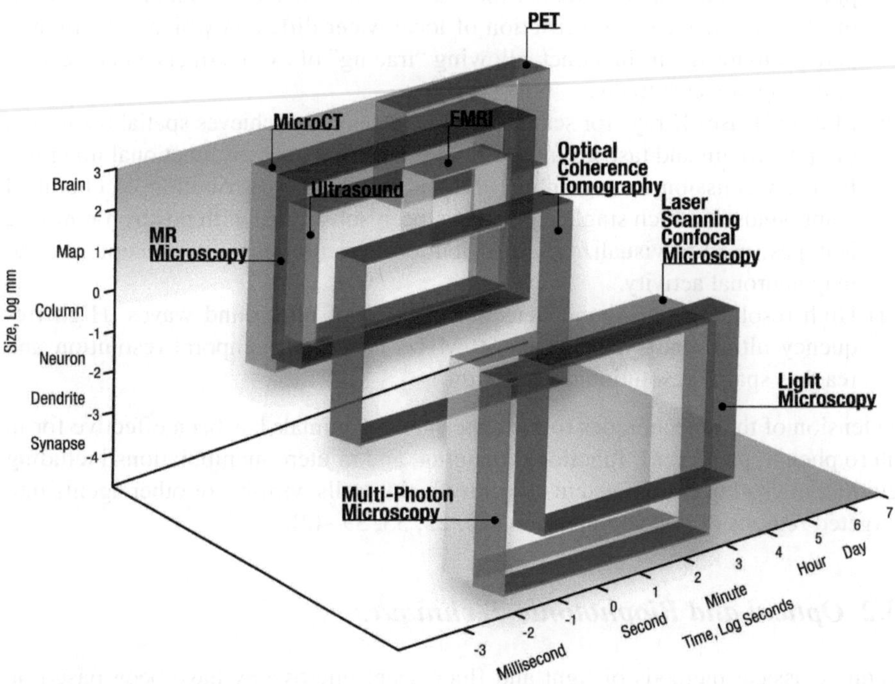

Fig. 1 Temporal and spatial resolution of imaging techniques in developmental neurobiology (PET-positron emission tomography, MR-magnetic resonance, FMRI-functional magnetic resonance imaging, CT-computerized tomography) [23, 33]

be immobilized while microscopy techniques are limited in depth of penetration into the specimen. Further improvements and development of methods that will use simultaneous combination of existing technologies are expected to overcome these limitations [29].

3.1 Non-optical Imaging Techniques

Several technological breakthroughs in the 1970s have revolutionized medicine and research in developmental neurobiology [15, 23, 33]. These include:

(a) Magnetic resonance imaging (MRI) detects the relaxation properties of hydrogen atoms in magnetic fields to generate 3D views of the inside of the brain. Because of the smaller size of the experimental animals, stronger magnetic fields can be used and higher resolution (micro-MRI) can be achieved. Contrast-enhanced MRI uses compounds with a high MRI contrast such as gadolinium or super-paramagnetic iron. In addition, manganese can be introduced to neurons and can replace calcium. It can be used to track axonal connections or map neuronal activity. Functional MRI detects changes in blood flow and hemoglobin oxygenation in response to altered neuronal activity and thus, allows mapping of the functional centers of the brain. Diffusion tensor imaging is a variant of MRI that detects the restriction of local water diffusion which is facilitated parallel to the axons in a tract, allowing "tracing" of axonal tracts and detection of damage to such tracts.
(b) Micro-CT uses X rays for scanning. This technology achieves spatial resolution of up to 10 μm and fast scanning velocity with potential for functional imaging.
(c) Positron emission tomography (PET) measures the distribution of chemical compounds in which single atoms have been substituted with positron-emitting isotopes; used to visualize the distribution of neurotransmitter receptors or to map neuronal activity.
(d) High resolution ultrasound detects reflection of ultrasound waves. High frequency ultrasound systems (40–60 MHz) have high temporal resolution and reaches spatial resolution up to 40 μm.

Extension of these techniques to the experimental animals has been effective for in utero phenotype screens, functional imaging, and in utero manipulations including guiding micro-injections systems for introducing cells, viruses, or other agents into targeted regions of developing brain [15, 23, 33, 39–42].

3.2 Optical and Biophotonic Techniques

While classical methods of light and fluorescent microscopy have been based on simultaneous illumination of the entire field of view, new techniques move a focus beam of light across the field of view to detect the signal of each spot sequentially and reconstruct the entire picture. Techniques used in developmental neurobiology for in vivo microscopy include the following:

(a) Confocal laser scanning microscopy (CLSM) is a technique for optical sectioning which uses a laser to focus light down to a cone through the specimen [23, 29, 33, 43–45]. The emitted fluorescence light is then sent back through a pinhole, which corresponds to the focal point in the specimen, thus blocking out light from other focal planes. By scanning the focal point within a plane, an image of a single plane within the tissue can be recreated (Fig. 2). Stepping the focus through the specimen builds up a 3D image. It requires high-intensity laser excitation to generate enough photons that can escape from the tissue which results in the damage both to the fluorophores, resulting in photobleaching, and to the viable tissue, resulting in production of free radicals and phototoxicity. These significantly limit the signal intensity, length of time-lapse imaging, and

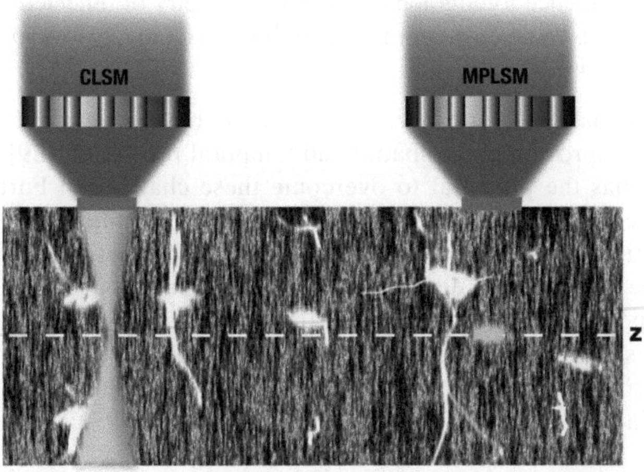

Fig. 2 Confocal laser scanning microscopy (CLSM) utilizes visible lasers to focus light down to a cone through the specimen. The emitted fluorescence is then sent back through a pinhole, which corresponds to the focal point in the specimen, thus blocking out fluorescence from other focal planes. An optical section can be created by scanning the focal points within the single plane while collection of consecutive planes builds up a three-dimensional image. Although the laser beam is focused only in the focal plane, cells in the path of the beam are irradiated and fluorofores above and below are excited. Continuing excitation and absorption of high energy photons by cell organelles led to photobleaching and phototoxic damage to living neurons. In the multiphoton laser scanning microscopy (MPLSM), fluorophores are excited by ultrafast high-intensity lasers in the near infra red spectrum of the wavelengths. This means that each photon carries approximately half of the energy of the photons of the wave lengths in the visible spectrum. The fluorophore absorbs two or more photons simultaneously and emits fluorescence. Because the probability of such an event depends on the square of the intensity, excitation drops off rapidly away from the focal point of the laser. The three dimensional image of the specimen can be reconstructed in a process similar to confocal imaging. The major advantages are (1) longer wave length photons can penetrate deeper in the tissue then photons from visible wave lengths, (2) since the image is generated by the non linear excitation which is limited to the focus, all the emitted photons that come from the focus, meaning that even if the fluorescence is scattered on the way out of the tissue, it can be collected to form an image (3) since the excitation is restricted to the focal point there is no photodamage to the parts of the specimen through which the laser beam passes or near focal point [23, 24, 29, 33]

viability of the tissue. The solution in overcoming these problems is the development of multiphoton microscopy based on excitation by simultaneous absorption of two or more comparably lower energy photons [24, 29, 33].

(b) Multiphoton laser scanning microscopy (MPLSM) microscopy is a scanning technique which uses ultra fast pulse lasers to achieve optical sectioning by limiting multiphoton excitation to the small volume of perifocal plane. It became dominant in vivo microscopy technique since high wave length lasers can penetrate deep into the tissue and the phototoxicity is limited only to the focal plane (Fig. 2) [24, 29–31, 36, 45, 46].

(c) Second-harmonic generation microscopy uses near-infrared lasers to generate frequency-doubled signals from the scattering light as it interacts with macromolecules allowing imaging of the intracellular structures such as the cytoskeleton [15, 23, 33].

(d) Optical coherence tomography uses interferometry of reflected light from biological tissues to reconstruct the image. It has a relatively low spatial resolution [23, 45, 47–49].

The main challenges for in vivo microscopy have been reduction of phototoxicity and improvement of spatial and temporal resolution [29]. Multiphoton microscopy has the potential to overcome these challenges. Further improvements of the ultra fast pulse lasers, applications of higher wave lengths, and miniaturization of objective lenses that can be inserted into the living brain will increase the penetration depth and scanning speed [46]. Additionally, the increasing power of image processing, highly sensitive detectors, introduction of objectives with high working distance and numerical aperture, and automation of microscope controls will continue to facilitate acquisition of video sequences in microscopy.

3.3 Advances in Labeling Methods

In addition to the development of the new imaging technologies, in vivo microscopy has depended on the advent of new tools to label neurons. New labeling techniques increased the ability to distinguish between structures and provide information about gene expression or the physiological state of the tissues. New labeling tools are developing to monitor neuronal cell structure, molecular biology and activity [23, 29, 50, 51].

3.3.1 Fluorescent Membrane Dyes

Vital dyes can stain living cells or extracellular components of the nervous tissue. They allowed some of the early in vivo imaging studies to be carried out. These dyes incorporate into and diffuse laterally within plasma membranes of neurons. The rapid and complete surface labeling is especially useful for determining the morphology of cells such as neurons that have very long and elaborate branching processes. A useful property of membrane dyes is that they can label cells in both living and fixed tissues. In brain slices, adequate levels of staining can be achieved in neuronal processes over 1 mm away from the labeling site within a few hours. Cell bodies of projection neurons can be back-labeled by injecting a tracer into target regions. Alternatively, axonal projections can be labeled by applying dye to the region of cell bodies, to dendrites, or along known axonal tracts. Vital fluorescent dyes have been extensively utilized for in vivo imaging in brain slices [1, 23, 27, 29].

A new approach to obtain a Golgi-like labeling of neurons has been developed. This technique is a modification of a gene-gun transfection technique. Tiny gold or tungsten particles coated with one or more fluorescent lipid dyes are propelled into brain tissues using a "gene gun." Neurons whose surfaces are contacted by dye-coated particles become labeled including the entire axonal and dendritic arbors, together with synaptic spines. Using particles coated with different dyes, it is possible to label and distinguish many neurons within a well defined tissue region [29, 52, 53]. In addition, fluorescent lipophilic dyes enable us to observe cells in vivo under the multiphoton microscope and subsequently examine the same fixed cells in the electron microscope at the subcellular level [52].

3.3.2 Immunofluorescent Techniques

Antibodies are powerful tools for labeling specific populations of cells and subcellular structures in neural tissues, but their use is typically limited to the fixed specimens. The availability of a wide spectrum of fluorescent probes combined with multiphoton or confocal imaging systems with multiple laser lines permits double- and triple-label immunohistochemical analyses of brain tissues. Use of longer wavelength fluorophores may improve the detectable fluorescence signal from deeper portions of the specimen. Combination of immunohistochemical staining combined with fluorescent membrane dyes in brain slices opens new possibilities for in vivo imaging [15, 23, 29, 33, 54].

3.3.3 Fluorescent Quantum Dots

The development of fluorescent quantum dots (semi-conductor nanocrystals) shows great promise for in vivo microscopy [23, 55, 56]. Available fluorescent dyes fade quickly and are subject to photobleaching from laser excitation, while multi-label preparations must be excited at multiple wavelengths to produce multicolored emission spectra. Quantum dots can persist for weeks without harming living tissue by phototoxicity. They are resistant to photobleaching, and different-sized crystals (1–10 nm) produce a range of colors with excitation at just one wavelength. Quantum dots are an especially interesting tool for in vivo imaging since they can be also hybridized with specific DNA sequences [23].

3.3.4 Voltage Sensitive Dyes

Recently developed dyes for cellular physiology imaging include both fluorescent and non-fluorescent voltage sensitive probes of intracellular calcium, magnesium, and several other molecules. Fluorescent calcium probes are especially useful since large changes in intracellular calcium are associated with neural electrical activity [23, 45, 57–62].

The most reliable way of introducing the physiological probes into neurons is direct intracellular injection. Microinjection allows selection of individual cells of interest and control of the intracellular dye concentration. This approach has been successfully used to image intracellular calcium at high spatial and temporal resolution within neuronal cell bodies, dendrites, and individual dendritic spines thus enabling studies of brain plasticity in vivo [23, 57, 58, 63, 64]. A novel technique which stain large populations of neurons with calcium dyes in vivo, called multicell bolus loading, has been recently developed to visualize calcium transients and monitor activity within specifically labeled regions of developing brain [23, 59, 62, 65, 66].

3.3.5 Molecular Genetic Tools

The complexity of neural tissues makes it difficult to map the organization of synaptic structures or label specific cell populations by using classical fluorescent labeling methods [6, 15, 23, 33]. Molecular genetic tools allow us to label defined populations of cells or specific subcellular structures. Genes derived from marine organisms that produce fluorescent proteins are essential for these techniques. A variety of approaches have been used to introduce foreign genes into the cells of brain slices in vitro or embryonic brain in vivo. These include viral constructs and non-viral transfection methodologies such as particle-mediated biolistics, liposome-mediated transfection, and electroporation. Transgenic techniques make it possible to coexpress a label (reporter) gene along with the gene of interest [29, 39, 51, 67, 68]. These fusion genes express combination of the two proteins in which the fluorescent protein works as a "tag" that makes the tagged protein visible in fluorescent light [29, 69, 70]. Under the control of cell type- or region-specific regulatory genes, fluorescent proteins can be used as a reporter system to label specific neuronal populations or subcellular targeting sequences [15, 23, 29, 45, 50, 51, 68, 69, 71–73]. Important advantage of this approach is that even proteins such as actin, which is highly sensitive to changes in its chemical structure, appear to work normally inside the nerve cells when joined to fluorescent proteins [25, 74]. For example, green fluorescent protein fused to a synaptic protein results in specific labeling of pre- or postsynaptic structures. This offered the possibility of mapping development of synaptic structures at the single-cell level [15, 33, 69, 70, 72, 73].

The breakthrough for in vivo imaging was the generation of transgenic mice that express high levels of fluorescent proteins in neurons [29, 70, 71]. Furthermore, by using mice lines with conditional expression, fluorescent neurons can be restricted to specific regions of the developing brain, and their projection tracts can be selectively marked. Currently, there are transgenic models for tracing many structures in the developing nervous system, including specific neuronal phenotypes, axon tracts, glial cells, and brain vasculature [29, 70].

4 In Vivo Imaging Models

Separate analysis of the biological mechanisms underlying each step in the brain development provides a reductionist approach for which well-defined, simplified, in vitro systems are crucial. However, more complex and closer to in vivo experimental models are necessary to fully understand neurogenesis [6, 15, 23, 33]. The advances in imaging technologies, neuronal labeling techniques and mouse genetics enabled us to explore neurogentic processes in living models at the organ, cellular, and molecular level. In vivo and in vitro imaging models include dissociated neuronal cultures, acute brain slices, and organotopic brain slice cultures. In vivo imaging in living animals include imaging of the brain in the early postnatal living mice and in vivo in utero technique of imaging microsurgically exposed brain of the living fetus in pregnant anesthetized mouse [1, 6, 29].

4.1 Dissociated Embryonic Brain Cell Cultures

In developmental neurobiology, cultured neurons have been used in a wide range of in vivo imaging studies, including growth cone dynamics, and growth factor signaling pathways [6, 53, 74–77].

The advantage of neuronal cultures over the other systems is complete control of the neuronal environment, which allows specific manipulations including the use of factors that regulate neuronal function and development. In addition, there are several methods for transfecting DNA into cultured neurons which allow genetic manipulations [68, 78]. Cultured neurons are optically easily accessible and can be fixed for subsequent immunofluorescent and electron microscopy analyses after in vivo and in vitro imaging [33, 53]. However, cultured neurons cannot be expected to provide definitive answers how neurons differentiate and grow in vivo. Since the neurons have been removed from their natural surroundings, dissociated, and grown on the artificial substrate of a two-dimensional glass coverslip with an externally supplied medium, many processes that depend on normal molecular gradients, cell-cell adhesion, early synaptic activity, or simply the presence of a three-dimensional extracellular matrix are disturbed in vitro. Intrinsically, inevitable problems with dissociated embryonic brain cell cultures are cellular diversity with subsequent problems identifying the phenotype of studied cells and distortion of cell axons and dendritic processes [28, 29, 33, 45].

4.2 Embryonic Brain Slices

Tissue slices derived from embryonic or early postnatal brain become a golden standard for the studies of neurogenesis and development of neuronal circuits.

Since they preserve the complex cytoarchitecture of the embryonic brain, display excellent cellular differentiation, and are relatively easy to prepare, brain slices have become the dominant model for in vivo microscopy [6, 28, 29, 33, 79, 80].

Since acute slices are short-lived, the need for in vitro preparations that can be used for long-term studies led to development of organotypic slice cultures. The initial steps in the preparation of acute slices and cultures involve microsurgical isolation and cutting of the embryonic brain into sections of 100–400 μm in thickness. Techniques for preparation of the organotopic brain slice cultures differ with respect to how the cultures are embedded and maintained, thickness of the slices and the time that slices survive in culture. In addition, organotopic slice cultures offer unique opportunities for developmental studies of molecular gradients and culturing the brain tissue derived from mutant animals with limited survival time. Fetal brain slices survive well on cultures, but the organotypic organization of tissue often becomes distorted since the majority of nerve cells are still in their migratory phase. The early postnatal period is ideal for preparation of organotopic slice cultures because the essentials of the cytoarchitecture are already established, the brain is larger and easier to dissect, and the nerve cells survive explantation more readily [1, 28, 29, 33, 45]. The major advantages of tissue slice preparations are greater optical and physiological accessibility over in vivo in living animal conditions and the maintenance of structural and functional integrity over dissociated cell culture preparations. Another advantage of the slice preparation is the possibility of long (>48 h) culturing times. Experiments based on immunohistological markers of cell death indicated minimal apoptosis in organotopic slice cultures up to 72 h in controlled in vitro conditions. The main disadvantages are structural and molecular gradients distortion due to the tissue isolation procedure, physiologically loss of extrinsic connections which cause isolated tissues to cease normal activity, and a finite period of time before tissue deteriorates [1, 28, 29].

4.3 Imaging in the Living Experimental Animals

The emerging strategy in developmental neurobiology is to study neurogenetic events in the brain of the living experimental animals [1, 6, 15, 23, 29, 33, 42]. Recently, several groups started exploring the brain of living embryos or early postnatal rodents with multi photon microscopy. The breakthroughs were in vivo imaging of structural dynamics of dendritic spines in the brain of early post natal animals and in vivo in utero imaging of neuronal migration in the murine embryonic cortex. In these studies minimally invasive microsurgical procedures under anesthesia were used to expose brains of living experimental animals to the multiphoton microscope objective [1, 42].

The main disadvantages of in vivo imaging in live animals are lack of consistent models, complexity of surgical procedures and anesthesia during the imaging, complex adjustment of existing multiphoton microscopes for imaging in vivo, and imaging artifacts due to physiological motions. The main advantages are preserved structural and molecular gradients including extrinsic neuronal connections, the possibility to simultaneously apply different modalities of imaging and analyze neurogenetic events on a cellular and organ level [1, 23, 33, 42].

5 The Impact of In Vivo Imaging on Studies of Cortical Development and Pathogenesis of Neurological Disorders

Early studies on human patients and animal models of neurological diseases have shown that disruption in brain development leads to abnormal behavior. This is the fundamental concept of the neurodevelopmental hypothesis of neurological disorders [2, 6, 7, 10, 13, 15, 18, 19, 29, 81, 82]. Subsequently, studies of neurological disorders started to focus on brain development.

More precisely, a basic structural unit of the cerebral cortex is the minicolumn, a radial array of cells through the thickness of the cortex which contains approximately constant number of cells in mammals. Neuronal proliferation, migration, and synaptogenesis are basic processes in cortical columns formation and maturation. Several recent studies demonstrated disturbances in the cytoarchitecture of the cortical columns in neurological disorders. This realization that the relatively subtle structural abnormalities of cortical columns which can not be detected by MRI, CT, ultrasound or standard pathohistological analyses might underlie neuropsychiatric disorders further propelled morphological and physiological microscopic studies of the brain development in vivo [3, 6, 16, 18, 54, 83–85]. The inaccessibility of the embryo to the spatial and temporal requirements adequate for in vivo microscopy was a major obstacle for the morphological and functional studies of neurogenesis in the complex environment of the developing brain in living experimental animals. This barrier has been overcome by new imaging technologies [1, 15, 29].

Future studies will eventually be able to correlate abnormalities in neuronal proliferation, migration, and synaptogenesis with specific neurological abnormalities [15, 33].

5.1 Neuronal Proliferation

The initial phase of brain development is the exponential proliferation of neuronal stem cells in the ventricular zone (pseudostratified neuroepithelium that surround prospective lateral ventricles) beginning at the neural plate stage and lasting until the beginning of neurogenesis. Neurogenesis begins as the first neurons generated from the ventricular zone start to migrate to the developing cortex. The relatively long duration of neurogenesis and complexity of produced subclasses of neuronal progenitors contribute to increased cortical surface and cognitive function in primates. During these initial stages of brain development there are two main modes of cell division: symmetrical which gives two similar daughter cells with the same proliferative fate and asymmetrical division which give one daughter cell which remains to proliferate and another that differentiates into a neuron and migrates to cortex. Developmental regulation of the rate and ratio of symmetrical and asymmetrical divisions control the size and complexity of the cerebral cortex [2, 3, 6, 10, 27, 29, 69, 86–91]. Despite the critical role of proliferate events in generating cortical grow, the mechanisms regulating the rate and transition between modes of division have been unexplored.

Several recent studies based on multiphoton microscopy in viable brain slices focused on mechanisms of neuronal proliferation and established parameters to which results from mutant animal models or manipulated systems can be compared [27, 86, 92–94]. More precisely, since the orientation of the mitotic plate before the cleavage is one mechanism for specifying the mode of division, initial studies of neuronal progenitor cells have correlated changing cleavage plane angles with cell division mode outcomes. They showed that vertical division cleavage planes, perpendicular to the ventricular surface, identify symmetrical divisions, while horizontal cleavage planes parallel to the ventricular surface, characterize asymmetrical divisions. Towards the end of the neurogenesis prevailing pattern of cell division become asymmetrical divisions while symmetrical divisions significantly decrease. In addition, in vivo multiphoton imaging has enabled us to analyze sequences of rapid molecular and cytoskeleton rearrangements that regulate onset, rate, and duration of proliferation of neuronal progenitor cells [27, 29, 33, 87, 88, 92, 93]. These studies provided the first video sequences of these processes in vivo and opened new avenues of research to reveal abnormalities of neuronal proliferation in animal models of neurological disorders [15, 27].

5.2 Neuronal Migration

The new imaging technologies have led to several breakthroughs in understanding the sequences of intra- and extracellular processes that control neuronal migration [1, 4, 26, 28, 29, 34, 35, 95–99]. Considering place of origin and directionality of movement, neuronal migration to the developing cortex can be classified into radial (proceeding radial from the ventricular to the pial surface) and tangential (running parallel to the brain surface). As new neurons arrive, developing cortex expands into mature six-layered cortex in an inside outside manner: the earliest generated neurons form the deepest layers and each new round of young neurons pass older neurons to occupy progressively superficial layers. Cortical neurons generated at the same point of time eventually occupy the same cortical layer regardless of mode of their migration or neuronal phenotype. Contact interaction between migrating neurons and the surfaces of neighboring cells play a decisive role in this highly precise process, selecting the migratory pathway and determining their final position. Misplacement of neurons leads to impaired differentiation and synaptic development and may possibly result in neurological deficits. For example, repeated exposure of pregnant rodents and primates to environmental agents, such as alcohol, drugs, neurotrophic viruses, ionizing irradiation or occurrence of specific genetic mutations have well documented potential to cause disruptions in neuronal migration and eventually cause cognitive impairment [1, 4, 6, 20, 22, 29, 90, 97, 99].

Early applications of confocal microscopy in rodent brain slices have elucidated the role of spontaneous intracellular calcium fluctuations in regulation of the neuronal migration [100, 101]. The potential of in vivo microscopy have been recognized and subsequent in vivo studies have described several different modes of migration and showed that the mode and rate of neuronal migration change throughout the time course of cortical development. Further studies based on multiphoton imaging in viable brain slices established normal parameters of neuronal migration dynamics. The latest studies focused on how tangentially migrating interneurons, which have to pass much larger distances then radialy migrating neurons, find their position in the developing cortex. Study by Ang et al. based on multiphoton microscopy, was the first documented in vivo in utero imaging of tangential neuronal migration in the murine embryonic cortex. These video sequences confirmed that areal and laminar positioning of interneurons is a highly organized process. They also revealed that interneurons' growth cones sense local molecular and morphological clues in the local environment during the final phase of tangential migration which implicated that the molecular coordinates for the positioning of the interneurons are contained within the appropriate layer of developing cortex [1, 29, 34, 35, 102].

The future expectation of in vivo studies is to focus on abnormalities of neuronal migration in animal models of neurological diseases and elucidate the underlying pathogenetic mechanisms [6, 15, 29].

5.3 Development of Neuronal Connections

The intensive growth of axonal projections and formation of synapses starts relatively early in brain development while positioning of the migrating neurons is still ongoing. Some of these projections play a regulatory role in the cortical development [54, 60, 103–107]. Newly arrived neurons in the developing cortex continue to express their phenotypes which have been determined by place and time of origin, and mode of migration. They start to extend dendrites and spines to receive connections and extend their axons towards appropriate targets and subsequently make synapses. Disturbances in the processes of neuronal proliferation or migration lead to malpositioning of the neurons within the cortex which might prevent formation of the appropriate connections and adversely affect mature brain function [16–18, 54, 105–110].

Studies of synaptogenesis were the earliest applications of in vivo microscopy [15, 33, 111]. They revealed significant discrepancies among the results obtained in vitro in cell cultures, in embryonic or early postnatal brain slices, and results obtained in brains of living animals [33, 63, 103, 109, 111–113, 115]. Although previous in vitro studies implicated that action potentials propagate into dendritic tree, studies based on confocal laser scanning microscopy and voltage sensitive dyes in brain slices revealed that action potentials remained more localized to the cell soma [32, 37, 116]. Subsequent multiphoton studies focused on synaptic development and maintenance or neuronal networks in the brain of living animals. Contrary to the previously dominant hypothesis that the synaptic contacts and network are relatively stable [117], these studies showed that the dendritic spines in the brain of early postnatal animals are constantly reconfiguring, forming de novo or disappearing within minutes [37, 63, 107, 109, 111, 112, 114, 118–120]. They clearly showed that spine morphology rapidly changes with the level of target innervations and stimulation and that sensory deprivation significantly decrease the dynamic of spine and dendrites formation [37, 38, 64, 66, 104, 121, 123, 124]. These studies thus confirmed that synaptogenesis is a highly dynamic process, and indicated that sensory experience drives structural plasticity of dendrites.

More recently studies based on multifoton microscopy started to simultaneously study neuronal morphology and physiology at the whole neuron level, at the level of dendrites, and dendritic spines in vivo [50, 63, 65, 77, 107, 109, 112, 113, 115, 118, 120, 125]. These studies might be crucial for our understanding of early neuronal circuits development and memory organization.

5.4 Advanced Microscopic Imaging in Animal Models of Neurological Diseases

Several mutant and pharmacological animal models have been implemented in the studies of genetic, molecular, and cellular causes of developmental neurological disorders. Although it is difficult to recapitulate the complex neurological deficits of human patients, these models are nevertheless powerful tools to analyze pathogenetic mechanisms at any stage of brain development [13, 15, 19, 29, 42, 126, 127].

Since many of these animal models are characterized by abnormal cytoarchitecture of cortical columns, recent in vivo studies based on multiphoton microscopy have been focusing on dynamic and morphological parameters of neuronal precursor expansion, migration and synaptogenesis which are essential for columnar formation [1, 15, 23, 29, 34, 35, 41, 54]. Recent findings of consistent structural columnar abnormalities in the several developmental neurological disorders in human patients and in their corresponding animal models additionally bolstered this research [6, 15–18, 21, 42, 54].

5.5 Functional Imaging in Developmental Neurobiology

Unlike computers, which are assembled before they are ever turned on, the nervous system is active at early stages of development. New advances in functional imaging enabled us to study early neuronal networks whose pattern of activity change with respect to the changing structure of the differentiating neuron or circuit. In fact, it appears that such incipient activity play an important role in shaping patterns of connectivity and cortical maturation [23, 104, 122, 128]. Although the presence of spontaneous patterned neuronal activity has been observed in pioneering electrophysiological studies in embryonic brain slices, much of what we know about the origins of these receptor-independent electrical activity has recently come from functional studies based on combination of multiphoton microscopy and voltage sensitive dyes [32, 66, 104, 113, 122, 128, 129].

Numerous studies have utilized Ca^{2+} imaging in early postnatal brain slices to study the development of neural circuits in brain [32, 62, 66]. Additionally, astrocytes also exhibit sizable intracellular fluctuations in calcium and a variety of stimuli can induce waves of calcium activity within glial cell networks. Such transglial calcium signals have been proposed to have an important role in long range cellular signaling within the developing brain. Studies based on multiphoton microscopy and voltage sensitive dyes have definitely demonstrated that synaptic activity plays an important role in the expansion of the dendritic arbors while spontaneous calcium oscillations also have a regulatory role in neuronal differentiation and growth cone guidance [32, 58–60, 66, 104, 114, 122, 128, 129].

Elucidating the role of these processes in cortical development might be crucial for understanding how immature neuronal spontaneous activity patterns solidify into mature neuronal network and eventually lead to normal cognitive function and memory [15, 23]. Ultimately, modern optical functional imaging will lead to the integration of morphological and physiological studies of neuronal network assembly.

6 Future Directions

There are two challenges for in vivo imaging of brain development. The first is to elucidate molecular pathways that underlie neurogenesis. Molecules ultimately mediate the pathogenesis of developmental neurological diseases and offer promising targets for possible therapeutic interventions. The second challenge is to bridge the gap between in vivo microscopy and the macroscopic phenotype of experimental animals and translate these results into clinical setting [6, 15, 23, 29, 82, 127].

In the field of non optical imaging technologies, progress is expected from the new MRI and PET technologies. High-resolution MRI is reaching spatial resolution sufficient to capture cortical columnar organization. Further advances in physics will enable MRI to detect magnetic fields induced by dendritic currents and detect changes associated directly with neuronal activity. Functional MRI spectroscopy will allow us to investigate molecular pathways in vivo. Most recently, markers for the several cellular and molecular mediators, that can be detected by MRI or by PET, have been generated. Some of these agents are detectable by microscopy what opens the possibility to correlate macro and molecular events in vivo [15, 23].

Further development of in vivo microscopy requires a multidisciplinary approach combining advances in molecular biology, physics and chemistry. Multiphoton microscopy will remain the method of choice for in vivo microscopy with deeper penetration into tissue and reduced photodamage [30, 46]. High wave length ultra fast pulse lasers will enable multiphoton microscopy to use intrinsic autofluorescence signals and not to depend on fluorescent molecular markers [15, 23, 29]. In addition, fiber-optic-based head-mounted two-photon systems that allow imaging in freely moving animals have been recently developed [31]. These methods have the potential to analyze the neurogenesis in a completely non invasive manner. In order to simultaneously create image structure and function in living cells, new applications based on the photo bleaching effect of laser on the fluorescent proteins are developing [15, 29, 30, 33, 50].

New fluorophores are developing to complement advances in imaging. New labeling techniques which can be simultaneously used for light, multiphoton, and electron microscopy, will enable simultaneous analyses at organ, cellular and subcellular level. The development of the new fluorescent probes that can be photoactivated with multiphoton excitation will further elucidate the role of neurotransmitters in the development of neuronal circuits in the brain [15, 23, 29, 30, 33, 51, 61, 62, 73].

Another important aspect of live imaging in general is a massive amount of data in the form of three-dimensional images at many time points. Storage and analysis of these amounts of data are still challenging. Novel automated analysis techniques including mathematical modeling and self learning pattern recognition software are developing to enable us to extract needed information [23, 29, 43, 74, 114].

7 Conclusion

In vivo imaging continues to be an increasingly important aspect of the multidisciplinary endeavor to merge and understand the morphological, physiological, and regulatory processes of neurogenesis. Refinement of the spatial and temporal resolution of in vivo imaging techniques and interpretation of collected data will require further integration of neuroscience, mathematics and physics in order to develop refined and plausible model of brain development [15, 29, 33].

The real challenge for in vivo imaging of brain development will eventually be to adjust available methods for work with primates and to translate results towards the clinical domain [15]. New approaches need to be developed to monitor how the behavior of groups of cells causes the clinical manifestations of neurological diseases. Further advances will allow us to longitudinally monitor how the neuronal proliferation and migration abnormalities affect synaptic wiring and ultimately the brain function.

In vivo imaging technologies definitely have the potential to elucidate the pathogenesis of developmental neurological disorders and make critical steps of brain development clinically detectable and therapeutically approachable. Ultimately, the future holds imaging technologies capable of detecting early neurodevelopmental disorders in children, identifying underlying biological substrates, and determining successful pharmacological or behavioral interventions.

References

1. Ang ES Jr, Haydar TF, Gluncic V, Rakic P (2003) Four-dimensional migratory coordinates of GABAergic interneurons in the developing mouse cortex. J Neurosci 23:5805–5815
2. Caviness VS Jr, Takahashi T, Nowakowski RS (1995) Numbers, time and neocortical neuronogenesis: a general developmental and evolutionary model. Trends Neurosci 18:379–383
3. Rakic P (1995) A small step for the cell, a giant leap for mankind: a hypothesis of neocortical expansion during evolution. Trends Neurosci 18:383–388
4. Rakic P (2000) Molecular and cellular mechanisms of neuronal migration: relevance to cortical epilepsies. Adv Neurol 84:1–14
5. Rakic P (2003) Elusive radial glial cells: historical and evolutionary perspective. Glia 43:19–32
6. Rakic PA (2006) Century of progress in corticoneurogenesis: from silver impregnation to genetic engineering. Cereb Cortex 16(Suppl):3–17
7. Ang ES Jr, Gluncic V, Duque A et al (2006) Prenatal exposure to ultrasound waves impacts neuronal migration in mice. Proc Natl Acad Sci USA 103:12903–12910
8. Bassell GJ, Kelic S (2004) Binding proteins for mRNA localization and local translation, and their dysfunction in genetic neurological disease. Curr Opin Neurobiol 14:574–581
9. Bond J, Roberts E, Mochida GH et al (2002) ASPM is a major determinant of cerebral cortical size. Nat Genet 32:316–320
10. Caviness VS Jr, Takahashi T, Nowakowski RS (2002) Neocortical malformation as consequence of nonadaptive regulation of neuronogenetic sequence. Ment Retard Dev Disabil Res Rev 6:22–33
11. Frints SG, Marynen P, Hartmann D et al (2003) CALL interrupted in a patient with non-specific mental retardation: gene dosage-dependent alteration of murine brain development and behavior. Hum Mol Genet 12:1463–1474
12. Gambello MJ, Darling DL, Yingling J et al (2003) Multiple dose-dependent effects of Lis1 on cerebral cortical development. J Neurosci 23:1719–1729
13. Guerrini R, Marini C (2006) Genetic malformations of cortical development. Exp Brain Res 173:322–333
14. Lo Nigro C, Chong CS, Smith AC et al (1997) Point mutations and an intragenic deletion in LIS1, the lissencephaly causative gene in isolated lissencephaly sequence and Miller-Dieker syndrome. Hum Mol Genet 6:157–164
15. Misgeld T, Kerschensteiner M (2006) In vivo imaging of the diseased nervous system. Nat Rev Neurosci 7:449–463
16. Casanova MF, Buxhoeveden D, Gomez J (2003) Disruption in the inhibitory architecture of the cell minicolumn: implications for autism. Neuroscientist 9:496–507
17. Casanova MF, Buxhoeveden DP, Switala AE et al (2002) Asperger's syndrome and cortical neuropathology. J Child Neurol 17:142–145
18. Casanova MF, Buxhoeveden DP, Switala AE et al (2002) Minicolumnar pathology in autism. Neurology 58:428–432
19. Fatemi SH (2001) Reelin mutations in mouse and man: from reeler mouse to schizophrenia, mood disorders, autism and lissencephaly. Mol Psychiatry 6:129–133
20. Kumada T, Lakshmana MK, Komuro H (2006) Reversal of neuronal migration in a mouse model of fetal alcohol syndrome by controlling second-messenger signaling. J Neurosci 26:742–756
21. Powell EM, Campbell DB, Stanwood GD et al (2003) Genetic disruption of cortical interneuron development causes region - and GABA cell type-specific deficits, epilepsy, and behavioral dysfunction. J Neurosci 23:622–631
22. Ross ME, Walsh CA (2001) Human brain malformations and their lessons for neuronal migration. Annu Rev Neurosci 24:1041–1070
23. Brainered EL, Hale ME (2006) In vivo and functional imaging in developmental physiology. In: Warburton S, Burggren W (eds) Comparative developmental physiology. Oxford University Press, New York, pp 21–40
24. Denk W, Svoboda K (1997) Photon upmanship: why multiphoton imaging is more than a gimmick. Neuron 18:351–357

25. Dent EW, Kalil K (2003) Dynamic imaging of neuronal cytoskeleton. Methods Enzymol 361:390–407
26. Hatanaka Y, Murakami F (2002) In vitro analysis of the origin, migratory behavior, and maturation of cortical pyramidal cells. J Comp Neurol 454:1–14
27. Haydar TF, Ang E Jr, Rakic P (2003) Mitotic spindle rotation and mode of cell division in the developing telencephalon. Proc Natl Acad Sci USA 100:2890–2895
28. Haydar TF, Bambrick LL, Krueger BK et al (1999) Organotypic slice cultures for analysis of proliferation, cell death, and migration in the embryonic neocortex. Brain Res Protoc 4:425–437
29. Haydar TF (2005) Advanced microscopic imaging methods to investigate cortical development and the etiology of mental retardation. Ment Retard Dev Disabil Res Rev 11:303–316
30. Helmchen F, Denk W (2002) New developments in multiphoton microscopy. Curr Opin Neurobiol 12:593–601
31. Helmchen F, Fee MS, Tank DW, Denk W (2001) A miniature head-mounted two-photon microscope. High-resolution brain imaging in freely moving animals. Neuron 31:903–912
32. Helmchen F, Waters J (2002) Ca^{2+} imaging in the mammalian brain in vivo. Eur J Pharmacol 447:119–129
33. Lichtman JW, Fraser SE (2001) The neuronal naturalist: watching neurons in their native habitat. Nat. Neurosci 4(Suppl):1215–1220
34. Nadarajah B, Alifragis P, Wong RO et al (2003) Neuronal migration in the developing cerebral cortex: observations based on real-time imaging. Cereb Cortex 13:607–611
35. Nadarajah B, Parnavelas JG (2002) Modes of neuronal migration in the developing cerebral cortex. Nat Rev Neurosci 3:423–432
36. Squirrell JM, Wokosin DL, White JG et al (1999) Long-term two-photon fluorescence imaging of mammalian embryos without compromising viability. Nat Biotechnol 17:763–767
37. Svoboda K, Denk W, Kleinfeld D et al (1997) In vivo dendritic calcium dynamics in neocortical pyramidal neurons. Nature 385:161–165
38. Trachtenberg JT, Chen BE, Knott GW, Feng G, Sanes JR et al (2002) Long-term in vivo imaging of experience-dependent synaptic plasticity in adult cortex. Nature 420:788–794
39. Gaiano N, Kohtz JD, Turnbull DH et al (1999) A method for rapid gain-of-function studies in the mouse embryonic nervous system. Nat Neurosci 2:812–819
40. Liu A, Joyner AL, Turnbull DH (1998) Alteration of limb and brain patterning in early mouse embryos by ultrasound-guided injection of Shh-expressing cells. Mech Dev 75:107–115
41. Olsson M, Campbell K, Turnbull DH (1997) Specification of mouse telencephalic and mid-hindbrain progenitors following heterotopic ultrasound-guided embryonic transplantation. Neuron 19:761–772
42. Strome EM, Doudet DJ (2007) Animal models of neurodegenerative disease: insights from in vivo imaging studies. Mol Imaging Biol 9:186–195
43. Megason SG, Fraser SE (2003) Digitizing life at the level of the cell: high-performance laser-scanning microscopy and image analysis for in toto imaging of development. Mech Dev 120:1407–1420
44. Rodriguez A, Ehlenberger D, Kelliher K, Einstein M, Henderson SC et al (2003) Automated reconstruction of three-dimensional neuronal morphology from laser scanning microscopy images. Methods 30:94–105
45. Yuste R, Lanni F, Konnerth A (eds) (2000) Imaging neurons. Cold Spring Harbor Lab, Cold Spring Harbor
46. Theer P, Hasan M, Denk W (2003) Two-photon imaging to a depth of 1000 μm in living brains by use of a TI:A1203 regenerative amplifier. Optics Lett 28:1002–1004
47. Kalatsky VA, Stryker MP (2003) New paradigm for optical imaging. Temporally encoded maps of intrinsic signal. Neuron 38:529–545
48. Obrig H, Villringer A (2003) Beyond the visible – imaging the human brain with light. J Cereb Blood Flow Metab 23:1–18
49. Rubin BD, Katz LC (1999) Optical imaging of odorant representations in the mammalian olfactory bulb. Neuron 23:499–511

50. Yuste R, Miller RB, Holthoff K, Zhang S, Miesenbock G (2000) Synapto-pHluorins: chimeras between pH-sensitive mutants of green fluorescent protein and synaptic vesicle membrane proteins as reporters of neurotransmitter release. Methods Enzymol 327:522–546
51. Zhang J, Campbell RE, Ting AY, Tsien RY (2002) Creating new fluorescent probes for cell biology. Nat Rev Mol Cell Biol 3:906–918
52. Gan WB, Grutzendler J, Wong WT, Wong RO, Lichtman JW (2000) Multicolor "DiOlistic" labeling of the nervous system using lipophilic dye combinations. Neuron 27:219–225
53. O'Brien JA, Lummis SC (2006) Diolistic labeling of neuronal cultures and intact tissue using a hand-held gene gun. Nat Protoc 1:1517–1521
54. Janusonis S, Gluncic V, Rakic P (2004) Early serotonergic projections to Cajal-Retzius cells: relevance for cortical development. J Neurosci 24:1652–1659
55. Larson DR, Zipfel WR, Williams RM, Clark SW, Bruchez MP et al (2003) Water-soluble quantum dots for multiphoton fluorescence imaging in vivo. Science 300:1434–1436
56. Michalet X, Pinaud FF, Bentolila LA, Tsay JM, Doose S, Li JJ, Sundaresan G, Wu AM, Gambhir SS, Weiss S (2005) Quantum dots for live cells, in vivo imaging, and diagnostics. Science 307:538–544
57. Djurisic M, Zochowski M, Wachowiak M, Falk CX, Cohen LB, Zecevic D (2003) Optical monitoring of neural activity using voltage-sensitive dyes. Methods Enzymol 361:423–451
58. Ellis-Davies GC (2003) Development and application of caged calcium. Methods Enzymol 360:226–238
59. Kettunen P, Demas J, Lohmann C, Kasthuri N, Gong Y et al (2002) Imaging calcium dynamics in the nervous system by means of ballistic delivery of indicators. J Neurosci Methods 119:37–43
60. Lohmann C, Myhr KL, Wong RO (2002) Transmitter-evoked local calcium release stabilizes developing dendrites. Nature 418:177–181
61. Politz JC (1999) Use of caged fluorochromes to track macromolecular movement in living cells. Trends Cell Biol 9:284–287
62. Rudolf R, Mongillo M, Rizzuto R, Pozzan T (2003) Innovation: looking forward to seeing calcium. Nat Rev Mol Cell Biol 4:579–586
63. Friedman HV, Bresler T, Garner CC, Ziv NE (2000) Assembly of new individual excitatory synapses: time course and temporal order of synaptic molecule recruitment. Neuron 27:57–69
64. Lendvai B, Stern EA, Chen B, Svoboda K (2000) Experience-dependent plasticity of dendritic spines in the developing rat barrel cortex in vivo. Nature 404:876–881
65. Brager DH, Luther PW, Erdelyi F et al (2003) Regulation of exocytosis from single visualized GABAergic boutons in hippocampal slices. J Neurosci 23:10475–10486
66. O'Donovan M, Ho S, Yee W (1994) Calcium imaging of rhythmic network activity in the developing spinal cord of the chick embryo. J Neurosci 14:6354–6369
67. Tsien RY (1998) The green fluorescent protein. Ann Rev Biochem 67:509–544
68. Washbourne P, McAllister AK (2002) Techniques for gene transfer into neurons. Curr Opin Neurobiol 12:566–573
69. Chen BE, Lendvai B, Nimchinsky EA et al (2000) Imaging high-resolution structure of GFP-expressing neurons in neocortex in vivo. Learn Mem 7:433–441
70. Spergel DJ, Kruth U, Shimshek DR, Sprengel R, Seeburg PH (2001) Using reporter genes to label selected neuronal populations in transgenic mice for gene promoter, anatomical, and physiological studies. Prog Neurobiol 63:673–686
71. Feng G, Mellor RH, Bernstein M, Keller-Peck C, Nguyen QT et al (2000) Imaging neuronal subsets in transgenic mice expressing multiple spectral variants of GFP. Neuron 28:41–51
72. Miyasaka N, Arimatsu Y, Takiguchihayashi K (1999) Foreign gene expression in an organotypic culture of cortical anlage after in vivo electroporation. Neuroreport 10:2319–2323
73. Patterson GH, Lippincott-Schwartz J (2002) A photoactivatable GFP for selective photolabeling of proteins and cells. Science 297:1873–1877
74. Ponti A, Vallotton P, Salmon WC, Waterman-Storer CM, Danuser G (2003) Computational analysis of f-actin turnover in cortical actin meshwork using fluorescent speckle microscopy. Biophys J 84:3336–3352

75. Mochizuki N, Yamashita S, Kurokawa K, Ohba Y, Nagai T et al (2001) Spatio-temporal images of growth-factor-induced activation of Ras and Rap1. Nature 411:1065–1068
76. Qian X, Shen Q, Goderie SK et al (2000) Timing of CNS cell generation: a programmed sequence of neuron and glial cell production from isolated murine cortical stem cells. Neuron 28:69–80
77. Schaefer AW, Kabir N, Forscher P (2002) Filopodia and actin arcs guide the assembly and transport of two populations of microtubules with unique dynamic parameters in neuronal growth cones. J Cell Biol 158:139–152
78. Brecht M, Fee MS, Garaschuk O et al (2004) Novel approaches to monitor and manipulate single neurons in vivo. J Neurosci 24:9223–9227
79. Gahwiler BH, Capogna M, Debanne D et al (1997) Organotypic slice cultures: a technique has come of age. Trends Neurosci 20:471–477
80. Stoppini L, Buchs PA, Muller D (1991) A simple method for organotypic cultures of nervous tissue. J Neurosci Methods 37:173–182
81. Alvarez-Buylla A, Garcia-Verdugo JM, Tramontin AD (2001) A unified hypothesis on the lineage of neural stem cells. Nat Rev Neurosci 2:287–293
82. Belmonte MK, Cook EH Jr, Anderson GM et al (2004) Autism as a disorder of neural information processing: directions for research and targets for therapy. Mol Psychiatry 9:646–663
83. Courchesne E, Carper R, Akshoomoff N (2003) Evidence of brain overgrowth in the first year of life in autism. JAMA 290:337–344
84. Hong SE, Shugart YY, Huang DT et al (2000) Autosomal recessive lissencephaly with cerebellar hypoplasia is associated with human RELN mutations. Nat Genet 26:93–96
85. Noctor SC, Flint AC, Weissman TA et al (2001) Neurons derived from radial glial cells establish radial units in neocortex. Nature 409:714–720
86. Gorski JA, Talley T, Qiu M et al (2002) Cortical excitatory neurons and glia, but not GABAergic neurons, are produced in the Emx1-expressing lineage. J Neurosci 22(15):6309–6314
87. Lyons DA, Guy AT, Clarke JD (2003) Monitoring neural progenitor fate through multiple rounds of division in an intact vertebrate brain. Development 130:3427
88. McCarthy M, Turnbull DH, Walsh CA et al (2001) Telencephalic neural progenitors appear to be restricted to regional and glial fates before the onset of neurogenesis. J Neurosci 21:6772–6781
89. Takahashi T, Nowakowski RS, Caviness VS Jr (1996) The leaving or Q fraction of the murine cerebral proliferative epithelium: a general model of neocortical neuronogenesis. J Neurosci 16:6183–6196
90. Tan SS, Kalloniatis M, Sturm K et al (1998) Separate progenitors for radial and tangential cell dispersion during development of the cerebral neocortex. Neuron 21:95–304
91. Walsh C, Cepko CL (1993) Clonal dispersion in proliferative layers of developing cerebral cortex. Nature 362:632–635
92. Das T, Payer B, Cayouette M, Harris WA (2003) In vivo time-lapse imaging of cell divisions during neurogenesis in the developing zebrafish retina. Neuron 37:597–609
93. Feng Y, Walsh CA (2004) Mitotic spindle regulation by Nde1 controls cerebral cortical size. Neuron 44:279–293
94. Noctor SC, Martinez-Cerdeno V, Ivic L et al (2004) Cortical neurons arise in symmetric and asymmetric division zones and migrate through specific phases. Nat Neurosci 7:136–144
95. Demyanenko GP, Schachner M, Anton E et al (2004) Close homolog of L1 modulates area-specific neuronal positioning and dendrite orientation in the cerebral cortex. Neuron 44:423–437
96. Gongidi V, Ring C, Moody M et al (2004) SPARC-like 1 regulates the terminal phase of radial glia-guided migration in the cerebral cortex. Neuron 41:57–69
97. Hirotsune S, Fleck MW, Gambello MJ et al (1998) Graded reduction of Pafah1b1 (Lis1) activity results in neuronal migration defects and early embryonic lethality. Nat Genet 19:333–339
98. Letinic K, Zoncu R, Rakic P (2002) Origin of GABAergic neurons in the human neocortex. Nature 417:645–649

99. Wichterle H, Turnbull DH, Nery S et al (2001) In utero fate mapping reveals distinct migratory pathways and fates of neurons born in the mammalian basal forebrain. Development 128: 3759–3771
100. Komuro H, Rakic P (1993) Modulation of neuronal migration by NMDA receptors. Science 260:95–97
101. Komuro H, Rakic P (1998) Orchestration of neuronal migration by activity of ion channels, neurotransmitter receptors, and intracellular Ca^{2+} fluctuations. J Neurobiol 37:110–130
102. Koster RW, Fraser SE (2001) Direct imaging of in vivo neuronal migration in the developing cerebellum. Curr Biol 11:1858–1863
103. Cline HT (2001) Dendritic arbor development and synaptogenesis. Curr Opin Neurobiol 11:118–126
104. Feller MB (1999) Spontaneous correlated activity in developing neural circuits. Neuron 22:653–656
105. Hutson LD, Chien CB (2002) Pathfinding and error correction by retinal axons: the role of astray/robo2. Neuron 33:205–217
106. Jin X, Hu H, Mathers PH et al (2003) Brain-derived neurotrophic factor mediates activity-dependent dendritic growth in nonpyramidal neocortical interneurons in developing organotypic cultures. J Neurosci 23:5662–5673
107. Trachtenberg JT, Trepel C, Stryker MP (2000) Rapid extragranular plasticity in the absence of thalamocortical plasticity in the developing primary visual cortex. Science 287:2029–2032
108. Greenough WT, Klintsova AY, Irwin SA et al (2001) Synaptic regulation of protein synthesis and the fragile X protein. Proc Natl Acad Sci USA 98:7101–7106
109. Jones JD, Buchanan J, Smith SJ (2000) Growth cone and dendrite dynamics in zebrafish embryos: early events in synaptogenesis imaged in vivo. Nat Neurosci 3:231–237
110. Kaufmann WE, Moser HW (2000) Dendritic anomalies in disorders associated with mental retardation. Cereb Cortex 10:981–991
111. Purves D, Lichtman JW (1987) Synaptic sites on reinnervated nerve cells visualized at two different times in living mice. J Neurosci 7:1492–1497
112. Gleason MR, Higashijima S, Dallman J, Liu K, Mandel G, Fetcho JR (2003) Translocation of CaM kinase II to synaptic sites in vivo. Nat Neurosci 6:217–218
113. Gomez TM, Spitzer NC (1999) In vivo regulation of axon extension and pathfinding by growth-cone calcium transients. Nature 397:350–355
114. Koh IY, Lindquist WB, Zito K, Nimchinsky EA, Svoboda K (2002) An image analysis algorithm for dendritic spines. Neural Comput 14:1283–1310
115. Walsh MK, Lichtman JW (2003) In vivo time-lapse imaging of synaptic takeover associated with naturally occurring synapse elimination. Neuron 37:67–73
116. Cox CL, Denk W, Tank DW et al (2000) Action potentials reliably invade axonal arbors of rat neocortical neurons. Proc Natl Acad Sci USA 97:9724–9728
117. Grutzendler J, Kasthuri N, Gan WB (2002) Long-term dendritic spine stability in the adult cortex. Nature 420:812–816
118. Alsina B, Vu T, Cohen-Cory S (2001) Visualizing synapse formation in arborizing optic axons in vivo: dynamics and modulation by BDNF. Nat Neurosci 4:1093–1101
119. Nedivi E, Wu GY, Cline HT (1998) Promotion of dendritic growth by CPG15, an activity-induced signaling molecule. Science 281:1863–1866
120. Wong WT, Wong RO (2000) Rapid dendritic movements during synapse formation and rearrangement. Curr Opin Neurobiol 10:118–124
121. Crair MC, Gillespie DC, Stryker MP (1998) The role of visual experience in the development of columns in cat visual cortex. Science 279:566–570
122. Katz LC, Shatz CJ (1996) Synaptic activity and the construction of cortical circuits. Science 274:1133–1138
123. Shepherd GM, Pologruto TA, Svoboda K (2003) Circuit analysis of experience-dependent plasticity in the developing rat barrel cortex. Neuron 38:277–289

124. Sin WC, Haas K, Ruthazer ES, Cline HT (2002) Dendrite growth increased by visual activity requires NMDA receptor and Rho GTPases. Nature 419:475–480
125. Gaietta G, Deerinck TJ, Adams SR et al (2002) Multicolor and electron microscopic imaging of connexin trafficking. Science 296:503–507
126. Mochida GH, Walsh CA (2001) Molecular genetics of human microcephaly. Curr Opin Neurol 14:151–156
127. Murcia CL, Gulden F, Herrup K (2004) A question of balance: a proposal for new mouse models of autism. Int J Dev Neurosci 23:265–275
128. Gu X, Olson EC, Spitzer NC (1994) Spontaneous neuronal calcium spikes and waves during early differentiation. J Neurosci 14:6325–6335
129. Wong RO, Chernjavsky A, Smith SJ, Shatz CJ (1995) Early functional neural networks in the developing retina. Nature 374:716–718

Congenital, Non-inheritable Chromosomal Abnormalities Responsible for Neurological Disorders

Riccardo Bianchi and Patrizia D'Adamo

Abstract Chromosomal abnormalities include aberrations of chromosome number and/or structure, such as trisomy, translocation, duplication, deletion. These abnormalities are caused by errors in chromosome disjunction during meiosis and result in conditions that range from prenatal death to relatively mild syndromes in adulthood. Common traits of syndromes due to chromosomal abnormalities that are compatible with postnatal life are altered development of the nervous system and various degrees of intellectual dysfunction (mental retardation). Recent advances in genome mapping and research on animal models have provided new insights into the mechanisms underlying mental retardation. Growing evidence indicates important roles of developmental processes that establish the "wiring" of brain networks and of signaling pathways that mediate activity-dependent synaptic plasticity. Understanding how these processes are altered by the chromosomal abnormalities may indicate therapeutic strategies to boost intellectual functions and, ultimately, to improve the quality of life of affected individuals.

Keywords Chromosome disorders • Mental retardation • Brain development • Altered synaptic plasticity • Transcription factor • Down syndrome

1 Introduction

Brain development begins shortly after conception and continues throughout the growth of a fetus. A complex genetic program coordinates the formation, growth, and migration of billions of neurons and their development into discrete, interacting brain regions. Interruption of this program, especially early in development, can

R. Bianchi
Robert F. Furchgott Center for Neural and Behavioral Science, and Department of Physiology and Pharmacology, State University of New York Downstate Medical Center, Box 29, 450 Clarkson Avenue, Brooklyn, NY 11203, USA
e-mail: rbianchi@downstate.edu

D. Janigro (ed.), *Mammalian Brain Development*, Contemporary Neuroscience,
DOI 10.1007/978-1-60761-287-2_10,

cause structural and functional defects in the brain. In addition, normal brain formation requires proper development of the surrounding skull, and skull defects may lead to brain malformation. Gross brain malformations can be identified through direct physical examination or with the use of imaging studies, such as computer tomography (CT) and magnetic resonance imaging (MRI). Common forms of brain malformations include anencephaly, encephalocele, spina bifida, Dandy-Walker malformations, holoprosencephaly, lissencephaly, schizencephaly, micrencephaly, megalencephaly, and ventriculomegaly. The diagnosis of these congenital brain defects may be confounded by subtle gradations of clinical presentations, variability in the phenotype, or phenotypic overlap between distinct syndromes.

Congenital brain defects may be caused by inherited genetic defects, spontaneous mutations within the genes of the embryo, or effects on the embryo due to the mother's infection, trauma or drug use. Abnormal number or structure of chromosomes in the embryo may lead to congenital, non-inheritable brain defects associated with disorders of the nervous system. Chromosome full or partial trisomies, monosomies, duplications, deletions, insertions, inversions, and translocations that are not necessarily lethal for the fetus affect postnatal development with different degrees of severity. Table 1 lists the most frequent syndromes caused by chromosomal abnormalities and the associated neurological disorders. The cytogenetic basis of many of these syndromes was established with the advent of chromosomal banding techniques that facilitated the detection of aneuploidies and of large structural chromosome rearrangements [1]. Powerful approaches to the investigation of specific chromosomal abnormalities are the fluorescent in situ hybridization (FISH) [2] and the spectral karyotyping (SKY) [3]. The development of comparative genomic hybridization (CGH), particularly CGH using microarrays, has further broadened the scope and resolution of microdeletions and microduplications, leading to the identification of new pathological conditions [4]. Comparative analysis also reveals critical chromosomal regions and narrows the search for genes and mechanisms underlying the brain abnormalities observed in these syndromes.

In this review, we consider structural and functional alterations of brain neuronal circuits in congenital chromosomal abnormalities that compromise the normal development of the brain. Recent data, both from humans and experimental animal models, suggest that abnormal formation and function of synaptic connections in the brain may underlie the associated cognitive disabilities. Thus, research focused on identifying mechanisms of synaptic dysfunction may open the way to potential therapeutic treatments.

2 Chromosomal Abnormalities and Cognitive Deficits

Table 1 indicates that mental retardation appear as a major phenotype in all syndromes caused by chromosomal abnormalities and range from mild learning disabilities, as in sex chromosome aneuploidies, to severe mental retardation, as in autosomal trisomies or partial deletions. In most cases, such cognitive impairments are associated with macroscopic craniofacial and brain structural alterations. Also, neurological disorders

Table 1 Congenital chromosomal abnormalities and major phenotypes

Chromosomal abnormality[a]	Karyotype (syndrome)	OMIM annotation	Frequency at birth	Cognitive and neurological conditions	Other abnormalities
Autosomal trisomies	Trisomy 21 (Down s.)	#190685	1/700–800[b]	Mental retardation	Heart defects, characteristic facial features, motor skills, leukemia, AD
	Trisomy 18 (Edwards s.)	–	1/3000–1/8,000	Severe mental retardation	Many physical birth defects, often death before first birthday
	Trisomy 13 (Patau s.)	–	1/10,000	Severe mental retardation	Many physical birth defects, often death before first birthday
Autosomal deletions	Deletion on Chr. 4 (Wolf-Hirschhorn s.)	#194190	1/50,000	Severe mental retardation, seizures	Heart defects, high blood pressure, poor muscle tone, and other problems
	Deletion on Chr. 5 (Cri-du-Chat s.)	#123450	1/20,000–1/50,000	Mental retardation	Cat-like high-pitched cry during infancy, physical abnormalities
	Heterozygous microdeletion on Chr. 7 (7q11.23) (Williams s.)	#194050	1/7,500–1/20,000	Mental retardation, impaired visuospatial constructive abilities	Supravalvular aortic stenosis, elfin facies, transient hypercalcemia in infancy
	Deletion on Chr. 15 (15q13; maternal allele) , or uniparental disomy for paternal allele (Angelman s.)	#105830	1/15,000	Mental retardation (severe language impairment); epilepsy	Microbrachicephaly, macrostomia, ataxia
	Deletion on Chr. 15 (15q13; paternal allele), or uniparental disomy for maternal allele (Prader-Willi s.)	#176270	1/10,000–1/20,000	Mental retardation	Obesity, hypotonia, hernias, hypogonadism, short stature
	Deletion on Chr. 17 (17p11.2) (Smith-Magenis s.)	#182290	1/15,000–1/25,000	Mental retardation, speech and motor delay, self-injurious behaviors	Brachicephaly, midface hypoplasia, prognathism, hypotonia, hoarse voice
	Deletion on Chr. 22 (22q11.2) (DiGeorge s.)	#188400	1/4,000	Learning disabilities	Heart defects, cleft lip/palate, immune system abnormalities, characteristic facial features

(continued)

Table 1 (continued)

Chromosomal abnormality[a]	Karyotype (syndrome)	OMIM annotation	Frequency at birth	Cognitive and neurological conditions	Other abnormalities
Sex chromosome aneuploidies	X (Turner s.)	–	1/2,500 girls	Normal intelligence (some difficulties in math and spatial concepts)	Short stature, webbed neck, hand/feet edema in neonates, heart and kidney defects
	XXY, XXXY (Klinefelter s.)	278850	1/500–1,000 boys	Normal intelligence or learning disabilities	Low testosterone, gynecomastia, hypogonadism
	XYY (XYY s.)	–	1/900–2,000 boys	Normal intelligence or learning disabilities	Tall stature, high testosterone, facial acne, poor coordination ("super-males")
	XXX (Triple X s.)		1/1,000 girls	Normal intelligence or learning disabilities	From no phenotypes to minor dysmorphism and tall stature ("metafemales")

[a]Rare chromosomal abnormalities at birth are not included in this table
[b]Maternal-age-dependent

include deficits of the sensory and motor systems, and age-related neurodegenerative processes (e.g., Alzheimer's disease-like pathology in Down syndrome). Other systems of organs are affected by the different syndromes to various extents – with heart defects being frequently associated with chromosomal abnormalities. Thus, in addition to direct effects of chromosomal abnormalities on the nervous system, neurological conditions can be secondarily affected by cardiovascular, respiratory, metabolic, and hormonal syndromic dysfunction.

2.1 Mental Retardation

Mental retardation manifests as impairment of intellectual functions and is practically characterized by intelligence coefficient (IQ) below 70 (relative to a normal IQ of 100). IQ ranges vary from mild (50–70) to profound (< 25), depending mainly on the type of chromosomal abnormality. However, large variations of IQ values between individuals affected by the same syndrome are also observed, suggesting that complex interactions between genetic, epigenetic, and environmental factors converge on affecting brain function.

A variety of cognitive impairments have been documented in different syndromes. For example, in Down syndrome (DS; trisomy 21) – the most frequent chromosomal abnormality at birth (Table 1) and the most common condition of mental retardation – major cognitive deficits in young adults have been reported in language and verbal communication skills [5, 6], executive functions [7], spatial memory [8], short- and long-term memory [9–11].

Whereas cognitive impairments are evident in adults, behavioral tests applied to infants have not revealed clear deficits in DS babies compared to age-matched non-syndromic individuals during the first months of life [12]. Nadel [13] suggested that the most predominant cognitive impairments in DS patients depend on brain regions, such as the hippocampus and the cerebellum, that mature postnatally. Thus, the forms of learning and memory dependent on these structures would not be available until some time after birth. The earliest learning deficits have been detected at about 9 months of age: Ohr and Fagen [14] showed that a group of 9-month-old infants with DS were impaired in learning contingency between arm movement and reinforcement, whereas other DS infants behaved normally, and suggested a relative decline in conditionability in DS infants after 6 months of age. This observation is consistent with studies on brain maturation that indicate a relatively normal DS brain at birth and appearance of morphological abnormalities after age 6 months [15].

Later in life these cognitive problems in DS individuals are compounded by Alzheimer's Disease (AD) that develops after about 35–40 years of age [16]. AD has been linked to abnormal precipitation and deposition of the protein β-amyloid in various brain regions [17, 18]. Overexpression of the gene for the amyloid precursor protein (APP) – located on chromosome 21 – is likely to underlie the development of AD in DS [19, 20]. Although AD is a neurodegenerative condition associated with aging, rather than a disorder of brain development, it has been recently proposed that overexpression of genes that lead to developmental cognitive deficits in DS may also participate in partial compensatory processes that delay AD progression later in life [21].

Individuals with Williams syndrome (WS; caused by a microdeletion on chromosome 7; Table 1) have a primary impairment in visuo-spatial performances, but both productive and receptive language relatively preserved [22]. Moreover, visual constructive, visual motor, and visuo-spatial working memory, selective attention and saccadic eye movement abilities are particularly affected [23, 24].

Learning disabilities are also observed in Angelman syndrome (AS) and in Prader-Willi syndrome (PWS), caused by a microdeletion on chromosome 15 of maternal or paternal origin, respectively [25], although neurological and cognitive impairments are more severe in AS and include seizures, language difficulties, and ataxia [26], whereas in PWS difficult interpersonal behaviors predominate [27].

Thus, mental retardation is a common feature of most syndromes caused by chromosomal abnormalities, but specific cognitive deficits in different syndromes vary depending on the affected brain region(s) and on the parental origin of chromosomal abnormality.

2.2 Brain Malformations

Brain malformations may constitute the structural basis for the cognitive impairments observed in syndromes caused by congenital chromosomal abnormalities (Table 1).

A sizeable fraction of all brain malformations (from about 10% of all anencephalies and ventriculomegalies, to 50% of all holoprosencephalies and Dandy-Walker malformations) are caused by trisomies or triploidies [28].

In DS, the brain is reduced in size and weight and shows abnormal patterns such as immature cortical gyri, irregular laminar formation and delayed myelination [29–31]. Brain volume reduction and abnormalities of cortical gyri and/or lamination are found in all the other syndromes listed on Table 1 [26, 32–55]. Specific brain regions and types of cortical abnormalities may differ between the various syndromes. For example, in WS cerebral size is reduced but cerebellar size is normal, whereas in DS both brain structures are reduced [56], and relative preservation of frontal and temporal cortices in WS contrasts with relative preservation of basal ganglia and diencephalon in DS [57].

A number of studies suggest a close relationship between the types of cognitive deficits that predominate in a specific syndrome and structural alterations in the brain regions involved in the same cognitive processes. For example, learning and memory tasks dependent on hippocampal function are impaired in DS [8, 13] and the hippocampus is partially atrophic in these patients [31, 58–60]. Also, it was noted that the behavioral tests used in DS infants depend on hippocampal function and that the hippocampus continues to develop after birth, reaching maturity around 1–2 years of age [13]. This suggests that some learning deficits in DS depend on dysfunction of the hippocampus and become more evident after 2 years of age, when learning increasingly relies on this brain region. Deficits in episodic and executive functions as working memory, that depend on the prefrontal cortex, are paralleled by a reduced size of the prefrontal cortical lobes in DS [58]. A reduced cerebellar volume in DS [31] may underlie motor skills impairments, hypotonia in infants, and fine motor deficits in adults [61], as well as acquisition of conditioned responses [13].

More severe cranio-facial and brain malformations are observed in trisomy 13 (cleft palate and holoprosencephaly) [62, 63] and in trisomy 18 (gyral abnormalities, hippocampal dysplasia, corpus callosum and cerebellar hypoplasia) [33, 64], and are associated with a 90–95% mortality within the first year of life [65, 66]. Other autosomal full trisomies – e.g., of chromosome 8, 9, 16, and 22 – are lethal for the embryo or very rare at birth, and cause, in survivors, marked cranial and brain malformations and severe mental and motor deficiencies [28].

In WS abnormalities have been identified in parietal and occipital lobes [34, 67, 68], consistent with visuo-spatial deficits and enhanced emotionality and face processing [68–70]. In AS, language impairments have been related to anomalous Sylvian fissure [40] and movement disorders and ataxia may depend on cerebellar abnormalities [71]. PWS is characterized by behavioral and cognitive deficits similar to those observed in autistic spectrum disorders and lesional pathologies of the frontal lobe. Thus, Prader-Willi symptoms may relate to abnormal frontal cognitive processes such as attention, working memory, and executive function that could be ascribed to abnormal orbito-frontal cortex functions [72, 73]. Causative relationships between specific brain abnormalities and characteristic cognitive/behavioral impairments have been also proposed for DiGeorge [41, 74, 75], Turner [51, 76–78], and Klinefelter [53] syndromes.

Despite the clear association between brain malformations and MR, cognitive deficits observed in the absence of structural brain anomalies and non-syndromic MR suggest that macroscopic malformations are not a pre-requisite for cognitive impairments. Recent data indicate that microstructural and functional abnormalities of neural circuits may be responsible for MR (see below).

2.3 Neurological Dysfunctions

Non-invasive neurophysiological studies have detected anomalies in brain activities recorded from individuals affected by chromosomal abnormalities. Convulsive seizures or abnormal electroencephalographic (EEG; e.g., spike and wave complexes, wave slowing) activities affect a relatively high proportion of these individuals [79], such as infantile spasm in DS [80–84], myoclonic and absence seizures and EEG spikes in AS [85–92], febrile seizures in PWS [93–95], infantile spasms in WS [96–98], myoclonic epilepsy and EEG fast spikes and wave complexes in WHS [87, 88, 99–101], generalized epilepsy in DGS [102, 103], and complex partial or generalized tonic-clonic seizures in Turner, Klinefelter, and triple X syndromes [104–107].

Recordings of evoked potentials have also revealed a variety of abnormal responses in different motor and sensory brain regions of DS, AS, PWS, and WS. Abnormal responses include potentials related to movements or motor preparation [108–110], and somatosensory [111–114], visual [115–119], auditory [120–123], and olfactory [124] evoked-related potentials.

In addition, abnormalities of event-related potentials associated with behavioral tasks [125–129] suggest altered neuronal activity during cognitive processing.

Anomalous brain activities suggest that cognitive impairments may arise from altered synaptic functional properties of the underlying neural networks resulting in dysfunctional neuronal synchronization and information processing.

3 Animal Models of Chromosomal Abnormalities

Recent advancement in our knowledge on the mechanisms underlying the effects of chromosomal abnormalities on neuronal synaptic connections and plasticity has been fueled by the identification of genes in affected chromosomes and by the development of various mouse models.

The first complete sequence of a human chromosome [130] provided a complete map of genes in chromosome 21 [131]. Three copies of genes located on chromosome 21 (trisomy 21) cause DS [132, 133]. Based on gene homology, a number of trisomic mice expressing three copies of genes located on murine chromosome 16 (MMU16) have been generated to model DS [134–137]. DS mouse models range from full trisomy 16 (Ts16, lethal at birth) or segmental trisomy of different lengths of MMU16 (e.g., Ts65Dn, Ts1Cje, Ms1Ts65) to single gene overexpression (transgenic mice), or they carry an extra copy of the human chromosome 21 (ES#21, Tc1). All these mouse models, except for the Ts16, survive into adulthood, display cranio-facial and brain abnormalities, as well as physiological and behavioral deficits comparable to the DS phenotype [134, 138].

Mouse models of deletion syndromes have also been generated by knocking out genes that are homologous to those identified in the corresponding deleted portions of the human chromosomes (see below).

4 Chromosomal Abnormalities and Synaptic Connections

Studies in recent years have extended the analysis of brain abnormalities to the cellular and molecular levels. A unifying perspective is emerging whereby mental retardation results from altered cortical synaptic connections, function, and plasticity in syndromes caused by chromosomal abnormalities [139], as well as in single-gene mutations [140] and in non-syndromic conditions [140, 141]. Chromosomal abnormalities compromise the normal development of the central nervous system, leading to abnormal formation of neuronal networks. In addition to structural abnormalities of brain connectivity, altered plasticity of cortical synapses may hinder cognitive processes throughout the lifespan of affected individuals.

4.1 Development of Brain Neuronal Networks

Microscopic studies have revealed that brain malformations are associated with abnormal development of neurons and of synaptic connections, resulting in altered synaptic structures.

Decreased neuronal density and size in the neocortex, hippocampus, and cerebellum [142–147] may account for the reduced volume of these structures in the DS brain (see above). Altered cortical layers, hypocellularity and small neurons are already evident in the DS fetus [29, 31, 56]. Abnormal neuronal density, cell size, and distribution has been reported also in the visual cortex of WS individuals [35, 148].

DS adult brain also show specific alterations in cortical synaptic structures, such as reduction in dendritic branching and length, and in spine density and number [142, 143, 145, 149, 150], as well as abnormally elongated spines [143, 151]. In contrast to adults, dendritic branching and length are greater in DS infants younger than 6 months than in normal infants and then decline steadily to the lower levels of adult DS [142], suggesting abnormal development and maturation of postsynaptic structures. Indeed, the prenatal DS brain has fewer synapses [152, 153] and cortical neurons with reduced dendritic arborizations and spine number [139, 142, 154]. Quantitative Golgi studies also showed dendritic abnormalities in trisomy 13 [151, 155] and in trisomy 18 [156] fetal and infant brains.

Studies on mouse models of these syndromes have shown abnormal brain development and synaptic structures, and have suggested underlying molecular mechanisms. Altered craniofacial development and phenotype resembling DS have been reported in the Ts65Dn and Ts1Cje mice [157–159]. In these mice, overall smaller brain and specific reduction of hippocampal and cerebellar volumes are associated with low density of neurons [160–164] due to defects in embryonic neurogenesis, such as cell cycle alteration, decreased cell proliferation, and increased apoptotic cell death [165–168]. Decreased sensitivity to the protein sonic hedgehog (Shh)-mediated signaling pathway has been found to cause the mitotic deficit of cerebellar granule cell progenitors [169]. Cerebellar transcriptome analysis during postnatal development has shown dysregulated expression of genes involved in development (homeobox genes) and has suggested that development has a greater effect than trisomy on global gene expression [170]. Altered neurogenesis and synaptic defects occur early during embryogenesis [166]. Trisomic mice show that abnormalities in cortical synaptic structures similar to those observed in DS brains [171–173] and global alterations of cerebellar transcriptome [174] are then maintained throughout adulthood.

4.2 Molecular Mechanisms of Abnormal Brain Development

A variety of molecular mechanisms may underlie abnormal development of brain neuronal networks because chromosomal abnormalities alter the number of multiple genes. Some of these mechanisms have emerged by identifying altered levels of specific protein products of affected genes during development (primary gene dosage effect; Fig. 1) and by analyzing development of transgenic mice overexpressing, or lacking, single genes that are homologous to those of the affected human chromosomes. In addition, chromosomal abnormalities have been shown also to alter the levels of proteins coded by euploid genes (secondary gene dosage effect; Fig. 1) that are involved in key developmental processes. Overall, abnormal development appears to depend on alterations in neurotransmission and neurotrophism, in intracellular signaling, and in gene expression.

Biogenic amines (acetylcholine, serotonin, and catecholamines) act as neurotransmitters as well as mediate differentiation signals during development thereby playing important roles in the formation of the brain circuits involved in cognitive processes

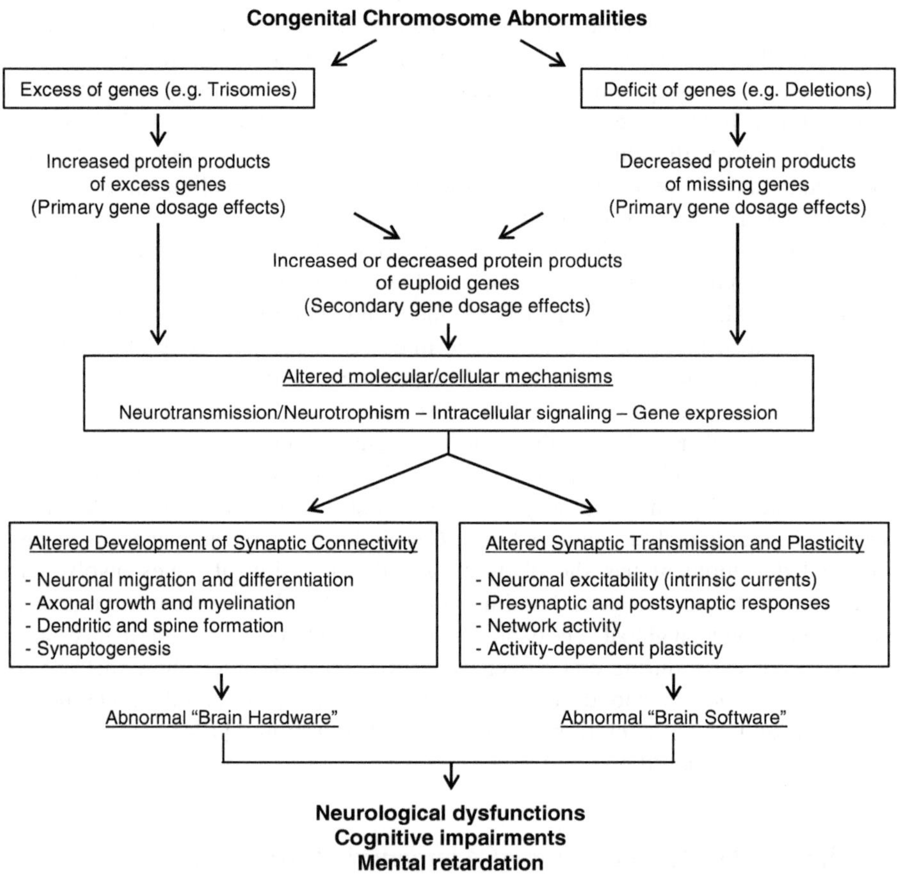

Fig. 1 Abnormal synaptic development, function, and plasticity mediates the effects of chromosomal abnormalities on brain dysfunction. Excess of genes includes full or partial trisomies and syndromes caused by supernumerary sex chromosomes. Deficit of genes includes deletions and monosomies (e.g., Turner syndrome)

[175–182]. In fetal (~4th–9th gestational month) DS brains, reduced levels of cholinergic muscarinic receptors [183], abnormal levels of serotonin 5-HT1A receptors [184], and reductions of serotonin and dopamine levels [185] have been reported. Immunoreactivity markers of cholinergic, monoaminergic, and serotonergic innervation in the fetal DS brain at the early second trimester did not differ from age-matched controls, suggesting a later appearance of neurotransmitters deficits in the DS fetus [186]. Fetal brains of the DS mouse model Ts16 showed a reduction in muscarinic receptor density and acetylcholine esterase-positive cells [187], decreased choline and increased myo-inositol levels [188], reduced neuronal uptake of choline [189], decreased number of tyrosine hydroxylase-positive cells [187]. These data suggest that altered levels of biogenic amines affect synaptogenesis and may contribute to developmental disabilities and mental retardation [179, 181, 182]. Furthermore, the levels of gamma-aminobutyric acid (GABA) and taurine in fetal DS frontal cortex are reduced [185], and glutamate-mediated survival [190] and NMDA-mediated migration [191] in cultured Ts16 neurons are compromised, suggesting that altered amino acid neurotransmitter action may also contribute to abnormal development in DS. Similar mecahnisms may be also relevant for AS, since a mouse model of AS harbors a $GABA_A$ receptor β3 subunit gene inactivation and shows reduced $GABA_A$ receptor density and function, hyperactivity, and epileptic seizures [192].

Neurotrophic factors direct cellular migration, axon targeting, and neuronal differentiation. In addition to neurodegeneration of basal forebrain cholinergic neurons in aging DS and trisomic mouse brains caused by deficit of the nerve growth factor (NGF) and the brain-derived neurotrophic factor (BDNF) [20, 193–196], altered levels of neurotrophic factors appear also to affect cell survival during development [182, 197, 198]. Failure of BDNF to promote survival of mouse Ts16 neurons in culture is associated with overexpression of a dysfunctional isoform of the tyrosine receptor kinase B, suggesting a neurotrophin signaling deficit in trisomic neurons [199, 200]. Interaction of neurotransmitters and neurotrophic factors may also be critical for normal brain development [182, 197, 198]. A defective glial-neuron interaction has been proposed to contribute to abnormal neuronal differentiation and development in DS [198]. The astroglial-derived calcium-binding and neurotrophic protein S100B, which gene is mapped to human chromosome 21, is present at high levels in DS brain [201, 202] and, when overexpressed in transgenic mice, promotes astrocytosis and neurite proliferation [203].

The levels or the function of a number of intracellular molecules mediating cellular responses to neurotransmitters and neurotrophic factors are altered by chromosomal abnormalities. The Rho family of GTPases stimulates neuronal outgrowth and spine formation [204, 205] and abnormal levels of proteins that interact directly with Rho GTPases or that regulate this signaling pathway may alter the organization and dynamics of cytoskeletal proteins involved in neuronal migration and synaptogenesis [141]. DS and WS have been recently associated with alterations of the Rho signaling pathway. The protein TTC3, encoded in the DSCR of chromosome 21, interacts with two effectors of the RhoA small GTPase and its overexpression inhibits neurite extension, whereas TTC3 knockdown stimulates neurite extension in PC12 cells [206]. The gene *Limk1* – deleted in WS – encodes for a neuronal serine/threonine kinase that acts downstream of both Rac and RhoA and controls actin dynamics by inactivating the actin depolymerization factor (ADF)/cofilin [207, 208]. *Limk1−/−* mice show alterations in ADF/cofilin phosphorylation, actin dynamics, spine morphology, synaptic plasticity, and spatial learning [209, 210].

Alterations in developing DS brain [211] also includes reduced levels in proteins of signaling pathways involved in neuronal differentiation and synaptic plasticity: Nck/Crk and 14-3-3 (adaptor proteins of the tyrosine receptor kinase/MAPK signaling) [212, 213], RACK1 (anchoring protein for translocation of protein kinase C to the membrane) [212], and Rab GDP-dissociation inhibitor (GDI) 2 and nucleoside diphosphate kinase (NDK)-B (modulators of the Rab-mediated recycling of synaptic vesicles) [153]. The finding that a mutation of the gene encoding GDI-1 causes X-linked mental retardation [214] indicates that a specific alteration in presynaptic function is sufficient to compromise cognitive abilities.

Various proteins associated with synaptic and cytoskeletal structures are dysregulated in fetal DS and trisomic mouse brains and are potentially responsible for abnormal synaptogenesis. Levels of synaptosome-associated proteins (SNAPs) and drebrin, an actin-binding proteins present in dendrites [215, 216], are reduced in trisomic fetal brains [153, 217]. On the other hand, proteins that are overexpressed include: intersectin, a scaffold protein that controls formation and branching of actin filaments [218, 219] and interacts with endocytotic/exocytotic synaptic processes [220, 221]; synaptojanin, a phosphatase involved in modulation of synaptic transmission and in synaptic vesicle endocytosis [222]; DSCAM, a family of immunoglobulin proteins critical for axon branching, dendritic arborization, and neuronal wiring [223–227]; DSCR1, inhibitor of calcineurin/nuclear factor of activated T cells (NFAT) signaling involved in synaptogenesis [228, 229]; DYRK1, a kinase that phosphorylates actin-binding proteins and may mediate spine membrane-actin cytoskeleton interaction [230]. It has been shown that overexpression of DSCR1 and of DYRK1 results in cooperative dysregulation of the NFAT-mediated signaling [228], indicating that interactions of proteins that converge on the same signaling pathways may compound the deleterious effects of chromosomal abnormalities.

Finally, evidence for altered transcription factors and regulators of translation in fetal brain tissue affected by chromosomal abnormalities indicates that abnormal gene expression plays a role in neurodevelopmental disorders. In fetal DS brain, the mRNA levels of different transcription factors critical for synapse formation and brain development are either reduced (REST, junD, nuclear factor κB) [231, 232] or increased (scleraxis) [231]. Substrates of DYRK1 include transcription factors of the MAPK signaling pathway, such as CREB [230], GLI1 [233], FKHR [234], ARIP4 [235]. Altered levels of various elongation factors and eukaryotic translation initiation factors [212] suggest that also post-transcriptional processes may be altered in the DS fetal brain [211].

Altered intrinsic excitability of Ts16 embryonic neurons in trisomic mice [236] may also contribute to abnormal brain development by affecting the extracellular levels of neurotransmitters and neurotrophic factors [198] and the activity of transcription factors [237].

4.3 Synaptic Transmission and Plasticity

Electrophysiological studies in adult mouse models have revealed abnormal synaptic function. In hippocampal slices of Ts65Dn mice, size and frequency of postsynaptic responses in CA3 pyramidal cells are altered [238] and the CA3 network shows abnormal synchronized epileptiform (interictal-like) activity [239, 240]. Abnormal synchronized neuronal activity – consisting of epileptic seizures and cerebellar fast oscillations in AS model mice [110, 192, 241, 242] and of unstable rhythm of respiratory neurons in PWS model mice [243] – suggests altered functional synaptic connectivity in these syndromes. Furthermore, abnormal increased synaptic long-term depression (LTD) in CA1 [244], decreased long-term potentiation (LTP) in CA1 [245–247] and in the dentate gyrus [173, 248], and increased propensity to develop prolonged synchronized (ictal-like) discharges [239] have been reported. Synaptic plasticity was also found altered in mouse models of WS (increased LTP in *Limk1−/−* mice) [209] and of AS (defective LTP in mice with maternal deficiency for Ube3a) [241, 249]. These data suggest that hippocampal activity-dependent plasticity is compromised by chromosomal abnormalities in congenital syndromes associated with mental retardation.

Recent studies have identified mechanisms underlying altered plasticity and have succeeded in reversing the deficits caused by the chromosomal abnormalities. In Ts65Dn mice, defective LTP in the dentate gyrus and hippocampus-dependent spatial learning were recovered by chronic administration of non-epileptic doses of $GABA_A$ receptor antagonists [250], and performance deficits on a fear conditioning test were rescued by acute intraperitoneal injection of the NMDA receptor antagonist memantine [251]. In addition to neurotransmission, additional data suggest that altered neuromodulation may also be responsible for deficits in synaptic plasticity and cognitive processes. Degeneration of basal forebrain cholinergic neurons was reversed by estrogen application to Ts65Dn adult female mice [252] and by NGF intracerebroventricular infusion [251]. Development of prolonged epileptiform discharges following blockade of $GABA_A$ receptor-mediated inhibition appeared to depend on synaptic activation of group I metabotropic glutamate receptor (mGluR) activation [239]. Chronic treatment with the serotonin uptake inhibitor fluoxetine rescues neurogenesis in Ts65Dn hippocampus [165].

An interesting parallelism can be drawn from studies on the mechanisms of Fragile X mental retardation syndrome (FXS) – the most common inherited mental retardation syndrome caused by epigenetic silencing of the gene *FMR1* that prevents the expression of the encoded protein, the fragile X mental retardation protein (FMRP) [253], a negative regulator of mRNA translation [254–258]. Data from FXS mouse models suggested exaggerated function of group I mGluR as the mechanism for a number of abnormal neural responses in FXS [259–262]. Thus, in addition to the proposed behavioral and structural similarities with DS [141, 263] and with other conditions of cognitive impairment [263], it is possible that altered activity-dependent synaptic plasticity mediated by metabotropic receptors coupled to protein synthesis is a mechanism underlying brain dysfunction in several syndromes associated with mental retardation.

5 Conclusions

Different hypothesis have been proposed to explain the phenotypic effects of chromosome abnormalities [264]. The hypothesis of gene dosage imbalance [265, 266] emphasizes the gene-specificity of chromosomal abnormalities and is consistent with the observation that different aneuploidies cause characteristic syndromes. On the other hand, non-specific common effects produced by different types of chromosomal imbalance [267, 268] suggest a generalized failure of the normal developmental processes (amplified developmental instability) [269, 270]. Additional mechanisms may include the parental origin of the affected chromosome that dramatically influences the phenotype of the syndrome as illustrated in AS and in PWS [271], and molecular misreading of normal DNA that leads to frameshifted proteins [272]. To discern between these mechanisms will require a complete delineation of differences and similarities between the complex and variable conditions of syndromic mental retardation. Combination to different extents of specific and non-specific effects of aneuploidies may account for the neural mechanisms of intellectual dysfunction.

In this context, we surmise that chromosomal abnormalities cause altered levels of distinct proteins that share key functions in synaptic development and plasticity. Thus, different aneuploidies would result in convergent effects on synaptic processes that are essential for normal learning abilities. Abnormal wiring of cortical networks and altered activity-dependent synaptic plasticity limit the computational capabilities of the brain and cause intellectual deficits in chromosomal syndromes. Current research into these mechanisms is providing windows of opportunity to develop pharmacological [250, 251, 273] and non-pharmacological (e.g., prophylactic use of nutrients [274] or enriched environment stimulation [275–278]) therapeutic treatments for prevention or compensation of neurological disorders in syndromes caused by chromosomal abnormalities.

References

1. Dave BJ, Sanger WG (2007) Role of cytogenetics and molecular cytogenetics in the diagnosis of genetic imbalances. Semin Pediatr Neurol 14:2–6
2. Liehr T, Starke H, Heller A et al (2006) Multicolor fluorescence in situ hybridization (FISH) applied to FISH-banding. Cytogenet Genome Res 114:240–244
3. Van den Veyver IB, Roa BB (1998) Applied molecular genetic techniques for prenatal diagnosis. Curr Opin Obstet Gynecol 10:97–103
4. Ylstra B, van den Ijssel P, Carvalho B, Brakenhoff RH, Meijer GA (2006) BAC to the future! or oligonucleotides: a perspective for micro array comparative genomic hybridization (array CGH). Nucleic Acids Res 34:445–450
5. Rohr A, Burr DB (1978) Etiological differences in patterns of psycholinguistic development of children of IQ 30 to 60. Am J Ment Defic 82:549–553

6. Ypsilanti A, Grouios G (2008) Linguistic profile of individuals with down syndrome: comparing the linguistic performance of three developmental disorders. Child Neuropsychol 14:148–170
7. Rowe J, Lavender A, Turk V (2006) Cognitive executive function in Down's syndrome. Br J Clin Psychol 45:5–17
8. Pennington BF, Moon J, Edgin J, Stedron J, Nadel L (2003) The neuropsychology of Down syndrome: evidence for hippocampal dysfunction. Child Dev 74:75–93
9. Carlesimo GA, Marotta L, Vicari S (1997) Long-term memory in mental retardation: evidence for a specific impairment in subjects with Down's syndrome. Neuropsychologia 35:71–79
10. Brown JH, Johnson MH, Paterson SJ, Gilmore R, Longhi E, Karmiloff-Smith A (2003) Spatial representation and attention in toddlers with Williams syndrome and Down syndrome. Neuropsychologia 41:1037–1046
11. Vicari S, Marotta L, Carlesimo GA (2004) Verbal short-term memory in Down's syndrome: an articulatory loop deficit? J Intellect Disabil Res 48:80–92
12. Ohr PS, Fagen JW (1991) Conditioning and long-term memory in three-month-old infants with Down syndrome. Am J Ment Retard 96:151–162
13. Nadel L (2003) Down's syndrome: a genetic disorder in biobehavioral perspective. Genes Brain Behav 2:156–166
14. Ohr PS, Fagen JW (1994) Contingency learning in 9-month-old infants with Down syndrome. Am J Ment Retard 99:74–84
15. Schmidt-Sidor B, Wisniewski KE, Shepard TH, Sersen EA (1990) Brain growth in Down syndrome subjects 15 to 22 weeks of gestational age and birth to 60 months. Clin Neuropathol 9:181–190
16. Teipel SJ, Hampel H (2006) Neuroanatomy of Down syndrome in vivo: a model of preclinical Alzheimer's disease. Behav Genet 36:405–415
17. Mehta PD, Dalton AJ, Mehta SP, Kim KS, Sersen EA, Wisniewski HM (1998) Increased plasma amyloid beta protein 1–42 levels in Down syndrome. Neurosci Lett 241:13–16
18. Mehta PD, Patrick BA, Dalton AJ et al (1999) Increased levels of tau-like protein in patients with Down syndrome. Neurosci Lett 275:159–162
19. Beyreuther K, Dyrks T, Hilbich C et al (1992) Amyloid precursor protein (APP) and beta A4 amyloid in Alzheimer's disease and Down syndrome. Prog Clin Biol Res 379:159–182
20. Isacson O, Seo H, Lin L, Albeck D, Granholm AC (2002) Alzheimer's disease and Down's syndrome: roles of APP, trophic factors and ACh. Trends Neurosci 25:79–84
21. Head E, Lott IT, Patterson D, Doran E, Haier RJ (2007) Possible compensatory events in adult Down syndrome brain prior to the development of Alzheimer disease neuropathology: targets for nonpharmacological intervention. J Alzheimers Dis 11:61–76
22. Vicari S, Carlesimo GA (2006) Short-term memory deficits are not uniform in Down and Williams syndromes. Neuropsychol Rev 16:87–94
23. Eckert MA, Galaburda AM, Mills DL, Bellugi U, Korenberg JR, Reiss AL (2006) The neurobiology of Williams syndrome: cascading influences of visual system impairment? Cell Mol Life Sci 63:1867–1875
24. Meyer-Lindenberg A, Mervis CB, Berman KF (2006) Neural mechanisms in Williams syndrome: a unique window to genetic influences on cognition and behaviour. Nat Rev Neurosci 7:380–393
25. Cassidy SB, Dykens E, Williams CA (2000) Prader-Willi and Angelman syndromes: sister imprinted disorders. Am J Med Genet 97:136–146
26. Clayton-Smith J, Laan L (2003) Angelman syndrome: a review of the clinical and genetic aspects. J Med Genet 40:87–95
27. Benarroch F, Hirsch HJ, Genstil L, Landau YE, Gross-Tsur V (2007) Prader-Willi syndrome: medical prevention and behavioral challenges. Child Adolesc Psychiatr Clin N Am 16:695–708
28. Siebert JR (2004) CNS manifestations of chromosomal change. Developmental neuropathology. Allen, Basel, Switzerland, pp 132–141
29. Golden JA, Hyman BT (1994) Development of the superior temporal neocortex is anomalous in trisomy 21. J Neuropathol Exp Neurol 53:513–520

30. Wisniewski HM, Silverman W, Wegiel J (1994) Ageing, Alzheimer disease and mental retardation. J Intellect Disabil Res 38(Pt 3):233–239
31. Raz N, Torres IJ, Briggs SD et al (1995) Selective neuroanatomic abnormalities in Down's syndrome and their cognitive correlates: evidence from MRI morphometry. Neurology 45:356–366
32. Nakamura Y, Hashimoto T, Sasaguri Y et al (1986) Brain anomalies found in 18 trisomy: CT scanning, morphologic and morphometric study. Clin Neuropathol 5:47–52
33. Inagaki M, Ando Y, Mito T et al (1987) Comparison of brain imaging and neuropathology in cases of trisomy 18 and 13. Neuroradiology 29:474–479
34. Chiang MC, Reiss AL, Lee AD et al (2007) 3D pattern of brain abnormalities in Williams syndrome visualized using tensor-based morphometry. Neuroimage 36:1096–1109
35. Galaburda AM, Bellugi UV (2000) Multi-level analysis of cortical neuroanatomy in Williams syndrome. J Cogn Neurosci 12(Suppl 1):74–88
36. Gaser C, Luders E, Thompson PM et al (2006) Increased local gyrification mapped in Williams syndrome. Neuroimage 33:46–54
37. Thompson PM, Lee AD, Dutton RA et al (2005) Abnormal cortical complexity and thickness profiles mapped in Williams syndrome. J Neurosci 25:4146–4158
38. Van Essen DC, Dierker D, Snyder AZ, Raichle ME, Reiss AL, Korenberg J (2006) Symmetry of cortical folding abnormalities in Williams syndrome revealed by surface-based analyses. J Neurosci 26:5470–5483
39. Guerrini R, Carrozzo R, Rinaldi R, Bonanni P (2003) Angelman syndrome: etiology, clinical features, diagnosis, and management of symptoms. Paediatr Drugs 5:647–661
40. Leonard CM, Williams CA, Nicholls RD et al (1993) Angelman and Prader-Willi syndrome: a magnetic resonance imaging study of differences in cerebral structure. Am J Med Genet 46:26–33
41. Campbell LE, Daly E, Toal F et al (2006) Brain and behaviour in children with 22q11.2 deletion syndrome: a volumetric and voxel-based morphometry MRI study. Brain 129:1218–1228
42. Chow EW, Zipursky RB, Mikulis DJ, Bassett AS (2002) Structural brain abnormalities in patients with schizophrenia and 22q11 deletion syndrome. Biol Psychiatry 51:208–215
43. van Amelsvoort T, Daly E, Henry J et al (2004) Brain anatomy in adults with velocardiofacial syndrome with and without schizophrenia: preliminary results of a structural magnetic resonance imaging study. Arch Gen Psychiatry 61:1085–1096
44. Tamraz J, Rethore MO, Lejeune J et al (1993) [Brain morphometry using MRI in Cri-du-Chat Syndrome. Report of seven cases with review of the literature]. Ann Genet 36:75–87
45. Tsao CY, Wenger GD, Bartholomew DW (2005) Cri du chat syndrome and complex karyotype in a patient with infantile spasms, hypsarrhythmia, nonketotic hyperglycinemia, and heterotopia. Am J Med Genet A 134:198–201
46. Moller RS, Hansen CP, Jackson GD et al (2007) Interstitial deletion of chromosome 4p associated with mild mental retardation, epilepsy and polymicrogyria of the left temporal lobe. Clin Genet 72:593–598
47. Gropman AL, Elsea S, Duncan WC Jr, Smith AC (2007) New developments in Smith-Magenis syndrome (del 17p11.2). Curr Opin Neurol 20:125–134
48. Brown WE, Kesler SR, Eliez S, Warsofsky IS, Haberecht M, Reiss AL (2004) A volumetric study of parietal lobe subregions in Turner syndrome. Dev Med Child Neurol 46:607–609
49. Molko N, Cachia A, Riviere D et al (2004) Brain anatomy in Turner syndrome: evidence for impaired social and spatial-numerical networks. Cereb Cortex 14:840–850
50. Kesler SR, Garrett A, Bender B, Yankowitz J, Zeng SM, Reiss AL (2004) Amygdala and hippocampal volumes in Turner syndrome: a high-resolution MRI study of X-monosomy. Neuropsychologia 42:1971–1978
51. Holzapfel M, Barnea-Goraly N, Eckert MA, Kesler SR, Reiss AL (2006) Selective alterations of white matter associated with visuospatial and sensorimotor dysfunction in turner syndrome. J Neurosci 26:7007–7013
52. Patwardhan AJ, Eliez S, Bender B, Linden MG, Reiss AL (2000) Brain morphology in Klinefelter syndrome: extra X chromosome and testosterone supplementation. Neurology 54:2218–2223

53. Itti E, Gaw Gonzalo IT, Pawlikowska-Haddal A et al (2006) The structural brain correlates of cognitive deficits in adults with Klinefelter's syndrome. J Clin Endocrinol Metab 91:1423–1427
54. Maymon R, Herman A, Reish O (2002) Brain anomalies associated with 47, XYY karyotypes detected on a prenatal scan. Prenat Diagn 22:490–492
55. Ehara H, Eda I (2001) Schizencephaly in triple-X syndrome. Pediatr Int 43:296–297
56. Jernigan TL, Bellugi U (1990) Anomalous brain morphology on magnetic resonance images in Williams syndrome and Down syndrome. Arch Neurol 47:529–533
57. Jernigan TL, Bellugi U, Sowell E, Doherty S, Hesselink JR (1993) Cerebral morphologic distinctions between Williams and Down syndromes. Arch Neurol 50:186–191
58. Pinter JD, Eliez S, Schmitt JE, Capone GT, Reiss AL (2001) Neuroanatomy of Down's syndrome: a high-resolution MRI study. Am J Psychiatry 158:1659–1665
59. Pinter JD, Brown WE, Eliez S, Schmitt JE, Capone GT, Reiss AL (2001) Amygdala and hippocampal volumes in children with Down syndrome: a high-resolution MRI study. Neurology 56:972–974
60. Kates WR, Folley BS, Lanham DC, Capone GT, Kaufmann WE (2002) Cerebral growth in Fragile X syndrome: review and comparison with Down syndrome. Microsc Res Tech 57:159–167
61. Latash ML, Corcos DM (1991) Kinematic and electromyographic characteristics of single-joint movements of individuals with Down syndrome. Am J Ment Retard 96:189–201
62. Moerman P, Fryns JP, van der Steen K, Kleczkowska A, Lauweryns J (1988) The pathology of trisomy 13 syndrome. A study of 12 cases. Hum Genet 80:349–356
63. Lin HY, Lin SP, Chen YJ et al (2007) Clinical characteristics and survival of trisomy 13 in a medical center in Taiwan, 1985–2004. Pediatr Int 49:380–386
64. Lin HY, Lin SP, Chen YJ et al (2006) Clinical characteristics and survival of trisomy 18 in a medical center in Taipei, 1988–2004. Am J Med Genet A 140:945–951
65. Rasmussen SA, Wong LY, Yang Q, May KM, Friedman JM (2003) Population-based analyses of mortality in trisomy 13 and trisomy 18. Pediatrics 111:777–784
66. Niedrist D, Riegel M, Achermann J, Schinzel A (2006) Survival with trisomy 18 – data from Switzerland. Am J Med Genet A 140:952–959
67. Reiss AL, Eckert MA, Rose FE et al (2004) An experiment of nature: brain anatomy parallels cognition and behavior in Williams syndrome. J Neurosci 24:5009–5015
68. Paterson SJ, Schultz RT (2007) Neurodevelopmental and behavioral issues in Williams syndrome. Curr Psychiatry Rep 9:165–171
69. Wang PP, Doherty S, Rourke SB, Bellugi U (1995) Unique profile of visuo-perceptual skills in a genetic syndrome. Brain Cogn 29:54–65
70. Farran EK, Jarrold C (2003) Visuospatial cognition in Williams syndrome: reviewing and accounting for the strengths and weaknesses in performance. Dev Neuropsychol 23:173–200
71. Reiss AL, Eliez S, Schmitt JE, Patwardhan A, Haberecht M (2000) Brain imaging in neurogenetic conditions: realizing the potential of behavioral neurogenetics research. Ment Retard Dev Disabil Res Rev 6:186–197
72. Jauregi J, Arias C, Vegas O et al (2007) A neuropsychological assessment of frontal cognitive functions in Prader-Willi syndrome. J Intellect Disabil Res 51:350–365
73. Walley RM, Donaldson MD (2005) An investigation of executive function abilities in adults with Prader-Willi syndrome. J Intellect Disabil Res 49:613–625
74. Zinkstok J, van Amelsvoort T (2005) Neuropsychological profile and neuroimaging in patients with 22Q11.2 Deletion Syndrome: a review. Child Neuropsychol 11:21–37
75. Glaser B, Schaer M, Berney S, Debbane M, Vuilleumier P, Eliez S (2007) Structural changes to the fusiform gyrus: a cerebral marker for social impairments in 22q11.2 deletion syndrome? Schizophr Res 96:82–86
76. Haberecht MF, Menon V, Warsofsky IS et al (2001) Functional neuroanatomy of visuo-spatial working memory in Turner syndrome. Hum Brain Mapp 14:96–107
77. Kesler SR, Blasey CM, Brown WE et al (2003) Effects of X-monosomy and X-linked imprinting on superior temporal gyrus morphology in Turner syndrome. Biol Psychiatry 54:636–646
78. Molko N, Cachia A, Riviere D et al (2003) Functional and structural alterations of the intraparietal sulcus in a developmental dyscalculia of genetic origin. Neuron 40:847–858

79. Yamanouchi H, Imataka G, Nakagawa E et al (2005) An analysis of epilepsy with chromosomal abnormalities. Brain Dev 27:370–377
80. Pueschel SM, Louis S, McKnight P (1991) Seizure disorders in Down syndrome. Arch Neurol 48:318–320
81. Stafstrom CE, Patxot OF, Gilmore HE, Wisniewski KE (1991) Seizures in children with Down syndrome: etiology, characteristics and outcome. Dev Med Child Neurol 33:191–200
82. Kaneko WM, Phillips EL, Riley EP, Ehlers CL (1996) EEG findings in fetal alcohol syndrome and Down syndrome children. Electroencephalogr Clin Neurophysiol 98:20–28
83. Silva ML, Cieuta C, Guerrini R, Plouin P, Livet MO, Dulac O (1996) Early clinical and EEG features of infantile spasms in Down syndrome. Epilepsia 37:977–982
84. Moller JC, Hamer HM, Oertel WH, Rosenow F (2001) Late-onset myoclonic epilepsy in Down's syndrome (LOMEDS). Seizure 10:303–306
85. Guerrini R, De Lorey TM, Bonanni P et al (1996) Cortical myoclonus in Angelman syndrome. Ann Neurol 40:39–48
86. Lalande M, Minassian BA, DeLorey TM, Olsen RW (1999) Parental imprinting and Angelman syndrome. Adv Neurol 79:421–429
87. Schinzel A, Niedrist D (2001) Chromosome imbalances associated with epilepsy. Am J Med Genet 106:119–124
88. Singh R, Gardner RJ, Crossland KM, Scheffer IE, Berkovic SF (2002) Chromosomal abnormalities and epilepsy: a review for clinicians and gene hunters. Epilepsia 43:127–140
89. Dan B, Boyd SG (2003) Angelman syndrome reviewed from a neurophysiological perspective. The UBE3A-GABRB3 hypothesis. Neuropediatrics 34:169–176
90. Laan LA, Vein AA (2005) Angelman syndrome: is there a characteristic EEG? Brain Dev 27:80–87
91. Williams CA (2005) Neurological aspects of the Angelman syndrome. Brain Dev 27:88–94
92. Wang PJ, Hou JW, Sue WC, Lee WT (2005) Electroclinical characteristics of seizures-comparing Prader–Willi syndrome with Angelman syndrome. Brain Dev 27:101–107
93. Varela MC, Kok F, Setian N, Kim CA, Koiffmann CP (2005) Impact of molecular mechanisms, including deletion size, on Prader-Willi syndrome phenotype: study of 75 patients. Clin Genet 67:47–52
94. Kumada T, Ito M, Miyajima T et al (2005) Multi-institutional study on the correlation between chromosomal abnormalities and epilepsy. Brain Dev 27:127–134
95. Thomson AK, Glasson EJ, Bittles AH (2006) A long-term population-based clinical and morbidity review of Prader-Willi syndrome in Western Australia. J Intellect Disabil Res 50:69–78
96. Trauner DA, Bellugi U, Chase C (1989) Neurologic features of Williams and Down syndromes. Pediatr Neurol 5:166–168
97. Tsao CY, Westman JA (1997) Infantile spasms in two children with Williams syndrome. Am J Med Genet 71:54–56
98. Mizugishi K, Yamanaka K, Kuwajima K, Kondo I (1998) Interstitial deletion of chromosome 7q in a patient with Williams syndrome and infantile spasms. J Hum Genet 43:178–181
99. Sgro V, Riva E, Canevini MP et al (1995) 4p(-) syndrome: a chromosomal disorder associated with a particular EEG pattern. Epilepsia 36:1206–1214
100. Valente KD, Freitas A, Fiore LA, Kim CA (2003) A study of EEG and epilepsy profile in Wolf-Hirschhorn syndrome and considerations regarding its correlation with other chromosomal disorders. Brain Dev 25:283–287
101. Battaglia A, Carey JC (2005) Seizure and EEG patterns in Wolf-Hirschhorn (4p-) syndrome. Brain Dev 27:362–364
102. Kao A, Mariani J, McDonald-McGinn DM et al (2004) Increased prevalence of unprovoked seizures in patients with a 22q11.2 deletion. Am J Med Genet A 129:29–34
103. El Tahir MO, Kerr M, Jones RG (2004) Two cases of generalized seizures and the Velocardiofacial syndrome – a clinically significant association? J Intellect Disabil Res 48:695–698
104. Tatum WO 4th, Passaro EA, Elia M, Guerrini R, Gieron M, Genton P (1998) Seizures in Klinefelter's syndrome. Pediatr Neurol 19:275–278

105. Grosso S, Pucci L, Farnetani M et al (2004) Epilepsy and electroencephalographic findings in pericentric inversion of chromosome 12. J Child Neurol 19:604–608
106. Tombini M, Marciani MG, Romigi A et al (2003) Bilateral frontal polymicrogyria and epilepsy in a patient with Turner mosaicism: a case report. J Neurol Sci 213:83–86
107. Vulliemoz S, Dahoun S, Seeck M (2007) Bilateral temporal lobe epilepsy in a patient with Turner syndrome mosaicism. Seizure 16:261–265
108. Chiarenza GA (1993) Movement-related brain macropotentials of persons with Down syndrome during skilled performance. Am J Ment Retard 97:449–467
109. Lalo E, Vercueil L, Bougerol T, Jouk PS, Debu B (2005) Late event-related potentials and movement complexity in young adults with Down syndrome. Neurophysiol Clin 35:81–91
110. Cheron G, Servais L, Wagstaff J, Dan B (2005) Fast cerebellar oscillation associated with ataxia in a mouse model of Angelman syndrome. Neuroscience 130:631–637
111. Ferri R, Del Gracco S, Elia M, Musumeci SA, Scuderi C, Bergonzi P (1994) Bit-mapped somatosensory evoked potentials in Down's syndrome individuals. Neurophysiol Clin 24:357–366
112. van Trotsenburg AS, Smit BJ, Koelman JH et al (2006) Median nerve conduction velocity and central conduction time measured with somatosensory evoked potentials in thyroxine-treated infants with Down syndrome. Pediatrics 118:e825–e832
113. Egawa K, Asahina N, Shiraishi H et al (2008) Aberrant somatosensory-evoked responses imply GABAergic dysfunction in Angelman syndrome. Neuroimage 39:593–599
114. Brandt BR, Rosen I (1998) Impaired peripheral somatosensory function in children with Prader-Willi syndrome. Neuropediatrics 29:124–126
115. Dustman RE, Callner DA (1979) Cortical evoked responses and response decrement in non-retarded and Down's syndrome individuals. Am J Ment Defic 83:391–397
116. John FM, Bromham NR, Woodhouse JM, Candy TR (2004) Spatial vision deficits in infants and children with Down syndrome. Invest Ophthalmol Vis Sci 45:1566–1572
117. Suttle CM, Turner AM (2004) Transient pattern visual evoked potentials in children with Down's syndrome. Ophthalmic Physiol Opt 24:91–99
118. Thompson DA, Kriss A, Cottrell S, Taylor D (1999) Visual evoked potential evidence of albino-like chiasmal misrouting in a patient with Angelman syndrome with no ocular features of albinism. Dev Med Child Neurol 41:633–638
119. Creel DJ, Bendel CM, Wiesner GL, Wirtschafter JD, Arthur DC, King RA (1986) Abnormalities of the central visual pathways in Prader-Willi syndrome associated with hypopigmentation. N Engl J Med 314:1606–1609
120. Diaz F, Zuron M (1995) Auditory evoked potentials in Down's syndrome. Electroencephalogr Clin Neurophysiol 96:526–537
121. Krecicki T, Zalesska-Krecicka M, Kubiak K, Gawron W (2005) Brain auditory evoked potentials in children with Down syndrome. Int J Pediatr Otorhinolaryngol 69:615–620
122. Chen YJ, Fang PC (2005) Sensory evoked potentials in infants with Down syndrome. Acta Paediatr 94:1615–1618
123. Gothelf D, Farber N, Raveh E, Apter A, Attias J (2006) Hyperacusis in Williams syndrome: characteristics and associated neuroaudiologic abnormalities. Neurology 66:390–395
124. Wetter S, Murphy C (1999) Individuals with Down's syndrome demonstrate abnormal olfactory event-related potentials. Clin Neurophysiol 110:1563–1569
125. Vieregge P, Verleger R, Schulze-Rava H, Kompf D (1992) Late cognitive event-related potentials in adult Down's syndrome. Biol Psychiatry 32:1118–1134
126. Kakigi R, Neshige R, Matsuda Y, Kuroda Y (1994) Do patients with Down's syndrome recognize Mickey Mouse? J Neurol Sci 121:22–26
127. Yoder PJ, Camarata S, Camarata M, Williams SM (2006) Association between differentiated processing of syllables and comprehension of grammatical morphology in children with Down syndrome. Am J Ment Retard 111:138–152
128. Stauder JE, Boer H, Gerits RH, Tummers A, Whittington J, Curfs LM (2005) Differences in behavioural phenotype between parental deletion and maternal uniparental disomy in Prader-Willi syndrome: an ERP study. Clin Neurophysiol 116:1464–1470

129. Grice SJ, Haan MD, Halit H et al (2003) ERP abnormalities of illusory contour perception in Williams syndrome. Neuroreport 14:1773–1777
130. Hattori M, Fujiyama A, Taylor TD et al (2000) The DNA sequence of human chromosome 21. Nature 405:311–319
131. Gardiner K, Davisson M (2000) The sequence of human chromosome 21 and implications for research into Down syndrome. Genome Biol 1:REVIEWS0002
132. Epstein CJ (1995) Epilogue: toward the twenty-first century with Down syndrome – a personal view of how far we have come and of how far we can reasonably expect to go. Prog Clin Biol Res 393:241–246
133. Antonarakis SE, Lyle R, Dermitzakis ET, Reymond A, Deutsch S (2004) Chromosome 21 and down syndrome: from genomics to pathophysiology. Nat Rev Genet 5:725–738
134. Dierssen M, Fillat C, Crnic L, Arbones M, Florez J, Estivill X (2001) Murine models for Down syndrome. Physiol Behav 73:859–871
135. Gardiner K, Fortna A, Bechtel L, Davisson MT (2003) Mouse models of Down syndrome: how useful can they be? Comparison of the gene content of human chromosome 21 with orthologous mouse genomic regions. Gene 318:137–147
136. Reeves RH (2006) Down syndrome mouse models are looking up. Trends Mol Med 12:237–240
137. Rachidi M, Lopes C (2007) Mental retardation in Down syndrome: from gene dosage imbalance to molecular and cellular mechanisms. Neurosci Res 59:349–369
138. Galdzicki Z, Siarey RJ (2003) Understanding mental retardation in Down's syndrome using trisomy 16 mouse models. Genes Brain Behav 2:167–178
139. Benavides-Piccione R, Ballesteros-Yanez I, de Lagran MM et al (2004) On dendrites in Down syndrome and DS murine models: a spiny way to learn. Prog Neurobiol 74:111–126
140. Zoghbi HY (2003) Postnatal neurodevelopmental disorders: meeting at the synapse? Science 302:826–830
141. Dierssen M, Ramakers GJ (2006) Dendritic pathology in mental retardation: from molecular genetics to neurobiology. Genes Brain Behav 5(Suppl 2):48–60
142. Becker LE, Armstrong DL, Chan F (1986) Dendritic atrophy in children with Down's syndrome. Ann Neurol 20:520–526
143. Marin-Padilla M (1976) Pyramidal cell abnormalities in the motor cortex of a child with Down's syndrome. A Golgi study. J Comp Neurol 167:63–81
144. Suetsugu M, Mehraein P (1980) Spine distribution along the apical dendrites of the pyramidal neurons in Down's syndrome. A quantitative Golgi study. Acta Neuropathol 50:207–210
145. Wisniewski KE (1990) Down syndrome children often have brain with maturation delay, retardation of growth, and cortical dysgenesis. Am J Med Genet Suppl 7:274–281
146. Zellweger H (1977) The prenatal diagnosis of anomalies and other pathological conditions (author's transl). Schweiz Rundsch Med Prax 66:836–841
147. Sylvester PE (1983) The hippocampus in Down's syndrome. J Ment Defic Res 27(Pt 3):227–236
148. Galaburda AM, Holinger DP, Bellugi U, Sherman GF (2002) Williams syndrome: neuronal size and neuronal-packing density in primary visual cortex. Arch Neurol 59:1461–1467
149. Takashima S, Ieshima A, Nakamura H, Becker LE (1989) Dendrites, dementia and the Down syndrome. Brain Dev 11:131–133
150. Takashima S, Iida K, Mito T, Arima M (1994) Dendritic and histochemical development and ageing in patients with Down's syndrome. J Intellect Disabil Res 38(Pt 3):265–273
151. Marin-Padilla M (1972) Structural abnormalities of the cerebral cortex in human chromosomal aberrations: a Golgi study. Brain Res 44:625–629
152. Petit TL, LeBoutillier JC, Alfano DP, Becker LE (1984) Synaptic development in the human fetus: a morphometric analysis of normal and Down's syndrome neocortex. Exp Neurol 83:13–23
153. Weitzdoerfer R, Dierssen M, Fountoulakis M, Lubec G (2001) Fetal life in Down syndrome starts with normal neuronal density but impaired dendritic spines and synaptosomal structure. J Neural Transm Suppl 61:59–70
154. Takashima S, Becker LE, Armstrong DL, Chan F (1981) Abnormal neuronal development in the visual cortex of the human fetus and infant with Down's syndrome. A quantitative and qualitative Golgi study. Brain Res 225:1–21

155. Ferrer I, Ribalta T, Fabregues I, Pineda M, Cusi V (1986) A Golgi study of cerebellar malformation in 13 trisomy. Clin Neuropathol 5:53–59
156. Jay V, Chan FW, Becker LE (1990) Dendritic arborization in the human fetus and infant with the trisomy 18 syndrome. Brain Res Dev Brain Res 54:291–294
157. Richtsmeier JT, Baxter LL, Reeves RH (2000) Parallels of craniofacial maldevelopment in Down syndrome and Ts65Dn mice. Dev Dyn 217:137–145
158. Richtsmeier JT, Zumwalt A, Carlson EJ, Epstein CJ, Reeves RH (2002) Craniofacial phenotypes in segmentally trisomic mouse models for Down syndrome. Am J Med Genet 107:317–324
159. Hill CA, Reeves RH, Richtsmeier JT (2007) Effects of aneuploidy on skull growth in a mouse model of Down syndrome. J Anat 210:394–405
160. Insausti AM, Megias M, Crespo D et al (1998) Hippocampal volume and neuronal number in Ts65Dn mice: a murine model of Down syndrome. Neurosci Lett 253:175–178
161. Baxter LL, Moran TH, Richtsmeier JT, Troncoso J, Reeves RH (2000) Discovery and genetic localization of Down syndrome cerebellar phenotypes using the Ts65Dn mouse. Hum Mol Genet 9:195–202
162. Kurt MA, Kafa MI, Dierssen M, Davies DC (2004) Deficits of neuronal density in CA1 and synaptic density in the dentate gyrus, CA3 and CA1, in a mouse model of Down syndrome. Brain Res 1022:101–109
163. O'Doherty A, Ruf S, Mulligan C et al (2005) An aneuploid mouse strain carrying human chromosome 21 with Down syndrome phenotypes. Science 309:2033–2037
164. Lorenzi HA, Reeves RH (2006) Hippocampal hypocellularity in the Ts65Dn mouse originates early in development. Brain Res 1104:153–159
165. Clark S, Schwalbe J, Stasko MR, Yarowsky PJ, Costa AC (2006) Fluoxetine rescues deficient neurogenesis in hippocampus of the Ts65Dn mouse model for Down syndrome. Exp Neurol 200:256–261
166. Chakrabarti L, Galdzicki Z, Haydar TF (2007) Defects in embryonic neurogenesis and initial synapse formation in the forebrain of the Ts65Dn mouse model of Down syndrome. J Neurosci 27:11483–11495
167. Contestabile A, Fila T, Ceccarelli C et al (2007) Cell cycle alteration and decreased cell proliferation in the hippocampal dentate gyrus and in the neocortical germinal matrix of fetuses with Down syndrome and in Ts65Dn mice. Hippocampus 17:665–678
168. Guidi S, Bonasoni P, Ceccarelli C et al (2008) Neurogenesis impairment and increased cell death reduce total neuron number in the hippocampal region of fetuses with Down syndrome. Brain Pathol 18(2):180–197
169. Roper RJ, Baxter LL, Saran NG, Klinedinst DK, Beachy PA, Reeves RH (2006) Defective cerebellar response to mitogenic Hedgehog signaling in Down [corrected] syndrome mice. Proc Natl Acad Sci USA 103:1452–1456
170. Dauphinot L, Lyle R, Rivals I et al (2005) The cerebellar transcriptome during postnatal development of the Ts1Cje mouse, a segmental trisomy model for Down syndrome. Hum Mol Genet 14:373–384
171. Kurt MA, Davies DC, Kidd M, Dierssen M, Florez J (2000) Synaptic deficit in the temporal cortex of partial trisomy 16 (Ts65Dn) mice. Brain Res 858:191–197
172. Belichenko PV, Masliah E, Kleschevnikov AM et al (2004) Synaptic structural abnormalities in the Ts65Dn mouse model of Down syndrome. J Comp Neurol 480:281–298
173. Belichenko PV, Kleschevnikov AM, Salehi A, Epstein CJ, Mobley WC (2007) Synaptic and cognitive abnormalities in mouse models of Down syndrome: exploring genotype-phenotype relationships. J Comp Neurol 504:329–345
174. Saran NG, Pletcher MT, Natale JE, Cheng Y, Reeves RH (2003) Global disruption of the cerebellar transcriptome in a Down syndrome mouse model. Hum Mol Genet 12:2013–2019
175. Haydon PG, McCobb DP, Kater SB (1984) Serotonin selectively inhibits growth cone motility and synaptogenesis of specific identified neurons. Science 226:561–564
176. Lankford KL, DeMello FG, Klein WL (1988) D1-type dopamine receptors inhibit growth cone motility in cultured retina neurons: evidence that neurotransmitters act as morphogenic growth regulators in the developing central nervous system. Proc Natl Acad Sci USA 85:4567–4571

177. Kasamatsu T (1991) Adrenergic regulation of visuocortical plasticity: a role of the locus coeruleus system. Prog Brain Res 88:599–616
178. Nguyen L, Rigo JM, Rocher V et al (2001) Neurotransmitters as early signals for central nervous system development. Cell Tissue Res 305:187–202
179. Okado N, Narita M, Narita N (2001) A biogenic amine-synapse mechanism for mental retardation and developmental disabilities. Brain Dev 23(Suppl 1):S11–S15
180. Hohmann CF (2003) A morphogenetic role for acetylcholine in mouse cerebral neocortex. Neurosci Biobehav Rev 27:351–363
181. Berger-Sweeney J (2003) The cholinergic basal forebrain system during development and its influence on cognitive processes: important questions and potential answers. Neurosci Biobehav Rev 27:401–411
182. Sodhi MS, Sanders-Bush E (2004) Serotonin and brain development. Int Rev Neurobiol 59:111–174
183. Florez J, del Arco C, Gonzalez A, Pascual J, Pazos A (1990) Autoradiographic studies of neurotransmitter receptors in the brain of newborn infants with Down syndrome. Am J Med Genet Suppl 7:301–305
184. Bar-Peled O, Gross-Isseroff R, Ben-Hur H, Hoskins I, Groner Y, Biegon A (1991) Fetal human brain exhibits a prenatal peak in the density of serotonin 5-HT1A receptors. Neurosci Lett 127:173–176
185. Whittle N, Sartori SB, Dierssen M, Lubec G, Singewald N (2007) Fetal Down syndrome brains exhibit aberrant levels of neurotransmitters critical for normal brain development. Pediatrics 120:e1465–e1471
186. Lubec B, Yoo BC, Dierssen M, Balic N, Lubec G (2001) Down syndrome patients start early prenatal life with normal cholinergic, monoaminergic and serotoninergic innervation. J Neural Transm Suppl 61:303–310
187. Kiss J, Schlumpf M, Balazs R (1989) Selective retardation of the development of the basal forebrain cholinergic and pontine catecholaminergic nuclei in the brain of trisomy 16 mouse, an animal model of Down's syndrome. Brain Res Dev Brain Res 50:251–264
188. Yao FS, Caserta MT, Wyrwicz AM (2000) In vitro 1H and 31P NMR spectroscopic evidence of multiple aberrant biochemical pathways in murine trisomy 16 brain development. Int J Dev Neurosci 18:833–841
189. Allen DD, Martin J, Arriagada C et al (2000) Impaired cholinergic function in cell lines derived from the cerebral cortex of normal and trisomy 16 mice. Eur J Neurosci 12:3259–3264
190. Bambrick LL, Yarowsky PJ, Krueger BK (1995) Glutamate as a hippocampal neuron survival factor: an inherited defect in the trisomy 16 mouse. Proc Natl Acad Sci USA 92:9692–9696
191. Behar TN, Colton CA (2003) Redox regulation of neuronal migration in a Down Syndrome model. Free Radic Biol Med 35:566–575
192. Homanics GE, DeLorey TM, Firestone LL et al (1997) Mice devoid of gamma-aminobutyrate type A receptor beta3 subunit have epilepsy, cleft palate, and hypersensitive behavior. Proc Natl Acad Sci USA 94:4143–4148
193. Holtzman DM, Li YW, DeArmond SJ et al (1992) Mouse model of neurodegeneration: atrophy of basal forebrain cholinergic neurons in trisomy 16 transplants. Proc Natl Acad Sci USA 89:1383–1387
194. Salehi A, Delcroix JD, Belichenko PV et al (2006) Increased App expression in a mouse model of Down's syndrome disrupts NGF transport and causes cholinergic neuron degeneration. Neuron 51:29–42
195. Bimonte-Nelson HA, Hunter CL, Nelson ME, Granholm AC (2003) Frontal cortex BDNF levels correlate with working memory in an animal model of Down syndrome. Behav Brain Res 139:47–57
196. Seo H, Isacson O (2005) Abnormal APP, cholinergic and cognitive function in Ts65Dn Down's model mice. Exp Neurol 193:469–480
197. Hohmann CF, Berger-Sweeney J (1998) Cholinergic regulation of cortical development and plasticity. New twists to an old story. Perspect Dev Neurobiol 5:401–425

198. Nelson PG, McCune SK, Ades AM, Nelson KB (2001) Glial-neurotrophic mechanisms in Down syndrome. J Neural Transm Suppl 61:85–94
199. Dorsey SG, Bambrick LL, Balice-Gordon RJ, Krueger BK (2002) Failure of brain-derived neurotrophic factor-dependent neuron survival in mouse trisomy 16. J Neurosci 22:2571–2578
200. Dorsey SG, Renn CL, Carim-Todd L et al (2006) In vivo restoration of physiological levels of truncated TrkB.T1 receptor rescues neuronal cell death in a trisomic mouse model. Neuron 51:21–28
201. Griffin WS, Sheng JG, Mrak RE (1998) Senescence-accelerated overexpression of S100beta in brain of SAMP6 mice. Neurobiol Aging 19:71–76
202. Sen J, Belli A (2007) S100B in neuropathologic states: the CRP of the brain? J Neurosci Res 85:1373–1380
203. Reeves RH, Yao J, Crowley MR et al (1994) Astrocytosis and axonal proliferation in the hippocampus of S100b transgenic mice. Proc Natl Acad Sci USA 91:5359–5363
204. Luo L (2000) Rho GTPases in neuronal morphogenesis. Nat Rev Neurosci 1:173–180
205. Ramakers GJ (2002) Rho proteins, mental retardation and the cellular basis of cognition. Trends Neurosci 25:191–199
206. Berto G, Camera P, Fusco C et al (2007) The Down syndrome critical region protein TTC3 inhibits neuronal differentiation via RhoA and Citron kinase. J Cell Sci 120:1859–1867
207. Arber S, Barbayannis FA, Hanser H et al (1998) Regulation of actin dynamics through phosphorylation of cofilin by LIM-kinase. Nature 393:805–809
208. Yang N, Higuchi O, Ohashi K et al (1998) Cofilin phosphorylation by LIM-kinase 1 and its role in Rac-mediated actin reorganization. Nature 393:809–812
209. Meng Y, Zhang Y, Tregoubov V et al (2002) Abnormal spine morphology and enhanced LTP in LIMK-1 knockout mice. Neuron 35:121–133
210. Meng Y, Zhang Y, Tregoubov V, Falls DL, Jia Z (2003) Regulation of spine morphology and synaptic function by LIMK and the actin cytoskeleton. Rev Neurosci 14:233–240
211. Engidawork E, Lubec G (2003) Molecular changes in fetal Down syndrome brain. J Neurochem 84:895–904
212. Freidl M, Gulesserian T, Lubec G, Fountoulakis M, Lubec B (2001) Deterioration of the transcriptional, splicing and elongation machinery in brain of fetal Down syndrome. J Neural Transm Suppl 61:47–57
213. Peyrl A, Weitzdoerfer R, Gulesserian T, Fountoulakis M, Lubec G (2002) Aberrant expression of signaling-related proteins 14-3-3 gamma and RACK1 in fetal Down syndrome brain (trisomy 21). Electrophoresis 23:152–157
214. D'Adamo P, Menegon A, Lo Nigro C et al (1998) Mutations in GDI1 are responsible for X-linked non-specific mental retardation. Nat Genet 19:134–139
215. Asada H, Uyemura K, Shirao T (1994) Actin-binding protein, drebrin, accumulates in submembranous regions in parallel with neuronal differentiation. J Neurosci Res 38:149–159
216. Hayashi K, Shirao T (1999) Change in the shape of dendritic spines caused by overexpression of drebrin in cultured cortical neurons. J Neurosci 19:3918–3925
217. Kojima N, Shirao T (2007) Synaptic dysfunction and disruption of postsynaptic drebrin-actin complex: a study of neurological disorders accompanied by cognitive deficits. Neurosci Res 58:1–5
218. Pucharcos C, Fuentes JJ, Casas C et al (1999) Alu-splice cloning of human Intersectin (ITSN), a putative multivalent binding protein expressed in proliferating and differentiating neurons and overexpressed in Down syndrome. Eur J Hum Genet 7:704–712
219. Pucharcos C, Estivill X, de la Luna S (2000) Intersectin 2, a new multimodular protein involved in clathrin-mediated endocytosis. FEBS Lett 478:43–51
220. Okamoto M, Schoch S, Sudhof TC (1999) EHSH1/intersectin, a protein that contains EH and SH3 domains and binds to dynamin and SNAP-25. A protein connection between exocytosis and endocytosis? J Biol Chem 274:18446–18454
221. Sengar AS, Wang W, Bishay J, Cohen S, Egan SE (1999) The EH and SH3 domain Ese proteins regulate endocytosis by linking to dynamin and Eps15. EMBO J 18:1159–1171

222. Arai Y, Ijuin T, Takenawa T, Becker LE, Takashima S (2002) Excessive expression of synaptojanin in brains with Down syndrome. Brain Dev 24:67–72
223. Saito Y, Oka A, Mizuguchi M et al (2000) The developmental and aging changes of Down's syndrome cell adhesion molecule expression in normal and Down's syndrome brains. Acta Neuropathol 100:654–664
224. Hattori D, Demir E, Kim HW, Viragh E, Zipursky SL, Dickson BJ (2007) Dscam diversity is essential for neuronal wiring and self-recognition. Nature 449:223–227
225. Schmucker D (2007) Molecular diversity of Dscam: recognition of molecular identity in neuronal wiring. Nat Rev Neurosci 8:915–920
226. Fuerst PG, Koizumi A, Masland RH, Burgess RW (2008) Neurite arborization and mosaic spacing in the mouse retina require DSCAM. Nature 451:470–474
227. Yamagata M, Sanes JR (2008) Dscam and Sidekick proteins direct lamina-specific synaptic connections in vertebrate retina. Nature 451:465–469
228. Fuentes JJ, Genesca L, Kingsbury TJ et al (2000) DSCR1, overexpressed in Down syndrome, is an inhibitor of calcineurin-mediated signaling pathways. Hum Mol Genet 9:1681–1690
229. Arron JR, Winslow MM, Polleri A et al (2006) NFAT dysregulation by increased dosage of DSCR1 and DYRK1A on chromosome 21. Nature 441:595–600
230. Yang EJ, Ahn YS, Chung KC (2001) Protein kinase Dyrk1 activates cAMP response element-binding protein during neuronal differentiation in hippocampal progenitor cells. J Biol Chem 276:39819–39824
231. Labudova O, Kitzmueller E, Rink H, Cairns N, Lubec G (1999) Gene expression in fetal Down syndrome brain as revealed by subtractive hybridization. J Neural Transm Suppl 57:125–136
232. Bahn S, Mimmack M, Ryan M et al (2002) Neuronal target genes of the neuron-restrictive silencer factor in neurospheres derived from fetuses with Down's syndrome: a gene expression study. Lancet 359:310–315
233. Mao J, Maye P, Kogerman P et al (2002) Regulation of Gli1 transcriptional activity in the nucleus by Dyrk1. J Biol Chem 277:35156–35161
234. Woods YL, Rena G, Morrice N et al (2001) The kinase DYRK1A phosphorylates the transcription factor FKHR at Ser329 in vitro, a novel in vivo phosphorylation site. Biochem J 355:597–607
235. Sitz JH, Tigges M, Baumgartel K, Khaspekov LG, Lutz B (2004) Dyrk1A potentiates steroid hormone-induced transcription via the chromatin remodeling factor Arip4. Mol Cell Biol 24:5821–5834
236. Galdzicki Z, Siarey R, Pearce R, Stoll J, Rapoport SI (2001) On the cause of mental retardation in Down syndrome: extrapolation from full and segmental trisomy 16 mouse models. Brain Res Brain Res Rev 35:115–145
237. Cano E, Canellada A, Minami T, Iglesias T, Redondo JM (2005) Depolarization of neural cells induces transcription of the Down syndrome critical region 1 isoform 4 via a calcineurin/nuclear factor of activated T cells-dependent pathway. J Biol Chem 280:29435–29443
238. Hanson JE, Blank M, Valenzuela RA, Garner CC, Madison DV (2007) The functional nature of synaptic circuitry is altered in area CA3 of the hippocampus in a mouse model of Down's syndrome. J Physiol 579:53–67
239. Bianchi R, Zhao W, Chuang SC (2005) Epileptiform activity in the CA3 region of hippocampal slices from a Down syndrome mouse model. In: Meeting of the Society for Neuroscience, Washington, DC. 2005 Abstract Viewer/Itinerary Planner (Online)
240. Bianchi R, Chuang SC, Vela-Duarte D, Lee T, Hua F, Small S, Moreno H (2007) Abnormal metabolism and epileptiform activity in the hippocampus of a Down syndrome mouse model. In: Society for Neuroscience, San Diego, CA. 2007 Abstract Viewer/Itinerary Planner (Program No. 889.9)
241. Jiang YH, Armstrong D, Albrecht U et al (1998) Mutation of the Angelman ubiquitin ligase in mice causes increased cytoplasmic p53 and deficits of contextual learning and long-term potentiation. Neuron 21:799–811
242. Servais L, Cheron G (2005) Purkinje cell rhythmicity and synchronicity during modulation of fast cerebellar oscillation. Neuroscience 134:1247–1259

243. Pagliardini S, Ren J, Wevrick R, Greer JJ (2005) Developmental abnormalities of neuronal structure and function in prenatal mice lacking the prader-willi syndrome gene necdin. Am J Pathol 167:175–191
244. Siarey RJ, Carlson EJ, Epstein CJ, Balbo A, Rapoport SI, Galdzicki Z (1999) Increased synaptic depression in the Ts65Dn mouse, a model for mental retardation in Down syndrome. Neuropharmacology 38:1917–1920
245. Siarey RJ, Stoll J, Rapoport SI, Galdzicki Z (1997) Altered long-term potentiation in the young and old Ts65Dn mouse, a model for Down Syndrome. Neuropharmacology 36:1549–1554
246. Siarey RJ, Villar AJ, Epstein CJ, Galdzicki Z (2005) Abnormal synaptic plasticity in the Ts1Cje segmental trisomy 16 mouse model of Down syndrome. Neuropharmacology 49: 122–128
247. Costa AC, Grybko MJ (2005) Deficits in hippocampal CA1 LTP induced by TBS but not HFS in the Ts65Dn mouse: a model of Down syndrome. Neurosci Lett 382:317–322
248. Kleschevnikov AM, Belichenko PV, Villar AJ, Epstein CJ, Malenka RC, Mobley WC (2004) Hippocampal long-term potentiation suppressed by increased inhibition in the Ts65Dn mouse, a genetic model of Down syndrome. J Neurosci 24:8153–8160
249. Weeber EJ, Jiang YH, Elgersma Y et al (2003) Derangements of hippocampal calcium/ calmodulin-dependent protein kinase II in a mouse model for Angelman mental retardation syndrome. J Neurosci 23:2634–2644
250. Fernandez F, Morishita W, Zuniga E et al (2007) Pharmacotherapy for cognitive impairment in a mouse model of Down syndrome. Nat Neurosci 10:411–413
251. Costa AC, Scott-McKean JJ, Stasko MR (2008) Acute injections of the NMDA receptor antagonist memantine rescue performance deficits of the Ts65Dn mouse model of Down syndrome on a fear conditioning test. Neuropsychopharmacology 33(7):1624–1632
252. Granholm AC, Sanders L, Seo H, Lin L, Ford K, Isacson O (2003) Estrogen alters amyloid precursor protein as well as dendritic and cholinergic markers in a mouse model of Down syndrome. Hippocampus 13:905–914
253. O'Donnell WT, Warren ST (2002) A decade of molecular studies of fragile X syndrome. Annu Rev Neurosci 25:315–338
254. Aschrafi A, Cunningham BA, Edelman GM, Vanderklish PW (2005) The fragile X mental retardation protein and group I metabotropic glutamate receptors regulate levels of mRNA granules in brain. Proc Natl Acad Sci USA 102:2180–2185
255. Caudy AA, Myers M, Hannon GJ, Hammond SM (2002) Fragile X-related protein and VIG associate with the RNA interference machinery. Genes Dev 16:2491–2496
256. Ishizuka A, Siomi MC, Siomi H (2002) A Drosophila fragile X protein interacts with components of RNAi and ribosomal proteins. Genes Dev 16:2497–2508
257. Jin P, Zarnescu DC, Ceman S et al (2004) Biochemical and genetic interaction between the fragile X mental retardation protein and the microRNA pathway. Nat Neurosci 7:113–117
258. Qin M, Kang J, Burlin TV, Jiang C, Smith CB (2005) Postadolescent changes in regional cerebral protein synthesis: an in vivo study in the FMR1 null mouse. J Neurosci 25:5087–5095
259. Bear MF, Huber KM, Warren ST (2004) The mGluR theory of fragile X mental retardation. Trends Neurosci 27:370–377
260. Huber KM, Gallagher SM, Warren ST, Bear MF (2002) Altered synaptic plasticity in a mouse model of fragile X mental retardation. Proc Natl Acad Sci USA 99:7746–7750
261. Yan QJ, Rammal M, Tranfaglia M, Bauchwitz RP (2005) Suppression of two major Fragile X Syndrome mouse model phenotypes by the mGluR5 antagonist MPEP. Neuropharmacology 49:1053–1066
262. Chuang SC, Zhao W, Bauchwitz R, Yan Q, Bianchi R, Wong RK (2005) Prolonged epileptiform discharges induced by altered group I metabotropic glutamate receptor-mediated synaptic responses in hippocampal slices of a fragile X mouse model. J Neurosci 25:8048–8055
263. Kaufmann WE, Moser HW (2000) Dendritic anomalies in disorders associated with mental retardation. Cereb Cortex 10:981–991
264. Gearhart JD, Oster-Granite ML, Reeves RH, Coyle JT (1987) Developmental consequences of autosomal aneuploidy in mammals. Dev Genet 8:249–265

265. Epstein CJ (1990) The consequences of chromosome imbalance. Am J Med Genet Suppl 7:31–37
266. Gardiner K (2004) Gene-dosage effects in Down syndrome and trisomic mouse models. Genome Biol 5:244
267. Vacik T, Ort M, Gregorova S et al (2005) Segmental trisomy of chromosome 17: a mouse model of human aneuploidy syndromes. Proc Natl Acad Sci USA 102:4500–4505
268. Olson LE, Roper RJ, Sengstaken CL et al (2007) Trisomy for the Down syndrome 'critical region' is necessary but not sufficient for brain phenotypes of trisomic mice. Hum Mol Genet 16:774–782
269. Shapiro BL (1983) Down syndrome – a disruption of homeostasis. Am J Med Genet 14:241–269
270. Pritchard MA, Kola I (1999) The "gene dosage effect" hypothesis versus the "amplified developmental instability" hypothesis in Down syndrome. J Neural Transm Suppl 57:293–303
271. Nicholls RD, Knepper JL (2001) Genome organization, function, and imprinting in Prader-Willi and Angelman syndromes. Annu Rev Genomics Hum Genet 2:153–175
272. van Leeuwen FW, Hol EM (1999) Molecular misreading of genes in Down syndrome as a model for the Alzheimer type of neurodegeneration. J Neural Transm Suppl 57:137–159
273. Seidl R, Tiefenthaler M, Hauser E, Lubec G (2000) Effects of transdermal nicotine on cognitive performance in Down's syndrome. Lancet 356:1409–1410
274. Thiel R, Fowkes SW (2005) Can cognitive deterioration associated with Down syndrome be reduced? Med Hypotheses 64:524–532
275. Martinez-Cue C, Baamonde C, Lumbreras M et al (2002) Differential effects of environmental enrichment on behavior and learning of male and female Ts65Dn mice, a model for Down syndrome. Behav Brain Res 134:185–200
276. Martinez-Cue C, Rueda N, Garcia E, Davisson MT, Schmidt C, Florez J (2005) Behavioral, cognitive and biochemical responses to different environmental conditions in male Ts65Dn mice, a model of Down syndrome. Behav Brain Res 163:174–185
277. Dierssen M, Benavides-Piccione R, Martinez-Cue C et al (2003) Alterations of neocortical pyramidal cell phenotype in the Ts65Dn mouse model of Down syndrome: effects of environmental enrichment. Cereb Cortex 13:758–764
278. Nithianantharajah J, Hannan AJ (2006) Enriched environments, experience-dependent plasticity and disorders of the nervous system. Nat Rev Neurosci 7:697–709

Index

A
Aromatase, 3, 4
Aastroglia, 5
ABCB1 gene, in humans, 69
Abnormal brain development, 201–204
Abnormal herniation, 140
Acepromazine, 85
Actin depolymerization factor (ADF), 204
Active chromatin, 23
S-Adenosyl methionine (SAM), 25
Adenylyl cyclase, 7
Adrenocorticotroph (ACTH), 30
Adult rats, hippocampal expression, 21
AED. *See* Anti-epileptic medications
A-esterase, 69
Alkylating agents, 66
Alzheimer's disease (AD), 65, 197
AMH gene, 4
γ-Amino butyric acid, 30, 112
  action in hippocampus, 92
  immunoreactivity in neuropil, 85
  receptor agonists, 86
  synthesis during SE, 89
AMPA receptors, 91
Amyloid precursor protein (APP), 197
Androgen, 5
Anencephaly, 66, 139
Angelman syndrome (AS), 198
Angiogenesis, 58, 59, 61, 67, 85
Animal models, of absence/generalized seizures, 114
Anomalous Sylvian fissure, 199
Anticonvulsant, 66
Antidepressants, 67
Antiepileptic drugs, 67
  candidate mechanisms for, 120–122
  cognitive effects of, 42
  naïve rats, effect on, 115
  rats with prior seizures, effect on, 119–120
  seizure-näive rodents, effect on, 116–118
Anti-epileptic medications, 42, 44–46. *See also* Antiepileptic drugs
  affects, on developing brain, 47
  in childhood, 48
  on cognition and learning, 47, 48
  for conditions other than epilepsy, 48
  exposure during gestation, 47
  as systemic teratogen, 47
Anti-neoplastic agents, 68
Antisense oligonucleotide (AS), 8
Anxiety disorders, 143–144
Anxiety-related behavior, 21
AP3 adaptor protein, 88
Apoptosis, 67, 111, 121, 137, 138, 180
Arched-back nursing (ABN), 19
Arnold Chiari II malformations, 140
Aromatization, of testosterone, 5
Asperger's syndrome, 143
Astrocyte precursor cells (APC), 62
Astrocytes, 6, 8, 60
Astroglial-derived calcium-binding, 203
Attention deficit hyperactivity disorder (ADHD), 143
Autism, 142
Autism spectrum disorders (ASD), 31
Autosomal dominant epilepsy with febrile seizure plus syndrome (ADEFS+), 114
Axon
  branching, immunoglobulin proteins for, 204
  delay myelinization, 67
  and dendritic outgrowth, 91
  distortion of, 179
  elongation, 3
  engulfment, 137

Axon (*cont.*)
growth, 6, 8
guidance, 2
hillock, 57
length, 4
from male neurons, 2
polarization, 3
projection, 57, 172, 177, 183
sprouting, 87
to track connections, 174
unmyelinated, 57
Axosomatic synapses, 6
5-Azacytidine, 32

B
Basal lamina, 64
BAX expression, 30
BBB. *See* Blood-brain barrier
B-cell lymphoma (BCL)-2 associated protein-X (BAX), 30
BCRP. *See* Breast cancer resistance protein
BDNF. *See* Brain derived neurotrophic factor
Becker, J. B., 2
Benzodiazepine, 120, 122
Betamethasone, 113, 114, 122
Bicuculline hydrochloride, 79
Bicuculline-induced ictal activity, 81
Bicuculline methiodide, 79
Biogenic amines, 202
Bisexuality, 2
Bjornaes, H., 45
Blood-borne substrates, 59
Blood-brain barrier, 59, 158
brain development and, 68–69
differentiation, 61
fetal-maternal unit along with developing, 166–167
functional characteristics, 62–64
induction and formation, 59–62
modulation tight junction, 64–65
role in neurological disease, 65–66
specific transport systems for, 64
Bone morphogenetic protein (BMP)-4, 54
Bourgeois, H., 45
B-Raf, 7
Brain derived neurotrophic factor, 9, 89, 110, 121, 203
Brain development, 54–56
developmental sequence, 55
effects of seizures on, 106–107
from inside-out, 56–57
nerve cells, maturation, 57
pre-natal pharmacological treatment and, 66–67
drug category, 68
stages in humans and rats, 107
VEGF in embryonic and adult brain, 58
Brain malformations, 43, 194, 198–199
Brain microvasculature, 59, 63
Brain neuronal networks, development, 201
Brain sexual differentiation, 2
Brain-specific promoter, 21
Brain trauma, 65
Breast cancer resistance protein, 163–164
8-Bromo-cAMP, 26
S-Bromowillardiine, 120

C
Ca^{2+}-dependent PKC inhibitor, 9
N-Cadherin, 54
Cahill, L., 2
Calcium mobilization, 7
N-CAM, 54
cAMP response element binding protein, 8, 9, 22, 27, 204
Carbamazapine, 47, 115, 120, 140
Carboxylesterase, 69
Carnitine transporter (OCTN2), 164–165
CD50 for induction of tonic-clonic seizures, 82
Cell damage, 89–90
Cell modulation, 62
Cellular polarization, 3
Central nervous system, 2, 20, 32, 47, 54, 57, 60, 69, 82, 84, 136, 137, 140, 147, 158
Cerebral palsy (CP), 147–148
Cerebrospinal fluid (CSF), 59, 139
Channelopathies, 43
Chemoconvulsant drugs, 78
susceptibility in immature animals
higher, 78–81
lower, 82
Chemokines, 66
Childhood absence epilepsy (CAE), 43
Children's cognitive development, 42
Children seizures, 43
Children's mean IQ, 48
ChIP. *See* Chromatin immunoprecipitation
ChIP assays, 21, 24
ChIP-based sequencing (ChIP-Seq), 32
ChIP-on-chip (ChIP-chip), 32
Chlorpyrifos (CPF), 67
Cholinergic system, 82
Chondroitin sulfate proteoglycan, 57

Chordin, 54
Chromatin immunoprecipitation, 21
Chromatin structure, 23
Chromosomal abnormality, 197, 199
animal models of, 200
and synaptic connections, 200
Cingulin, 65
Citron kinase gene (CitK), 87
Claudin-3 and-5, 65
CNS. *See* Central nervous system
CNS malformations, 47
Cognition, clinical studies on, 44–46
Coloboma (Cm) mouse, with microdeletion, 88
Comparative genomic hybridization (CGH), 194
Computer tomography (CT), 194
Confocal laser scanning microscopy (CLSM), 175
Congenital brain defects, 194
Congenital chromosomal abnormalities, 195–196
Congenital rubella infections, 146
Convulsions, 78
Cortical malformations, 84–86
Corticosterone plasma levels, 30
Corticotropin-releasing factor (CRF), 19
Corticotropin-releasing hormone (CRH), 90
Co-transfection, NGFI-A expression, 21
CpG dinucleotides, in GR exon 1_7 promoter sequence, 23–24
CREB. *See* cAMP response element binding protein
CRF synthesis, 20
Cyclic adenosine 3,5 monophosphate (cAMP), 8, 22, 24, 26, 65
Cyclooxygenase-2 (COX-2), 108
Cyclosporin A, 158
Cytokines, 32, 58, 65, 84, 89, 92, 147
5′-Cytosine-phosphodiester-guanine (CpG)-3′ dinucleotide sequences, 23

D

Dandy-Walker malformations, 194, 198
Davies, D. C., 2
DAX-1 gene, 4
Dementia, 65
Demethylases, 25
(^{14}C)2-Deoxyglucose (2-DG) autoradiographic studies, 81
Diazinon (DZN), 67
Diencephalon in DS, 198
DNA conformation, 24
DNA demethylation, 25–27
DNA methylation, 23–26, 28, 29, 32
persistent alterations in, 28–31
DNA synthesis, in astrocytes, 62
Down syndrome, 196–201, 203–205
Doxorubicin, 158
Drug-associated teratogenicity, 121
Drug-induced neurodegeneration, 115
Drugs in pregnancy, 68, 165
Drug transporters, 163
at blood-placental barrier, 162
DS. *See* Down syndrome
DSCAM, immunoglobulin proteins, 204
DSCR1, inhibitor of calcineurin, 204
DXC1 gene, 142
DYRK1 kinase, 204

E

E2. *See* Estradiol
EAA-related agents, 79
E2-albumin construct (E2-BSA), 7
Early childhood seizures, 42
EC. *See* Endothelial cells
E-cadherin, 54
EEG ictal activity, 80
E2-induced axogenesis, 9
E2-induced growth, of neuronal processes, 5
Eker mutant rat, 86
Electrically-induced seizures, 82–83
Electrophoretic mobility shift assay (EMSA), 24
Embryonic brain slices, 179–180
Endothelial cells, 59, 62, 65, 66
Environmental adversity, 19
Environment, influencing brain development, 19
Epigenetic changes, 23
Epigenetic programming, by maternal care, 25–26
Epigenetic reprogramming, 28
Epilepsy, 65
clinical studies on, 44–46
mothers with, 47
Epileptogenesis, 92–93
ER α, estrogen receptor, 7
ER-α expression, 28
ERα mRNA, 6
ER density, 4
ERK activation, 9
ERK pathway, by E2, 9
ERK pathway UO126, 9
ERK phosphorylation, 9
ER linked to intracellular signaling pathway (mER), 7
Erythropoietin, 121

E-selectin, 66
Estradiol, 2, 7, 8
17β-Estradiol, 121
Estradiol-sensitive receptor, 7
Estrogen receptor (ER)-α 1b promoter, 28
Estrogen receptors (ER), 3
Estrogen response element (ERE), 7
Ethanol, and myelinization, 67
Excitatory amino acids (EAA), 79
Extraneuronal monoamine transporter (OCT3), 164

F
Febrile seizures, 48, 83–84
Female GD19 embryos, respond to E2, 3
Female hypothalamus, 3
Fibroblast growth factor (FGF), 54
Fibronectin, 64
flathead mutant rat, 86
Fluorescent in situ hybridization (FISH), 194
Fluorescent membrane dyes, 176–177
Fluorescent quantum dots, 177
Fluorophores, 186
Fluoxetine, 205
Flurothyl, 107–109
Flurothyl-induced SE, in PN6 rats, 108
Flurothyl-induced seizures, 107–109
Folate, metabolic pathways, 141
Folic acid deficiency, 140
Follistatin, 54
Fragile X mental retardation protein (FMRP), 205
Fragile X mental retardation syndrome (FXS), 205
Freezing focus, 112
Freezing lesion, 113

G
GABA. *See* γ-Amino butyric acid
$GABA_A$ receptor, 80, 92, 109
  antagonists, 78
$GABA_A$ signaling, 121
GABAergic neurotransmission, 93
GABAergic system, 85
Gabapentin, 48
GAD67 expression, 30
GAD inhibitor, 82
GAERS rats, 114
Gamma-GTP, 60
Ganglioside GD3, 57
GAP-43 protein, 3
GD16 embryos, 3
GD19 embryos, 4
Gene expression, in developing brain, 23
Gene mutations, 43
Genetic models, of epilepsy, 86–89, 114–115
Genetic mutations, linked with human epilepsies, 114
Germ-line inheritance, 21
Glia cells, 5, 6
Glial cell line-derived neurotrophic factor (GDNF), 121
Gliogenesis, 60
Glucocorticoid (GC) negative feedback, in hippocampus, 20
Glucocorticoid receptor, 19
  and enhanced GC feedback sensitivity, 20
  gene expression, regulation of, 22, 27
  and HPA stress responses, 19, 21, 25, 26
  mRNA splice variants, 21
  promoter altering NGFI-A binding, 23
  promoter, resulting in DNA demethylation, 27
Glucose transporter type 1 (GLUT-1), 60
Glutamate decarboxylase (GAD), 30, 82
Glutamatergic receptors, 86
Glutamyl transpeptidase, 60
Glycoprotein, 64
Gonadal steroids, 5
Gonadotrophin releasing hormone, neurons, 7
GPR30, 7
G-protein-coupled chemokine receptors (GPCR), 66
G protein coupled receptors, 7
G-proteins, 65
GR. *See* Glucocorticoid receptor
Grabowski, T. J., 2
GR exon 1_7 promoter, 26
  in human genome, 32
  luciferase construct, 21
  NGFI-A binding to, 21
  sequence in hippocampus, methylation across, 23
GR mRNA splice variants, 21

H
Harvey, V. R., 4
Healthy diet, 43
Heparan sulphate proteoglycan, 64
High resolution ultrasound, 174
Hippocampal-dependent memory, 31
Hippocampal GR. *See also* Glucocorticoid receptor
  expression, 26
  function, 21

Histone acetylation, 24, 27, 28, 31
Histone acetyl transferase (HAT), 24, 31
Histone deacetylase (HDAC) inhibitors, 25, 121
Histones, 23
HIV-defining factor, in children, 147
H3K9 acetylation, 24, 25, 30
Holoprosencephaly, 194, 198
Homeobox (Hox) genes, 55
Homeostasis, 58
Homocysteic acid, 79
Homocysteine, 79
  metabolic pathways, 141
Hormonal treatment, 3
HPA. *See* Hypothalamic-pituitary-adrenal axis
Hsp90, in estrogen-induced activation, 7, 8
5-HT, 21, 26
5-HT_7 receptors, 26
Human equilibrative nucleoside transporters 1 and 2 (hENT1 and hENT2), 164
Hydrocephaly, 66, 140
Hypertension, 65
Hyperthermia-induced seizures, 83, 84
Hyperthermic seizures, 112
Hypomyelination, 121
Hypothalamic-pituitary-adrenal axis, 20, 31
  and behavioral responses, 18
  regulation, 21
  responses, to stress, 19–21
Hypothalamus, 1, 3, 4, 20, 90, 108
Hypotonia, in infants, 199

I

Ibotenate, 86
Ictogenesis, 91–92
Idiopathic epilepsy syndrome, 106
Immunofluorescent techniques, 177
Inositol triphosphate (IP3), 8
Insulin-like growth factor (IGF)-2, 29, 62
Insulin-like growth factor I (IGF-I) receptors, 3
Insulin-like growth factor receptor (IGF-I R_β), 8
Intelligence coefficient (IQ), 42, 197
Interleukin-6 (IL-6), 60
Interleukin (IL)-1β, 32
Intracellular free Ca^{2+}, 9
In vivo imaging
  models, 179
    dissociated embryonic brain cell cultures, 179
    embryonic brain slices, 179–180
    living experimental animals, imaging in, 180–181
  rational for, 172
  technologies, 173–174
    labeling methods, advancement in, 176–179
    non-optical imaging techniques, 174
    optical and biophotonic techniques, 174–176
In vivo multiphoton imaging, 182
IQ test, 44
Isoflavinoid genistein, 29
Isomorphic neural mechanisms, 2

J

JAM-1, junctional adhesion molecules, 65

K

KA. *See* Kainic acid
KA-induced status epilepticus (KA-SE), 109
Kainic acid, 79, 109–110
KA-SE, in immature rats, 109
KA seizures, 109–110
Ketamine, 81
Kimchi, T., 2
Kindling phenomenon, 82
Klinefelter syndromes, 199
knocking out (KO) genes, 86

L

Laminin, 64
Lamotrigine, 47, 48, 115
Language impairments, 199
Lethal phenotype, 114
Levetiracetam, 115
L-Glutamate-1-carboxylase, 30
Limbic automatisms, 91
Lipopolysaccharide (LPS), 84, 112
Lissencephaly, 194
Lithium-pilocarpine model, 91
Lithium/pilocarpine SE, 91
  on PN20 rat, 110
L-Methionine, 25
Long-term potentiation (LTP), 205
LTP phenotypes, 31

M

α2-Macroglobulin, 60
Magnetic resonance imaging (MRI), 148, 173, 174, 185, 194

Male-type brain circuitry, 5
MAM. *See* Methylazoxymethanol
MAM exposure, 85
MAM lesion, 113
MAP kinase cascade, 8–9
MAPK signaling pathway, 8, 204
Maternal behavior, in rat, 19
  LG-ABN mothers, 20, 21
    binding of NGFI-A protein to, 24
    decrease hypothalamic CRF expression and, 20
    epigenetic state, of GR exon 1_7 promoter, 25
    hippocampal GR expression, 21, 27
    hippocampal NGFI-A expression and, 21
    during postnatal period, 19
Maternal care, in health and stress reactivity, 19
Maternal-fetus interface, 156
Maternal programming, of hippocampal GR gene expression, 22
Maternal stress, on offspring, 120
MBD2 deficient mice, 28
MBD-2 silences methylated genes, 27
MECP2 gene, 31
MeCP2 knockout mice, 114
Medial hinge point (MHP), 54
Medial preoptic (MPOA) area, 28
Megalencephaly, 194
Meningocele, 139
Mental retardation, 197–198
Metalloproteinase, 154
Methionine
  metabolic pathways, 141
  pharmacological manipulation, 29
  treatment, 29
Methylazoxymethanol, 66, 84, 113
Methyl-CpG binding domain protein (MBD)-2, 27
Methyl CpG-binding protein 2 (MECP2), 31, 143
N-Methyl-D-aspartic acid (NMDA), 79
Methylenetetrahydrofolate reductase (MTHFR), 29
5,10-Methylene tetrahydrofolate reductase (MTHFR), 140
MET receptor tyrosine kinase genes, 142
Michaelis-Menten law, 160
Micrencephaly, 194
Micro-CT uses X rays for scanning, 174
Microencephaly, 87
Mitogen-activated protein kinase (MAPK), 7
MK-801, 121
  treatment, 81
MMPs inhibitor, 60
Molecular genetic tools, 178–179
Monoamine oxidase inhibitors, 67
Moshé, S. L., 42, 49
Mother-offspring interaction, 18
Mothers, with epilepsy, 47
MRI spectroscopy, 185
MTHFR deficiency, 140
Mullerian inhibiting hormone, 4
Multidrug resistance-associated proteins (MRPs), 161, 163
Multidrug resistance 1 *(MDR1)* humans, 69
Multiphoton laser scanning microscopy (MPLSM) microscopy, 176
Multiple sclerosis, 65
Mutation
  of 3′ CpG dinucleotide, 26
  of *TSC1* and *TSC2* genes, 87
Myelomeningocele, 139, 140

N

Näive brain, triggered seizures in
  chemical models, 107–111
    flurothyl, 107–109
    kainic acid, 109–110
    pilocarpine, 110–111
  by physical means, 111–112
    continuous hippocampal stimulation, 111
    electrical kindling, 111
    freezing focus, 112
    hyperthermic seizures, 112
Nck/Crk and 14-3-3, adaptor proteins, 204
Neonatal flurothyl-induced seizures, 119
Neoneurogenesis, 110
Neovascularization, of mammalian brain, 60
Network plasticity, 89–90
Neural migration, 136–137
  errors, 140–142
Neural network pruning, 137–138
Neural progenitor cells (NPC), 60
Neural tube defects (NTD), 138
  classification of, 139
Neural tube disorders (NTDs), 134
Neurite growth, 2
Neuritogenesis, 3
Neuritogenic effect, E2, 9
  on dopaminergic diencephalic cells, 4
Neuroectodermal factors, during embryogenesis, 61
Neuroendocrine systems, 18, 19
Neurogenesis, 92, 110
Neurological disorders, 65

impact of in vivo imaging on studies of, 65
functional imaging, in neurobiology, 184–185
microscopic imaging, in animal models, 184
neuronal connections, development of, 183–184
neuronal migration, 182–183
neuronal proliferation, 181–182
Neurological dysfunctions, 199–200
Neuron-glia interactions, 6
Neurons, in hippocampal cultures stain, 4
Neurophysiological functions, 18
Neurotoxicity
administration of anticonvulsant, 66–67
antibiotics, drugs, 67
antiepileptic drugs, 67
of ethanol, 67
industrial chemicals, 67
nitrogen mustard alkylating agents, 66
to organophosphorous insecticides, 67
Neurotrophic factors, 5
Neurotrophin-3 (NT-3), 121
Neurotrophins, 6
Neurulation, 134–135
Newer AEDs, 47
NFAT-mediated signaling, 204
NGFI-A, transcription factor
binding consensus sequence, 32
expression, in neonatal offspring, 22–23
in regulation of GR expression, 27
NIH Collaborative Perinatal project, 44, 45
NMDA-induced seizures, 79
NMDA receptors, 91, 108, 112, 121
antagonists, 81, 119, 121
Noggin, 54
Nonsteroidal receptor antagonist, 7
NR2B/NR2A ratio, 91
Nucleoside diphosphate kinase (NDK)-B, 204
Nutrient supply, 18

O

Obsessive compulsive disorders (OCD), 143
Occludin, 62
Occluding 1 and 5, 65
Oligodendrocytes, 57
Optical coherence tomography, 176
Oxcarbezapine, 47

P

Pavlov's conditioning models, 144
Pentylenetetrazole, 80, 89, 90
induced seizures, 81, 109
tonic-clonic seizures, 81
Peripheral nervous system (PNS), 57
Pervasive developmental disorders (PDD), 142–143
P-Glycoprotein (P-gp), 60, 69, 161
Pharmacodynamic, 68
Phenobarbital, 47, 48, 67, 115
Phenotypic plasticity, 18, 29, 33
Phenytoin, 47, 67, 115
Phospho-ERK signal, 9
Phosphorylation
alterations in ADF/cofilin, 204
of CREB, 8
of ERK1 and ERK2, 8, 9
estradiol treatment by, 8
Phytoestrogen, 29
Pilocarpine, 110–111
Pilocarpine-induced seizures, 82
Pituitary-adrenal system, 20
PKC activation, by Ro32-0432, 9
Placenta, 153
anatomy and development, 153–155
levels of protection at, 156–157
mechanisms, of drug passage across, 158–165
active transport, 160–161
BBB *vs.* placental barrier, 158–159
efflux transporters, 161, 163, 164
facilitated diffusion, 160
influx/efflux transporters, 164–165
passive diffusion, 160
placental permeability, affecting, 159
metabolic properties of, 157–158
xenobiotics percolation through, 165–166
Placental barrier
active transport of xenobiotics across, 160, 166
and cortical malformations, 84
drug disposition, 155
drug distribution, similarities between BBB, 158–159
drug transporters at, 162
levels of protection at, 156–157
PN7–9, and electrographic seizures, 91
p75^{NGFR}, receptor, 6
Polymorphism, 88
Pompolo, S., 4
Positron emission tomography (PET), 174
Post-translational modification, 7
Post traumatic stress disorders (PTSD), 143
Prader-Willi syndrome (PWS), 198, 199
Prenatal betamethasone exposure, 122
Prenatal corticosteroids, for seizures, 113–114

Prenatal infectious disease exposure, 144
commonly acquired fetal infections and role, 145
cytomegalovirus (CMV), 144, 146
human immunodeficiency virus infection, 147
rubella infection, 146
TORCH, 144
Prenatal stressors/antiepileptics, 122
Primate models, 18
Programmed cell death, 2
Protein kinase A (PKA), 7, 22, 26
P-selectin, 66
PTZ. *See* Pentylenetetrazole

R
Rab GDP-dissociation inhibitor (GDI) 2, 204
RACK1 (anchoring protein), 204
Rat hippocampal GR gene, 21
Recurrent pilocarpine-induced SE, 110
Reelin expression, 30, 140
Repetitive kindled seizures, 120
Retinoic acid, 54
Rett syndrome, 143
Rett syndrome model, 114
Reverse transcription-polymerase chain reaction (RT-PCR), 60
Rho signaling pathway, 203

S
Schizencephaly, 86, 194
Schwann cells, 57
Sedating agents, 85
SE-induced morphological damage, 89
Seizures, 42
in abnormal brain, 112–113
freezing lesion, 113
MAM lesion, 113
prenatal corticosteroids, 113–114
cognitive effects, 45
pattern in flurothyl seizure mode, 80
Serotonin, 21
uptake inhibitor, 205
5-HT1A receptor, 203
Serum protein, role in drug delivery, 165
Seven-transmembrane receptor (7TMR), 7
Sex steroid-dependent differentiation, 3
Sexual differences, in amount of GAP-43, 3
Sexual dimorphisms, in brain of vertebrates, 1, 2
SF-1 gene, 4
Singh, M., 7
Sonic hedgehog (Shh), 54
mediated signaling pathway, 201
SOX-9 gene, 4
Specific epilepsy syndrome, 106
Spectral karyotyping (SKY), 194
Spina bifida (SB), 139–140
Spontaneous recurrent seizures (SRS), 91
Spontaneous seizures, 86, 90–91, 110
SRY gene, 4
Status epilepticus (SE), 78, 107
induced by pilocarpine, 111
Stress-diathesis models, 18
Stress resilience, 18
Substantia nigra pars reticulata (SNR), 81, 92
Swann, J. W., 108
Synaptic long-term depression, 205
Synaptic plasticity, 30, 31, 110, 204–206
Synaptic reorganization in hippocampus, 89
Synaptic transmission, 205–206
Synaptogenesis, 2, 91
Synaptojanin, 204
Synaptosome-associated proteins (SNAPs), 204

T
Tamoxifen, 7
Taylor's type, cortical dysplasia, 86
Temporal pattern, of sexual differentiation, 3
Tenascin, 64
Teratogenic effects, AEDs, 47
Terminal deoxynucleotidyl transferase-mediated biotin-dUTP nick endlabeling (TUNEL), 30
Testosterone, 2, 5
Tetrandrine, 158
Thalidomide, 67
Therapeutic implications, 29
TH immunoreactive neurons, in mesencephalon, 4
Tight junctions (TJ), 62
Tish mutant rat, 86
Tissue plasminogen activator (tPA), 62
TJ associated protein 7H6, 65
TJ proteins ZO-1, 64
Tonic-clonic seizure, 80
Topiramate, 47, 115, 120, 121
in pediatric trials, 48
Transcription factor nerve-growth-factor-inducible-protein-A (NGFI-A), 21
Trans-endothelial electrical resistance (TEER), 60, 62
Transferrin receptor, 60
Transforming growth factor-β (TGF-β), 54

Transporters inhibitors, 158
Trichostatin A (TSA), 25, 27, 30
Tricyclic drugs, 67
Trisomy 16, 200
Trisomy 21, 197, 200
Trisomy 13 and 18, 199
TSC1 and *TSC2* genes, 86
TSC2 mutations, 43
T-type calcium channel gene CACNA1H, 88
Tuberous sclerosis, 43, 86
Turner syndromes, 199
Type I antiestrogen ICI182780, 7
Type IV collagen, 64
Tyrosine-expressing neurons, 4
Tyrosine hydroxylase (TH), 3, 4
Tyrosine kinases, 65
Tyrosine kinase type A receptors, 6
Tyrosine kinase type B (TrkB), 3, 7, 8
Tyrosine receptor kinase B, 203

V
Valproic acid (VPA), 47, 140
  induced neurodegeneration, 121
Vascular endothelial growth factor (VEGF), 58
Vascular endothelium, differentiation of, 62
Vasopressinergic fibers, 4
Ventriculomegaly, 194
Ventromedial hypothalamic neurons, 3
Verapamil, 158
Vernadakis, A., 78
Vining, E. P., 48
Viral teratogenic effect, 146
Voltage sensitive dyes, 178

W
WAG/Rij rats, 114
Wechsler Intelligence Test for Children–Revised (WISC-R), 45
Williams syndrome (WS), 198
Wilson, C. A., 2
Wingless int (Wnt3), 54
Wnt genes, 55
Women, with epilepsy, 47
Woodbury, D. M., 78
WT-1 gene, 4

X
Xenobiotics, 59, 69
  percolation through placenta, 165–166
  recognized as substrates by, 161
X-linked mental retardation, 204
Xylazine, 85

Y
Y chromosome, 4

Z
Zonisamide, 47

MIX
Papier aus verantwortungsvollen Quellen
Paper from responsible sources
FSC® C105338

If you have any concerns about our products,
you can contact us on
ProductSafety@springernature.com

In case Publisher is established outside the EU,
the EU authorized representative is:
Springer Nature Customer Service Center GmbH
Europaplatz 3, 69115 Heidelberg, Germany

Printed by Libri Plureos GmbH
in Hamburg, Germany